The Project Physics Course

Reader

UNIT 4 Light and Electromagnetism

A Component of the Project Physics Course

Published by
HOLT, RINEHART and WINSTON, Inc.
New York, Toronto

This publication is one of the many instructional materials developed for the Project Physics Course. These materials include Texts, Handbooks, Teacher Resource Books, Readers, Programmed Instruction Booklets, Film Loops, Transparencies, 16mm films and laboratory equipment. Development of the course has profited from the help of many colleagues listed in the text units.

Directors of Harvard Project Physics

Gerald Holton, Department of Physics, Harvard University

F. James Rutherford, Capuchino High School, San Bruno, California, and Harvard University

Fletcher G. Watson, Harvard Graduate School of Education

ISBN: 0-03-084561-0

89 039 9876

Picture Credits

Cover: *Current,* 1964, by Bridget Riley. Emulsion on composition board, 58⅜ x 58⅞". Courtesy of The Museum of Modern Art, New York City.

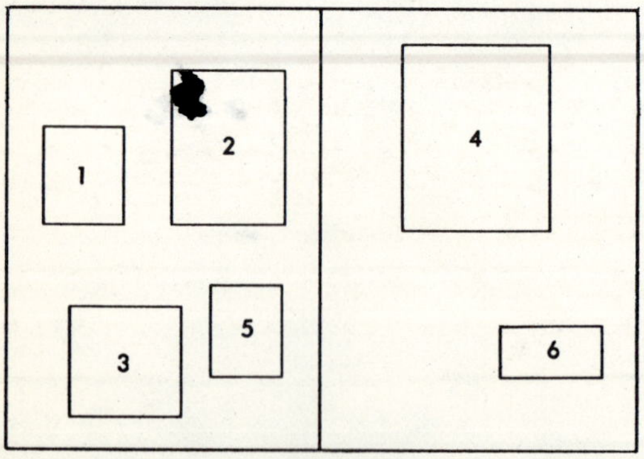

Picture Credits for frontispiece.

(1) Photograph by Glen J. Pearcy.

(2) *Jeune fille au corsage rouge lisant* by Jean Baptiste Camille Corot. Painting. Collection Bührle, Zurich.

(3) Harvard Project Physics staff photo.

(4) *Femme lisant* by Georges Seurat. Conté crayon drawing. Collection C. F. Stoop, London.

(5) *Portrait of Pierre Reverdy* by Pablo Picasso. Etching. Museum of Modern Art, N.Y.C.

(6) *Lecture au lit* by Paul Klee. Drawing. Paul Klee Foundation, Museum of Fine Arts, Berne.

Sources and Acknowledgments
Project Physics Reader 4

1. *Letter from Thomas Jefferson, June, 1799,* from *Scripta Mathematica,* Volume 1, 1932, pages 88–90. Reprinted with permission from the Manuscript Collection of Teachers College Library, Columbia University. Also in *Science and the Common Understanding* by J. Robert Oppenheimer. Reprinted with permission of Simon and Schuster, Inc.
2. *On the Method of Theoretical Physics* by Albert Einstein from *Essays in Science,* pages 12–21, Philosophical Library, New York, copyright 1934 by Estate of Albert Einstein. Reprinted with permission.
3. *Systems, Feedback and Cybernetics* from *Introduction to Natural Science* by V. Lawrence Parsegian, Alan S. Meltzer, Abraham S. Luchins and K. Scott Kinerson, copyright © 1968 by Academic Press, Inc. Reprinted with permission.
4. *Velocity of Light* from *Studies In Optics* by A. A. Michelson, copyright 1927 by The University of Chicago Press. Reprinted with permission.
5. *Popular Application of Polarized Light* from *Polarized Light* (Momentum #7) by William A. Shurcliff and Stanley S. Ballard, copyright © 1964 by Litton Educational Publishing Inc. Reprinted with permission of Van Nostrand Reinhold Company.
6. *Eye and Camera* by George Wald from *Scientific American,* August 1950, copyright © 1950 by Scientific American, Inc. Reprinted with permission. All rights reserved.
7. *The Laser—What It Is and Does* from *The Story of the Laser* by John M. Carroll, copyright © 1964 by E. P. Dutton & Co., Inc., Reprinted with permission of E. P. Dutton & Co., Inc., and Souvenir Press Ltd.
8. *A Simple Electric Circuit: Ohm's Law* from *The New College Physics: A Spiral Approach* by Albert V. Baez. W. H. Freeman and Company, copyright © 1967. Reprinted with permission.
9. *The Electronic Revolution* from *Voices from the Sky* by Arthur C. Clarke, Harper & Row, Publishers, New York, copyright © 1965. "The Electronic Revolution," copyright © 1962 by The New York Times Company. Reprinted with permission of the author and his agent, Scott Meredith Literary Agency, and David Higham Associates, Ltd .
10. *The Invention of the Electric Light* by Matthew Josephson from *Scientific American,* November

1959, copyright © 1959 by Scientific American, Inc. Reprinted with permission. All rights reserved.
11. *High Fidelity* from *Reproduction of Sound* by Edgar Villchur, copyright © 1962 by Acoustic Research, Inc., Cambridge, Massachusetts. Copyright © 1965 by Dover Publications, Inc. Reprinted with permission.
12. *The Future of Direct Current Power Transmission* by N. L. Allen from *Endeavor,* Volume CCVI, No. 97, Imperial Chemical Industries Limited, London. Reprinted with permission.
13. *James Clerk Maxwell* by James R. Newman, Part II, from *Scientific American,* June 1955, copyright © 1955 by Scientific American, Inc. Reprinted with permission. All rights reserved.
14. *On the Induction of Electric Currents* from *A Treatise on Electricity and Magnetism* by James Clerk Maxwell, Volume 2, 1873, The Clarendon Press, Oxford. Reprinted with permission.
15. *The Relationship of Electricity and Magnetism* from *Faraday, Maxwell, and Kelvin,* by D. K. C. MacDonald, copyright © 1964 by Educational Services Incorporated. Reprinted with permission of Doubleday & Company, Inc. (Science Study Series.)
16. *The Electromagnetic Field* from *The Evolution of Physics* by Albert Einstein and Leopold Infeld. Published by Simon and Schuster, copyright © 1961. Reprinted with permission.
17. *Radiation Belts Around the Earth* by James A. Van Allen from *Scientific American,* March 1959. Copyright © 1959 by Scientific American, Inc. Reprinted with permission. All rights reserved. Available separately at 20¢ each as Offprint No. 248 from W. H. Freeman and Company, 666 Market Street, San Francisco, California 94104.
18. *A Mirror for the Brain* from *The Living Brain* by W. Grey Walter, copyright 1953, © 1963 by W. W. Norton & Company, Inc. Reprinted with permission of W. W. Norton & Company, Inc., and Gerald Duckworth & Co. Ltd.
19. *Scientific Imagination* from *The Feynman Lectures on Physics* by Richard P. Feynman, Robert B. Leighton and Matthew L. Sands, copyright © 1964 by Addison-Wesley Publishing Company, Inc. Reprinted with permission.
20. *Lenses and Optical Instruments* from PSSC Physics, D. C. Heath and Company, Boston. Copyright © 1965 by Educational Services Incorporated. Reprinted with permission.
21. *Baffled,* by Keith Waterhouse from Punch, July 10, 1968, copyright © 1968 *Punch,* London.

LIBRARY HOURS

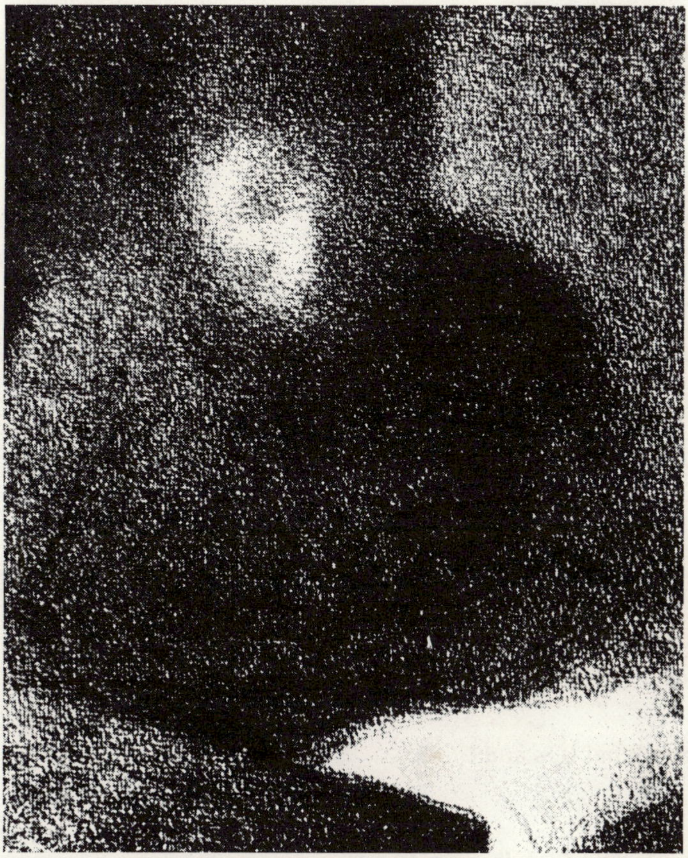

This is not a physics textbook. Rather, it is a physics reader, a collection of some of the best articles and book passages on physics. A few are on historic events in science, others contain some particularly memorable description of what physicists do; still others deal with philosophy of science, or with the impact of scientific thought on the imagination of the artist.

There are old and new classics, and also some little-known publications; many have been suggested for inclusion because some teacher or physicist remembered an article with particular fondness. The majority of articles is not drawn from scientific papers of historic importance themselves, because material from many of these is readily available, either as quotations in the Project Physics text or in special collections.

This collection is meant for your browsing. If you follow your own reading interests, chances are good that you will find here many pages that convey the joy these authors have in their work and the excitement of their ideas. If you want to follow up on interesting excerpts, the source list at the end of the reader will guide you for further reading.

Reader 4

Table of Contents

A great American writes about the significant role of science in the education of the individual and in the creation of American society.

1 Letter from Thomas Jefferson

June 1799

Monticello June 18. 99.

DEAR SIR,

I have to acknolege the reciept of your favor of May 14. in which you mention that you have finished the 6. first books of Euclid, plane trigonometry, surveying and algebra and ask whether I think a further pursuit of that branch of science would be useful to you. There are some propositions in the latter books of Euclid, and some of Archimedes, which are useful, and I have no doubt you have been made acquainted with them. Trigonometry, so far as this, is most valuable to every man, there is scarcely a day in which he will not resort to it for some of the purposes of common life; the science of calculation also is indispensible as far as the extraction of the square and cube roots; Algebra as far as the quadratic equation and the use of logarithms are often of value in ordinary cases; but all beyond these is but a luxury; a delicious luxury indeed; but not to be indulged in by one who is to have a profession to follow for his subsistence. In this light I view the conic sections, curves of the higher orders, perhaps even spherical trigonometry, Algebraical operations beyond the 2d dimension, and fluxions. There are other branches of science however worth the attention of every man: Astronomy, botany, chemistry, natural philosophy, natural history, anatomy. Not indeed to be a proficient in them; but to possess their general principles and outlines, so as that we may be able to amuse and inform ourselves further in any of them as we proceed through life and have occasion for them. Some knowlege of them is necessary for our character as well as comfort. The general elements of astronomy and of natural philosophy are best acquired at an academy where we can have the benefit of the instruments and apparatus usually provided there: but the others may well be acquired from books alone as far as our purposes require. I have indulged myself in these observations to you, because the evidence cannot be unuseful to you of a person who has often had occasion to consider which of his acquisitions in science have been really useful to him in life, and which of them have been merely a matter of luxury.

I am among those who think well of the human character generally. I consider man as formed for society, and endowed by nature with those dispositions which fit him for society. I believe also, with Condorcet, as mentioned in your letter, that his mind is perfectible to a degree of which we cannot as yet form any conception. It is impossible for a man who takes a survey of what is already known, not to see what an immensity in every branch of science yet remains to be discovered, and that too of articles to which our faculties seem adequate. In geometry and calculation we know a great deal. Yet there are some desiderata. In anatomy great progress has been made; but much is still to be acquired. In natural history we possess knowlege; but we want a great deal. In chemistry we are not yet sure of the first elements. Our natural philosophy is in a very infantine state; perhaps for great advances in it, a further progress in chemistry is necessary. Surgery is well advanced; but prodigiously short of what may be. The state of medecine is worse than that of total ignorance. Could we divest ourselves of every thing we suppose we know in it, we should start from a higher ground and with fairer prospects. From Hippocrates to Brown we have had nothing but a succession of hypothetical systems each having it's day of vogue, like the fashions and fancies of caps and gowns, and yielding in turn to the next caprice. Yet the human frame, which is to be the subject of suffering and torture under these learned modes, does not change. We have a few medecines, as the bark, opium, mercury, which in a few well defined diseases are of unquestionable virtue: but the residuary list of the materia medica, long as it is, contains but the charlataneries of the art; and of the diseases of doubtful form, physicians have ever had a false knowlege, worse than ignorance. Yet surely the list of unequivocal diseases and remedies is capable of enlargement; and it is still more certain that in the other branches of science, great fields are yet to be explored to which our faculties are equal, and that to an extent of which we cannot fix the limits. I join you therefore in branding as cowardly the idea that the human mind is incapable of further advances. This is precisely the doctrine which the present despots of the earth are inculcating, and their friends here re-echoing; and applying especially to religion and politics; ' that it is not probable that any thing better will be discovered than what was known to our fathers '. We are to look backwards then and not forwards for the improvement of science, and to find it amidst feudal barbarisms and the fires of Spital-fields. But thank heaven the American mind is already too much opened, to listen to these impostures; and while the art of printing is left to use, science can never be retrograde; what is once acquired of real knowlege can never be lost. To preserve the freedom of the human mind then and freedom of the press, every spirit should be ready to devote itself to martyrdom; for as long as we may think as we will, and speak as we think, the condition of man will proceed in improvement. The generation which is going off the stage has deserved well of mankind for the struggles it has made, and for having arrested that course of despotism which had overwhelmed the world for thousands and thousands of years. If there seems to be danger that the ground they have gained will be lost again, that danger comes from the generation your contemporary. But that the enthusiasm which characterises youth should lift it's parracide hands against freedom and science would be such a monstrous phaenomenon as I cannot place among possible things in this age and this country. Your college at least has shewn itself incapable of it; and if the youth of any other place have seemed to rally under other banners it has been from delusions which they will soon dissipate. I shall be happy to hear from you from time to time, and of your progress in study, and to be useful to you in whatever is in my power; being with sincere esteem Dear Sir

Your friend & servt
Th: Jefferson

Einstein discusses some of the factors that lead to a scientific theory.

2 On the Method of Theoretical Physics

Albert Einstein

An essay—1934.

IF YOU want to find out anything from the theoretical physicists about the methods they use, I advise you to stick closely to one principle: don't listen to their words, fix your attention on their deeds. To him who is a discoverer in this field the products of his imagination appear so necessary and natural that he regards them, and would like to have them regarded by others, not as creations of thought but as given realities.

These words sound like an invitation to you to walk out of this lecture. You will say to yourselves, the fellow's a working physicist himself and ought therefore to leave all questions of the structure of theoretical science to the epistemologists.

Against such criticism I can defend myself from the personal point of view by assuring you that it is not at my own instance but at the kind invitation of others that I have mounted this rostrum, which serves to commemorate a man who fought hard all his life for the unity of knowledge. Objectively, however, my enterprise can be justified on the ground that it may, after all, be of interest to know how one who has spent a life-time in striving with all his might to clear up and rectify its fundamentals looks upon his own branch of science. The way in which he regards its past and present may depend too much on what he hopes for the future and aims at in the present; but that is the inevitable fate of anybody who has occupied himself intensively with a world of ideas. The same thing happens to him as to the historian, who in the same way, even though perhaps unconsciously, groups actual events around ideals which he has formed for himself on the subject of human society.

Let us now cast an eye over the development of the theoretical system, paying special attention to the relations between the content of the theory and the totality of empirical fact. We are concerned with the eternal antithesis between the two inseparable components of our knowledge, the empirical and the rational, in our department.

We reverence ancient Greece as the cradle of western science. Here for the first time the world witnessed the miracle of a logical system which proceeded from step to step with such precision that every single one of its propositions was absolutely indubitable—I refer to Euclid's geometry. This admirable triumph of reasoning gave the human intellect the necessary confidence in itself for its subsequent achievements. If Euclid failed to kindle your youthful enthusiasm, then you were not born to be a scientific thinker.

But before mankind could be ripe for a science which takes in the whole of reality, a second fundamental truth was needed, which only became common property among philosophers with the advent of Kepler and Galileo. Pure logical thinking cannot yield us any knowledge of the empirical world; all knowledge of reality starts from experience and ends in it. Propositions arrived at by purely logical means are completely empty as regards reality. Because Galileo saw this, and particularly because he drummed it into the scientific world, he is the father of modern physics—indeed, of modern science altogether.

If, then, experience is the alpha and the omega of all our knowledge of reality, what is the function of pure reason in science?

A complete system of theoretical physics is made up of concepts, fundamental laws which are supposed to be valid for those concepts and conclusions to be reached by logical deduction. It is these conclusions which must correspond with our separate experiences; in any theoretical treatise their logical deduction occupies almost the whole book.

This is exactly what happens in Euclid's geometry, except that there the fundamental laws are called axioms and there is no question of the conclusions having to correspond to any sort of experience. If, however, one regard Euclidean geometry as the science of the possible mutual relations of practically rigid bodies in space, that is to say, treats it as a physical science, without abstracting from its original empirical content, the logical homogeneity of geometry and theoretical physics becomes complete.

We have thus assigned to pure reason and experience their places in a theoretical system of physics. The structure of the system is the work of reason; the empirical contents and their mutual relations must find their representation in the conclusions of the theory. In the possibility of such a representation lie the sole value and justification of the whole system, and especially of the concepts and fundamental principles which underlie it. These latter, by the way, are free inventions of the human intellect, which cannot be justified either by the nature of that intellect or in any other fashion *a priori*.

These fundamental concepts and postulates, which cannot be further reduced logically, form the essential part of a theory, which reason cannot touch. It is the grand object of all theory to make these irreducible elements as simple and as few in number as possible, without having to renounce the adequate representation of any empirical content whatever.

The view I have just outlined of the purely fictitious character of the fundamentals of scientific theory was by no means the prevailing one in the eighteenth or even the nineteenth century. But it is steadily gaining ground from the fact that the distance in thought between the fundamental concepts and laws on one side and, on the other, the conclusions which have to be brought into relation with our experience grows larger and larger, the simpler the logical structure becomes—that is to say, the smaller the number of logically independent conceptual elements which are found necessary to support the structure.

Newton, the first creator of a comprehensive,

workable system of theoretical physics, still believed that the basic concepts and laws of his system could be derived from experience. This is no doubt the meaning of his saying, *hypotheses non fingo*.

Actually the concepts of time and space appeared at that time to present no difficulties. The concepts of mass, inertia and force, and the laws connecting them seemed to be drawn directly from experience. Once this basis is accepted, the expression for the force of gravitation appears derivable from experience, and it was reasonable to hope for the same in regard to other forces.

We can indeed see from Newton's formulation of it that the concept of absolute space, which comprised that of absolute rest, made him feel uncomfortable; he realized that there seemed to be nothing in experience corresponding to this last concept. He was also not quite comfortable about the introduction of forces operating at a distance. But the tremendous practical success of his doctrines may well have prevented him and the physicists of the eighteenth and nineteenth centuries from recognizing the fictitious character of the foundations of his system.

The natural philosophers of those days were, on the contrary, most of them possessed with the idea that the fundamental concepts and postulates of physics were not in the logical sense free inventions of the human mind but could be deduced from experience by "abstraction"—that is to say by logical means. A clear recognition of the erroneousness of this notion really only came with the general theory of relativity, which showed that one could take account of a wider range of empirical facts, and that too in a more satisfactory and complete manner, on a foundation quite different from the Newtonian. But quite apart from the question of the superiority of one or the other, the fictitious character of fundamental principles is perfectly evident from the fact that we can point to two essentially different principles, both of which correspond with experience to a large extent; this proves at the same time that every attempt at a logical deduction of the basic concepts and postulates of mechanics from elementary experiences is doomed to failure.

If, then, it is true that this axiomatic basis of theoretical physics cannot be extracted from experience but must be freely invented, can we ever hope to find the right way? Nay more, has this right way any existence outside our illusions? Can we hope to be guided in the right way by experience when there exist theories (such as classical mechanics) which to a large extent do justice to experience, without getting to the root of the matter? I answer without hesitation that there is, in my opinion, a right way, and that we are capable of finding it. Our experience hitherto justifies us in believing that nature is the realization of the simplest conceivable mathematical ideas. I am convinced that we can discover by means of purely mathematical constructions the concepts and the laws connecting them with each other, which furnish the key to the understanding of natural phenomena. Experience may suggest the appropriate mathematical concepts, but they most certainly cannot be deduced from it. Experience remains, of course, the sole criterion of the physical utility of a mathematical construction. But the creative principle resides in mathematics. In a certain sense, therefore, I hold it true that pure thought can grasp reality, as the ancients dreamed.

In order to justify this confidence, I am compelled to make use of a mathematical conception. The physical world is represented as a four-dimensional continuum. If I assume a Riemannian metric in it and ask what are the simplest laws which such a metric system can satisfy, I arrive at the relativist theory of gravitation in empty space. If in that space I assume a vector-field or an anti-symmetrical tensor-field which can be inferred from it, and ask what are the simplest laws which such a field can satisfy, I arrive at Clerk Maxwell's equations for empty space.

At this point we still lack a theory for those parts of space in which electrical density does not disappear. De Broglie conjectured the existence of a wave field, which served to explain certain quantum properties of matter. Dirac found in the spinors field-magnitudes of a new sort, whose simplest equations enable one to a large extent to deduce the properties of the electron. Subsequently I discovered, in conjunction with my colleague, that these spinors form a special case of a new sort of field, mathematically connected with the four-dimensional system, which we called "semivectors." The simplest equations to which such semivectors can be reduced furnish a key to the understanding of the existence of two sorts of elementary particles, of different ponderable mass and equal but opposite electrical charge. These semivectors are, after ordinary vectors, the simplest mathematical fields that are possible in a metrical continuum of four dimensions, and it looks as if they described, in an easy manner, certain essential properties of electrical particles.

The important point for us to observe is that all these constructions and the laws connecting them can be arrived at by the principle of looking for the mathematically simplest concepts and the link between them. In the limited nature of the mathematically existent simple fields and the simple equations possible between them, lies the theorist's hope of grasping the real in all its depth.

Meanwhile the great stumbling-block for a field-theory of this kind lies in the conception of the atomic structure of matter and energy. For the theory is fundamentally non-atomic in so far as it operates exclusively with continuous functions of space, in contrast to classical mechanics, whose most important element, the material point, in itself does justice to the atomic structure of matter.

The modern quantum theory in the form associated with the names of de Broglie, Schrödinger, and Dirac, which operates with continuous functions, has overcome these difficulties by a bold piece of interpretation which was first given a clear form by Max Born. According to this, the spatial functions which appear in the equations make no claim to be a mathematical model of the atomic structure. Those functions are only supposed to determine the mathematical probabilities of the occurrence of such structures if measurements were taken at a particular spot or in a certain state of motion. This notion is logically unobjectionable and has important successes to its credit. Unfortunately, however, it compels one to use a continuum the number of whose dimensions is not that ascribed to space by physics hitherto (four) but rises indefinitely with the number of the particles constituting the system under consideration. I cannot but confess that I attach only a transitory importance to this interpretation. I still believe in the possibility of a model of reality—that is to say, of a theory which represents things themselves and not merely the probability of their occurrence.

On the other hand it seems to me certain that we must give up the idea of a complete localization of the particles in a theoretical model. This seems to me to be the permanent upshot of Heisenberg's principle of uncertainty. But an atomic theory in the true sense of the word (not merely on the basis of an interpretation) without localization of particles in a mathematical model, is perfectly thinkable. For instance, to account for the atomic character of electricity, the field equations need only lead to the following conclusions: A portion of space (three-dimensional) at whose boundaries electrical density disappears everywhere, always contains a total electrical charge whose size is represented by a whole number. In a continuum-theory atomic characteristics would be satisfactorily expressed by integral laws without localization of the formation entity which constitutes the atomic structure.

Not until the atomic structure has been successfully represented in such a manner would I consider the quantum-riddle solved.

One process can cause another; that one in turn, can be the cause of a further sequence of events—including the modification of the original process itself. This article is a primer to basic ideas in applied science, engineering, and information theory.

3 Systems, Feedback, Cybernetics

V. Lawrence Parsegian, Alan S. Meltzer, Abraham S. Luchins, K. Scott Kinerson

From the textbook, *Introduction to Natural Science*, 1968.

THE READER will recall that following the quotation from Teilhard de Chardin in Chapter 1, we proposed extending the scope of our interests to include analysis of *relationship* and *interrelationship of natural phenomena* to each other. We have come to a point that requires a more formal development of such interrelationships.

6.1 Extension of "systems"

One of the accomplishments of the Newtonian period was the strengthening of the concept that in material or physical situations at least, things do not happen without a causing force. A stone does not begin to move or come to a stop of its own volition. In this chapter we shall utilize that concept, but with three extensions.

The first extension takes into account the fact that in most situations surrounding an event (such as the hurling of a stone), the immediate event is itself part of a larger situation or *system* that includes various other articulating parts or related events. (That is, there is a person who throws the stone, and the throwing has relation to some cause or purpose.)

The second extension may perhaps be thought of as related to the action-reaction principle, namely, that within the context of the system involving an event (a stone is thrown) there is often a *feedback* effect (for example, the one at whom the stone is thrown may hurl it back).

The third extension includes in the system both material things (stones) and human beings along with biological processes and the less tangible thought processes.

What do we mean by the term *system*? We might refer to the weight suspended from a spring as a system that executes simple harmonic motion. The governor that controls the speed of an engine is a control system. We also speak of a *system of highways,* the *economic system* of a nation, a *system of thought,* and of many others. The combination locks that protect the vault of a savings bank make up a protective system, but this can also be said to be only a subsystem of the banking institution. The banking institution is itself only a subsystem within the larger community economics, and the latter is a subsystem of national economics. The chain of larger and larger subsystems, or the nesting of subsystems within larger subsystems, may lead to very complex assemblies and relationships.

While an accurate, all-encompassing definition for the term is not easy to give, we can note a few of the characteristics that are usually present in what we call a system:

(1) A system is likely to have two or more parts, elements, or aspects, which tend to have some functional relation to each other (like the bolt and key of the lock, or the president and staff of the bank).

(2) Because systems are usually subsystems of larger units it is usually helpful (and often necessary) to confine one's study to the smallest unit that encompasses the particular functional elements and interrelationships that are under study. (For example, the locksmith can quite properly repair a fault in the lock system of the bank vaults without considering the question of the merits of socialism for the nation's banking system.)

(3) A *control system* has within itself regulatory functions for control of variables such as speed of a motor, the temperature of a room, the price of commodities, or international trade in narcotics.

(4) It is usually possible to identify an "input" and an "output" portion (or aspect) of a system. For example, a key placed in a lock and turned (input) will cause the bolts to move (output); or an order from a president of an industrial firm (input) can double the selling price of its commercial products (output). We shall find, however, that most systems have more than one form of input, as well as a variety of functional relationships that produce quite varied output.

(5) Usually (nearly always in systems that include regulatory functions) there is some form of *feedback* from the output to the input, which may greatly modify the net output of the system. [For example, when the selling prices of the commercial products of paragraph (4) were doubled, the consumers could have initiated strong feedback by refusing to buy the products; and the industry's board of directors could have exerted even stronger feedback by firing the president and hiring another who would hold the prices at a more acceptable level.] The role of feedback will be given considerable attention in the discussion that follows.

We shall now turn to a more detailed introduction to systems, feedback, and control.

6.2 *Cyclic character of natural phenomena*

In Chapter 5 we learned that a mass suspended from a spring executes simple harmonic motion when displaced slightly from its equilibrum position. When the motion was recorded on a moving sheet of paper (to illustrate the motion as a function of time), the oscillations were recorded as sine or cosine waveforms. It was shown that the motion was initiated when *potential energy was added to the system* of weight and spring (by manually raising the weight from its rest position, against the pull of gravity, or by pulling it down and extending the spring). In either case, the pull of gravity or the pull of the spring alternately introduced a *restoring force,* which tended to return the displaced *mass* to its original position (Fig. 5.24). But since force applied to mass accelerates the mass and thereby increases its velocity (Eq. 5-1), by the time the mass reached the "zero" or initial position it had acquired so much *velocity* (because the *potential energy* we added manually had become kinetic energy at that point) that the mass moved past the zero point to the other extreme. There would have been few or no oscillations at all, on the other hand, if the weight had been subjected to so much frictional drag that the added (potential) energy was lost as heat.† (This might have been the case if the weight moved in a viscous liquid.)

What about cyclic behavior in other phenomena of nature? A very common form can be demonstrated in electric circuits in which the electric energy rapidly passes back and forth between parts of an oscillating circuit until the electric energy dissipates as heat or radiates away from the circuit (as in the transmission of radio waves).

We shall find that there can be many forms of oscillatory behavior when a "disturbance" changes the energy level of a system and introduces a restoring force that causes the energy to convert to another form rather than completely dissipate into the heat energy of the environment. The term *energy* may apply not only to mechanical, electrical, or chemical characters in physical systems, but also to institutional and personal pressures in social situations.

Let us now go to other phenomena that show cyclic or periodic variation. (See Figs. 6.1(a) through 6.1(d), for graphical examples of such cyclic variations.) We might utilize various sensing devices to record changes in the temperature of an air-conditioned room as a function of time, the height of the tides of the sea, wind velocity, the automobile traffic on a road, rainfall, the movements of a tall building or of the long span of a bridge, or the temperature of the earth. We might also look up past statistics on wheat production, the stock market, attendance at church, tourist travel, populations of animals, or the length of women's skirts, and plot these in graph form as function of time. We would find that many phenomena in nature and in animal or social activity have variations of an oscillating character (Fig. 6.1). It can be demonstrated that in all these situations which show oscillations about some average point, there is present a *restoring force* that comes into play whenever there is energy change in a system. To be sure, the magnitude and shapes of these oscillations and waves vary considerably from the sine waves we observe with a weight on a spring. The periods may vary from 10^{-15} sec in the case of light waves, to several hours for the period of the tides, and to many years in the case of other cycles of nature and of some social customs. Nevertheless, all are subject to some common influence, not the least of

† We shall learn in Chapter 10 that the kinetic energy of the system goes into faster, random motion of the molecules that make up the parts of the system. The increased molecular motion raises the temperature of the parts of the system, as though it were heated by a flame. There is therefore a correspondence or equivalence between the energy in a flame and mechanical motion of the system.

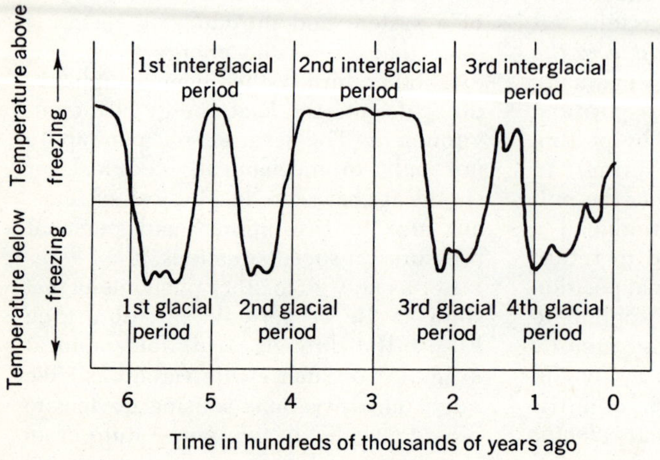

Fig. 6.1. (*a*) *Cyclic temperature variations during the ice ages. Current theory attributes these long, slow temperature variations to relatively minor changes in the atmospheric carbon dioxide content (see Chapter 15, Sec. 2). (Adapted from graph in G. H. Drury,* The Face of the Earth, *Penguin (Pelican book), pg. 157.)*

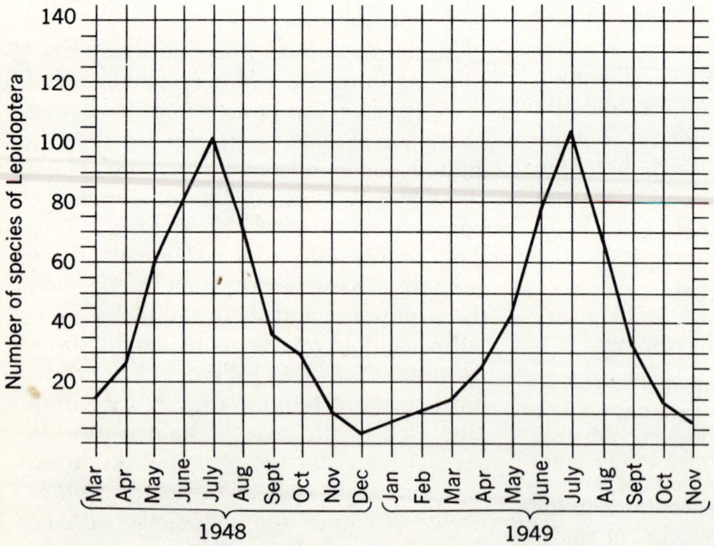

Fig. 6.1. (*b*) *Cyclic variations in numbers of species of Lepidoptera (butterflies and moths) captured in light traps at Woking, Surrey in 1948–49. The number of different species of captured reveals seasonal cyclic variations that are obviously related to weather conditions. Note peaks in successive Julys, when Lepidoptera conditions are ideal, and low values in winter when conditions are poor. (From C. B. Williams,* Patterns in the Balance of Nature, *Academic Press, 1964, pg. 159.)*

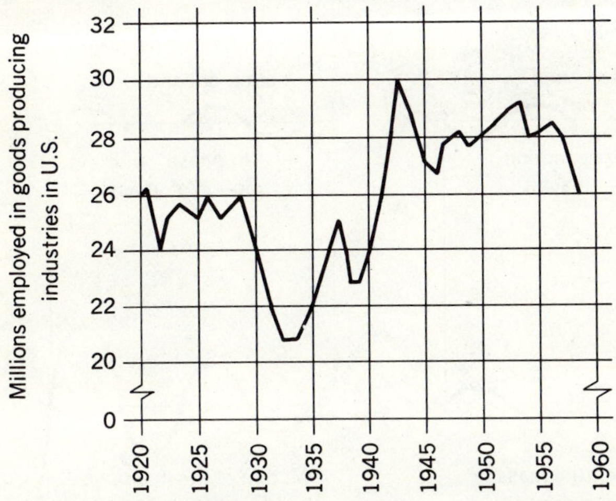

Fig. 6.1. (*c*) *Cyclic character of employment levels in U.S. goods producing industries, 1920–1960. Note large amplitude cycles superimposed on more normal fluctuations as a result of the depression in the early 1930's and of World War II during the early 1940's.*

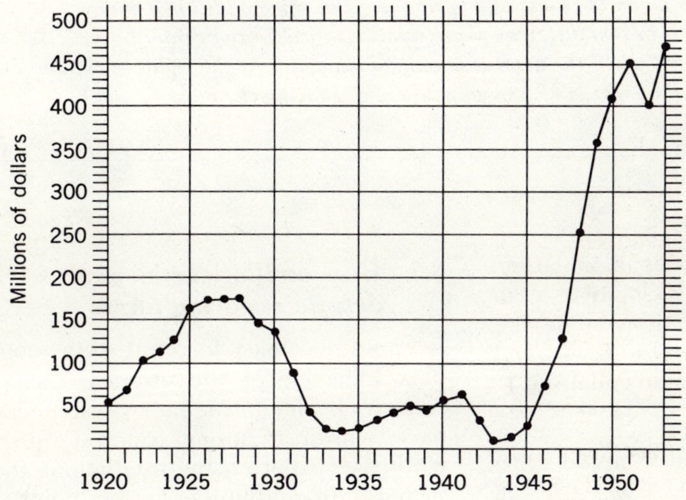

Fig. 6.1. (*d*) *Cyclic variation in value of new construction of religious buildings.*

which is the fact that nature is dynamic and in a state of continuous change, and indeed that static situations represent special and almost trivial aspects of nature and of man.

Is the presence of some restoring force sufficient assurance that a system will experience only moderate oscillations without going to extremes? Indeed it is not, as we can learn from the dramatic example of the failure of the Tacoma Narrows suspension bridge of Tacoma, Washington. When the bridge was opened to traffic on July 1, 1940, there were observed, in addition to the ordinary oscillations of the bridge, some unexpected

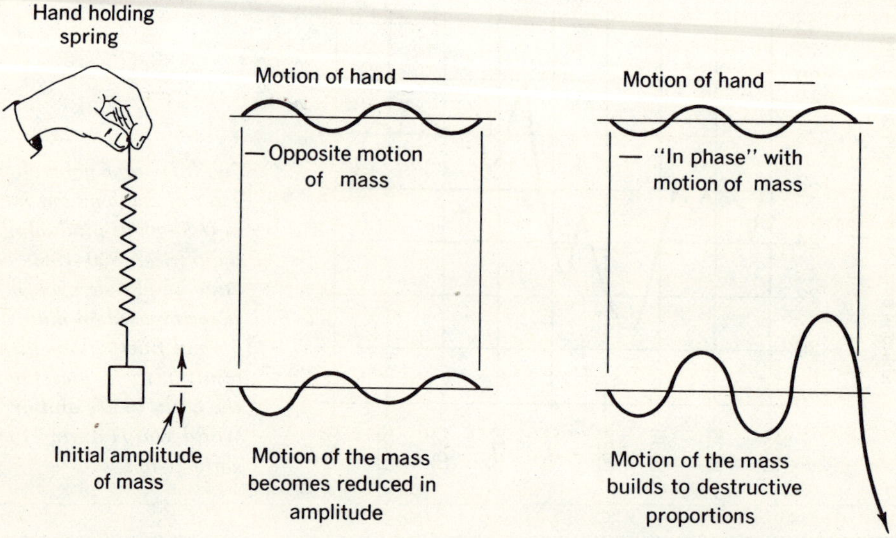

Fig. 6.2. What happens when the mass of a spring is given some additional energy by movement of the hand in two different phase relationships? In the center figure the hand is moved upward when the mass is moving downward. At the right, the hand is moved upward when the mass is also moving upward, causing the mass to take large swings.

transverse (vertical) modes of vibration. On November 7 a wind velocity of 40 to 45 mph made the vibrations so severe that the bridge was closed to traffic, and by 11:00 A.M. the main span collapsed.†

† A 4-minute film produced by the Ohio State University and distributed by The Ealing Corporation of Cambridge, Massachusetts, gives the very dramatic story of the final oscillation of the bridge prior to its collapse. Every reader should see this film and the variation it offers of "simple harmonic motion" involving the twisting and turning of this huge span of steel and concrete. The new bridge that was built on the original anchorages and tower foundations included deep stiffening trusses instead of girders, and has been entirely successful.

6.3 *How oscillations increase despite restoring forces*

It is not necessary to resort to the complex behavior of the original Tacoma Narrows bridge to see how a system may have within it strong restoring forces while yet experiencing oscillations that increase in amplitude to the point of destruction. The reader can duplicate the phenomenon with the simple weight on a spring as follows (Fig. 6.2):

Choose a weight and spring combination that gives an oscillatory period between $\frac{1}{5}$ and $\frac{1}{3}$ sec. Hold the spring firmly and steady in your hand, and observe that the weight executes the usual simple harmonic motion, eventually coming to a stop. Now prepare to move your holding hand up or down in synchronism with the motion of the weight and with two alternative movements.

First, *raise* your hand (about a half-inch will do) whenever the weight is moving *downward,* and lower it an equal distance whenever the weight is *moving up.* With a little analysis you can see that the weight tends to *reduce* amplitude because the movements of your hand *increase the restoring force* on the weight. Note that the movement of your hand is *180 deg out of phase* with the motion of the weight.

Next, repeat the experiment with the same up-and-down motion of your hand, but now change the timing to be *in phase* with the motion of the weight. That is, move your hand upward when the weight is moving upward, and downward when the weight is moving downward. There still is restoring force, and the weight continues to oscillate up and down; but now the amplitude of oscillations *increases* until it becomes dangerous to continue the experiment.

Why did the same *amount* of motion of your hand have such opposite effects, depending only on its phase relationship to the motion of the weight? The reason is that in the second case the increments of energy that were introduced by each *in-phase* motion of your hand tended to *add to and increase* the energy of the system represented by the spring and weight. Conversely, the hand motion that was completely out of phase with the motion of the weight detracted from the energy of the system.†

We can now extend this experiment to apply to the early Tacoma Narrows bridge experience. Obviously, the energy of the wind became converted to energy of oscillation of the bridge. Why did the wind energy not become absorbed in the concrete and steel of the bridge? Undoubtedly much of it did become absorbed and changed to heat energy, but not all of it. Apparently when the wind blew to produce a movement of the span at some point along the bridge, the conditions were just right to cause this movement to act as a traveling wave, which on backward reflection returned to the same point in just the right phase to support (rather than oppose) a new movement at that point, caused by the continued blowing of the wind. Had the physical structure of the bridge been different in length or mass, the returning wave could have opposed (out of phase with) any new movement at *A*, and thus would have added to the stability of the system.

We see, therefore, that for a system to be *stable,* the relationship of the forces and time characteristics must be such that *the amplitude and energy of the system will not increase.* This calls for special attention with respect to the *phase relationships* that obtain between feedback of energy from one part of the system to another part. When the *feedback opposes* the direction of the initial change that produced the feedback, the system tends to be *stable.* In contrast, when the returning *feedback* of energy *supports* the direction of initial change, the system tends to add to the initial energy gain and to be *unstable.* This means we must delve into the theory of system control.

† The reader is urged to perform this experiment and to attempt a careful analysis of the various factors (energies and forces) that become involved in the two cases. For example, in the second case the increments of energy are added to the spring-weight system. Where does the hand energy go in the first case?

6.4 *Modifying cyclic changes: controls*

While most fluctuations of nature go their own way without inviting human concern, there are some important cases in which it becomes necessary to interfere, that is, to modify the natural pattern or to control or hold the fluctuations to smaller changes. For example, the farmer may not want to depend entirely on natural rainfall to assure a good crop, so he intervenes by irrigating the fields when there is not enough rainfall. Because in the course of the year there are wide fluctuations in the temperature of the earth, he installs a control system in his home to keep the temperature within comfortable limits.

Many types of controls are involved in our daily life. We shall learn that the human body has a remarkable control system to maintain its own temperature within very close limits. The body's motor functions, by which we move our arms and legs in an accurate and determined manner, are possible only because of the operation of fine control systems. Industrial production relies heavily on control of temperature, pressure, chemical composition, and similar factors. The application of control principles extends to community and national life. Despite their variety, we shall find that there are some common characteristics among them. Also, within a specific control system there can be intermixed a wide variety of elements of widely different types. Take, for example, the very common experience of driving an automobile. Here, the steering control allows the driver to follow the curvature of the road effectively, and many other electromechanical parts as well in the motor and transmission systems affect the driving operation. But we shall learn before long that nearly every aspect of the driver's being—his metabolism, muscle and nerve action, his thinking process—and the life of his community are all parts of the system that encompasses the simple driving experience.

6.5 *Introduction to on-off control*

We return to the harmonic motion of the weight suspended from a spring and note that, so far, we have neither tried to restrict the amplitude of the motion nor put the movement to some useful application. In each assembly the added energy is converted and reconverted from kinetic energy to potential energy and then back again to kinetic energy. (If there were no frictional losses, the motion would continue forever, since the system would then be self-contained, that is, a closed or isolated system that neither receives energy from nor gives energy to the outside.) Such systems have limited value except as one may use them in a clock or metronome to tell time from the oscillations.†

If there were no frictional or other loss of energy from the system, the motion would have a periodicity of T seconds. Since friction is present, the oscillations become continually smaller in magnitude, and the period of each cycle becomes slightly longer ($T + \Delta T$) until the mechanical energy dissipates as heat energy and the movement ceases altogether (Fig. 6.3). In general, friction or damping is likely to make a system more stable.

We can design an oscillator to do some-

† Of course, as any such device requires periodic additions of energy to the driving springs, and therefore the person who winds the spring becomes part of the system.

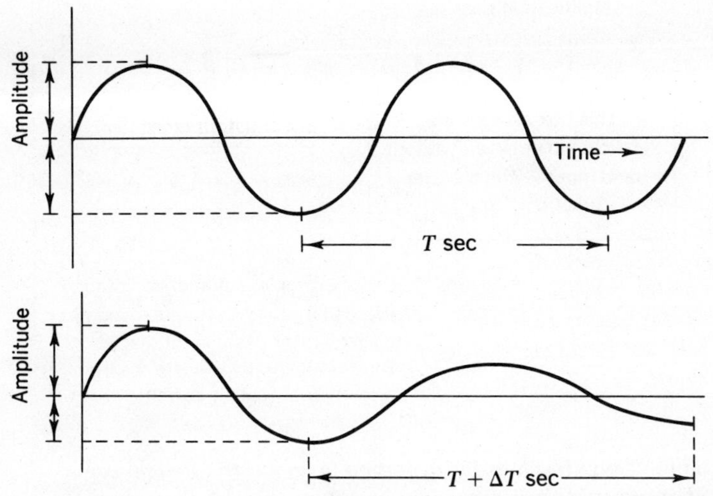

Fig. 6.3. How the period of simple harmonic motion changes when there is friction in the system. (The period of seconds increases to $T + \Delta T$ sec, while the amplitude of motion decreases.)

thing more by adding an electric switch so that the dropping weight sends an electrical signal to some device. As we know from common experience, the simple operation of an electric switch can initiate (or trigger) many motor or relay functions that bring into play the vast energy resources of electric power-generating stations. Figure 6.4 illustrates the relationship between input and output, with a transform function that relates the two along with a source of energy.

Suppose that we incorporate such an electric switch as part of a control system for automatically filling a bucket with water. Figure 6.5 illustrates how the dropping pail signals that the pail is full and also turns off the stream of water. This becomes a simple on-off control system in which the electrical signal provides a *feedback* function as part of the control system. (Later we shall introduce the idea that the feedback also represents *information*.)

We examine this process of filling the bucket in a little more detail. When the water flows into the bucket at a very slow rate, the bucket settles slowly and the signal switch has time to stop the flow of water and bring the bucket to a gentle stop. This is shown as curve *A* of Fig. 6.6, which shows very little dropping of the bucket below the desired level (that is, there is very little overshoot beyond the desired control point). The behavior becomes quite different when the water flows into the bucket at a rapid rate, however. The switch operates as it did before, but the rapid dropping of the bucket develops enough momentum to overshoot the desired final position by a substantial amount. The bucket will oscillate violently above and below the desired control height for some time and the switch will open and close erratically (curve *B*). In fact, if the response rates and delays in the switching and valve devices should turn out to be particularly

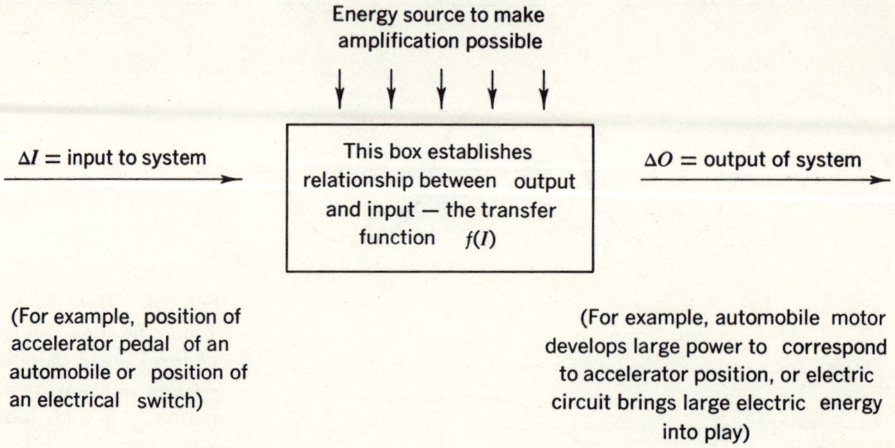

Fig. 6.4. *How a small* input *change (such as the operation of an electric switch) can bring into play sources of energy and thereby produce an* output *that may be quite different in form and magnitude from the input. Each such conversion can be referred to as involving a transformation (transfer function or transform function).*

unsuitable, the water would be turned on and off in such erratic manner as to recall the sad fate of the Tacoma Narrows bridge; see curve *B*, dotted line, Fig. 6.6.

In the case of room-temperature control, the thermostat is likely to be kept at one temperature (for comfort), say, around 72°F. In the case of the baking oven, the temperature setting will vary with the requirements for baking a cake or roasting meat. In either case, the temperature will vary (or hunt) around the set control point. The hunting or oscillations can be decreased if the rate of heat input is slow. But this would increase the time needed to bring the room or oven to the desired temperature. With on-off control, the heating unit becomes fully hot whenever the control switch turns it on. By the time the temperature at the thermostat reaches the desired temperature to turn off the heat, the region of the heater units becomes much hotter than necessary, and this excess heat drives the temperature well above the desired temperature. A similar delay in reactivating the heating unit as the temperature drops below the desired level causes continual hunting above and below the desired temperature.

We shall appreciate more and more, as we examine more cases, that the "control" of a variable rarely results in an exact holding of the variable to the desired control value. *Nearly always, the variable will hunt or vary about that control value.* Therefore, the function of a successful control system is to hold the variable *within acceptable departures from the desired control value.*

6.6 *Negative versus positive feedback*

In all the examples given above, while it is clear that control at a point usually ends up as hunting around that point, even this

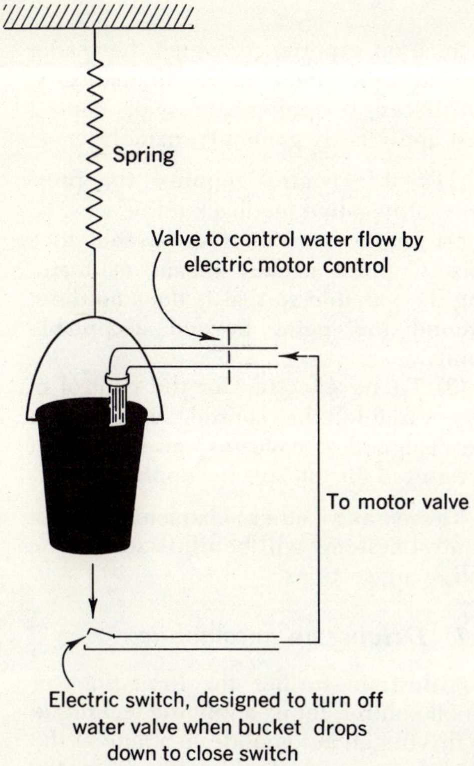

Fig. 6.5. *A simple system for controlling the filling of a bucket.*

Fig. 6.6. *How the bucket of Fig. 6.5 behaves: Bucket A is filled slowly and settles gradually to its final level after switch cuts off water flow. Bucket B (solid line) is filled rapidly and overshoots final position, rebounds, and hunts for an equilibrium position that is lower than that for bucket A because extra water was added after the first rebound above the switch-off level. With a different spring tension for bucket B (dotted line), the hunting may cause addition of sufficient extra water on each cycle so that the amplitude steadily increases until the system collapses.*

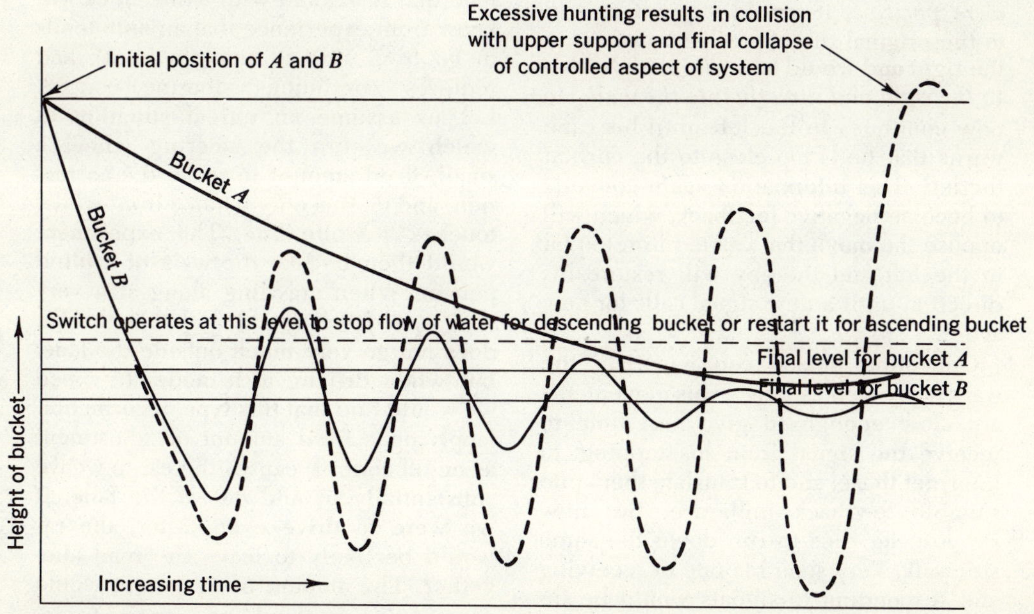

degree of control is achieved only when *negative feedback* is present. Thus, in the case of the full bucket, the switch turns off the water (since it was the "water-on" condition that filled the bucket). In the case of room-temperature control (which we shall discuss presently in detail), the heaters must be turned *on* when the room temperature is too low, and *off* when the temperature is too high.

The examples of feedback, as well as the limitations of on-off (sometimes called bang-bang) control can be illustrated further by the example of a blind person walking down a street with his cane. As he progresses along the sidewalk the tapping of his cane tells him when he is too close to the buildings on the right. This *information,* when processed through his brain and muscle system, serves as *feedback* to change his direction. Since his movements have taken him too far to the right, now he must move to the left and therefore the *feedback must be negative.* If the influence of feedback *were positive,* it would support or add to the original direction that took him to the right and would take *him even farther to the right and directly into the wall.* He now continues to the left until his cane warns that he is too close to the curb at the left. This information again converts to become negative feedback, which will oppose the move that carried him too far to the left and thereby will restore his direction until a new signal calls for new action.

Our blind person can negotiate the walk fairly well as long as his movements are slow enough to give him time to receive the signal from his tapping, to interpret these, and to translate them into suitable feedback influence. But now suppose he tries to run down the same sidewalk. Very soon his rate of receiving and responding to signals would be inadequate, and he would be running in a zigzag or colliding with obstacles.

Such an experience, which the reader can himself check rather dramatically, illustrates several features of control that apply fairly generally, namely:

(1) Stable control requires the presence of negative feedback influences.

(2) Stable control of a variable to a "fixed" point usually means maintaining the variable so that is does not hunt around the point beyond acceptable limits.

(3) To be effective for the control of any variable, the control system must be designed to have response rates that are suited for the specific application.

These and other characteristics of control systems will be illustrated in the following sections.

6.7 *Driving an automobile*

To illustrate further the limitations of on-off control, let us apply the technique to driving an automobile in a lane of the road that is marked with white lines. We know from experience that an auto tends to go from side to side (to hunt), and requires continuous steering control. Let us assume an unreal situation in which we turn the steering wheel a small, fixed amount to make the correction, and do this only when a front wheel touches a white line. The experiment would then be like the walk of a blind person. When crawling along at a very slow speed we would find that the car does not go very much outside the lane, but when driving at a moderate speed we would find that this type of correction (applying a fixed amount of adjustment as on-off control) causes the car to weave substantially in and out of the lane. If we were to drive even faster, the car would be likely to leave the road altogether. The amount of overshoot would depend on how slowly we respond to visual signals and take action (see Fig. 6.7).

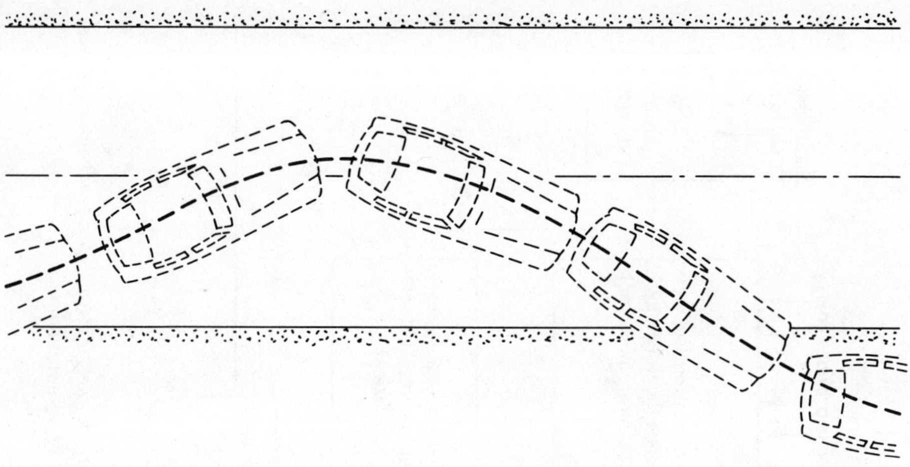

Fig. 6.7. Difficulty of driving an auto by on-off control technique.

Fortunately not many people drive in this manner because control of an automobile utilizes a much more sophisticated system of elements than is possible with on-off control. In fact, not many automatic industrial processes can compare with the sophistication and effectiveness of good auto driving, since human judgment enters this operation to a remarkable degree. To begin with, as the auto moves to a new position or direction, the driver is kept continually informed of the nature of each new situation through his sense of sight and general physical awareness. That is, there is continuous feedback, or information, reaching him to guide his next move. The element of judgment or experience also enters. He can vary the sharpness of turn of the steering wheel to conform to the sharp right turn. This is called *proportional control*. In addition, he can see a curve in the road ahead long before the auto has reached the curve. He can therefore anticipate the move (*anticipatory control*) and thus reduce delay in his action (Fig. 6.8).

The driver of an automobile is aware of several elements that make control more difficult. If the steering wheel has looseness or "play" in the shaft or gear system, the steering wheel must be turned several degrees of angle before there is any effect on the front wheel directions. This play, or region of no response, is sometimes called the *dead zone* of the system. The driver himself may be a little slow in judging the situation and taking action. This "lag or slowness of response together with looseness in the steering system, can make for wider overshoot in the movement of the car. If the throttle sticks, the motor hesitates, or the brakes seize, the driver will not be able to assure smooth "feel" and ride. Finally, roughness of the road can introduce random fluctuations that add uncertainty to the normal small feedback of information. A driver is not likely to give delicate

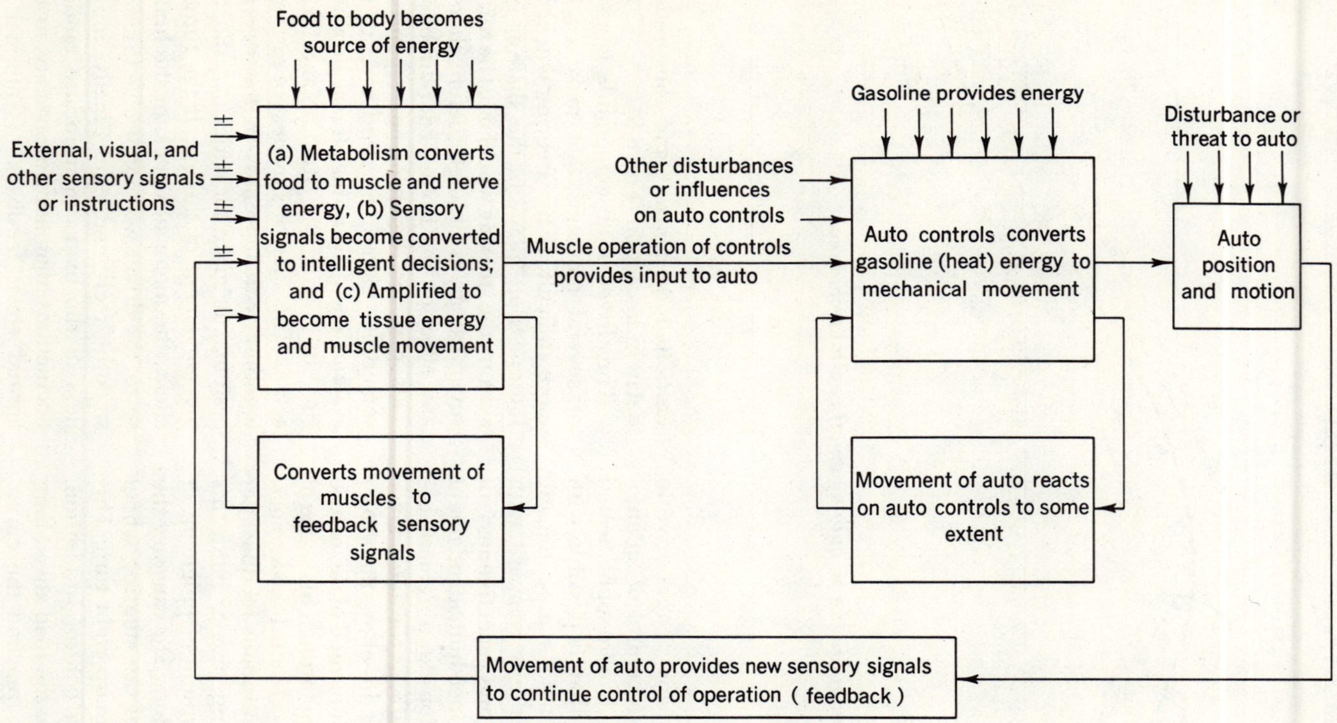

Fig. 6.8. *The input, output, energy source, and feedback influences that bear on the driving of an automobile.*

guidance to the auto when his whole body is being shaken. This background confusion is often called *noise* or *static* when one is referring to transmission of signal or of information. It exists in almost every type of control circuit, sometimes in the form of vibration of an automobile or plant equipment. It occurs in the normal radioactivity background of the environment, which disturbs radiation measurements. In very sensitive electronic circuitry it shows up in the random movements of the electrons. Similar phenomena are present in social situations and in biological organisms that maintain balance in their internal functions and with their environment.

Because cyclic and control aspects of nature are exceedingly important, we must consider control principles and nomenclature in a little more detail before looking at the several types of systems that are common.

6.8 *Some control principles—nomenclature*

Before beginning detailed discussion we need some convenient terminology and symbols for representing the elements and functions that make up systems. When the driver of an automobile engages the gears and steps on the accelerator pedal, the motor races and the car moves forward with an expenditure of energy that is vastly greater than the energy applied to the pedal. The power is amplified. We may represent this by a diagram such as Fig. 6.4. The input, ΔI in this case, appears to be simply the change in position of the pedal and the small energy required to make the change. The box in Fig. 6.4 represents the change or transformation (of function) that the input ΔI initiates or experiences; in this case the function produces motor power at a level that is related (and perhaps proportional) to the position of the throttle or accelerator pedal. We can refer to the box as representing a transfer function $f(I)$, or converter, which produces ΔO.

What is the source of the energy that makes this conversion possible? In this case, the energy source is the chemical energy in the tank of gasoline. The pedal, therefore, is nothing more than a lever device for controlling the use of this chemical energy. When one includes the tank of gasoline and the driver along with the automobile, the system becomes a *closed* (or *conservative*) *system.*† Without either one, the system would be incomplete. (It is common practice to omit the *sources* of energy from block diagrams of control systems and to indicate only the energy input and output for a system.)

The operator has freedom to depress the accelerator pedal quickly or slowly, as he wishes. A question that is frequently important for analyzing the behavior of control systems is the following: What is the nature of the output response wher the input is given a quick change? A quick increment of input change, which we may represent by ΔI, is usually called an input *step function.*‡ Figure 6.9 illustrates what might happen. Usually there is some lag in the rise of motor power, as shown by the curved rise and fall of the output. This lag will not be serious in the case of the automobile, since the input is not likely to be reversed rapidly very often. In general it is desirable to have as much lag as one can tolerate, consistent with adequate control; otherwise the system will be too ready to "jump" and probably to overshoot the

† We neglect the fact that as gasoline is used up it must be replaced, bringing the entire petroleum industry into our system. Likewise, food for the driver is neglected.

‡ The term *step function* is often associated with *on-off* changes because the change of power or direction assumes the form of a sudden change. This is illustrated by the shape of the heat input as it is turned on and off in Figs. 6.9 and 6.10.

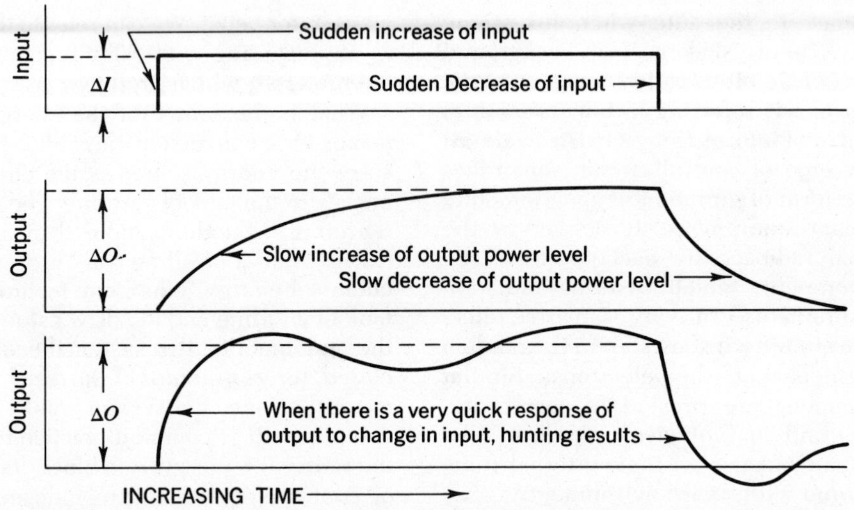

Fig. 6.9. When the input is given a quick change (step function), the response of the output may be designed to be slow or rapid. In general, a slow response of output produces less hunting than does rapid response.

mark and hunt badly before settling down. The lower curve of Fig. 6.9 illustrates the nature of the "hunting" that results when the system is made to respond too quickly to change in input. How the output will respond depends on the characteristics of the system and on the features incorporated in the transform function box of Fig. 6.4.

If the system of moving parts includes large, heavy components such as the flywheel and other parts of the automobile motor, we can appreciate that quicker response is possible only if the motor is designed to have adequate extra power to give the desired acceleration. But excessive power can make control less smooth and more "jumpy," not to mention excessively costly in gasoline and in the complexity of the motor itself. The goal for design of most systems is to find a happy compromise that makes the system *adequately responsive and yet stable against excessive hunting, and which is not too expensive in dollars or in use of energy.*

The system we have been discussing has the features of *proportional control.* That is, the accelerator pedal may be depressed to give large or small change, and the motor power level will respond with some proportional relationship. We backtrack a little to discuss *on-off control* before proceeding further.

6.9 *More on on-off control*

Earlier we discussed how difficult it would be to hold an automobile within the lane of the road if we applied on-off control principles to adjust the steering. Despite certain limitations, on-off control devices are used very commonly in homes and in industry because of their simplicity. It is very easy to design an electric iron, an oven, or room-temperature control, to operate an electric switch to turn on (or off) the electric power whenever the temperature falls below (or rises above) set values. Figure 6.10 illustrates how this might apply to the thermostat controls for heating a room in the winter time.

As shown in Fig. 6.10(a), 72°F is the temperature desired for this room. But all thermostats and switching devices require a differential zone of temperature change in which to go on and off; otherwise they would act too frequently and probably erratically because of vibration conditions and momentary temperature fluctuations in the immediate neighborhood of the thermostat. We start with the temperature dropping in the upper curve of this figure. When the temperature reaches the lower edge of the differential

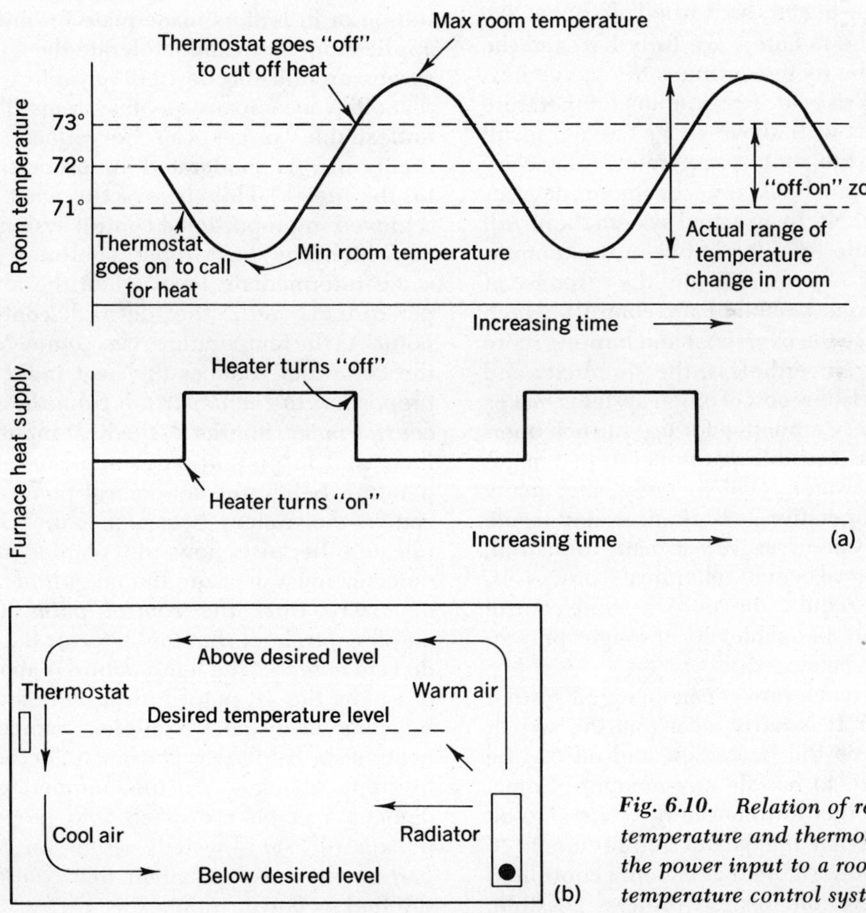

Fig. 6.10. Relation of room temperature and thermostat to the power input to a room-temperature control system.

temperature zone (71°F in Fig. 6.10(a)), the thermostat switch turns on the heater. This assumes that there are no significant time delays in the response of the thermostat or the heater controls. (In actual experience there are always some delays.) The radiators around the room take much more time to heat up, and the temperature of the air in the room continues to drop until it reaches some point which is well below the lower limit of the control range (about 70° in Fig. 6.10(a)).

As the hot radiators heat the air in the room, the temperature at the thermostat starts to climb again, and at the 73°F level the heaters are turned off. But at that point the radiators are fully hot, and the air in the room continues to receive heat and to rise to a maximum temperature which is well above 73°F. The net result is that the room temperature may vary by as much as four or more degrees Fahrenheit. In an actual system there will be a little time lag between the temperature at 71°F or 73°F and the response of the thermostat and heater controls, which can make the overshoot and hunting more severe. Nevertheless, the simplicity and relatively low cost of on-off systems makes them very attractive for use in such operations as controlling temperatures, maintaining water level in tanks, and many other operations. Biological and some social systems, as well as many industrial, mechanical, and chemical processes, usually require the more accurate control that can be achieved through proportional-type systems.

How much power can an on-off system control? It is fairly clear that the switch that turns the heater on and off can be designed to handle any amount of electric or other form of energy. The amount depends on the power requirements to keep the variable that is being controlled as close to the desired value as possible. A general rule might be to design the power level so that the controller calls for heat about half the time, and the heater remains off half the total time. Sometimes the control is improved by supplying a portion of the power continuously at a low, fixed level, and allowing the control system to add or subtract a smaller increment of power as needed.

6.10 Characteristics of proportional control

The on-off type of temperature control, in which the power is usually turned full-on or full-off, is inadequate for many applications that cannot tolerate the wide surges around the desired control point that often accompany on-off systems. The undesirable surges can be reduced if the power is moderated in proportion to the need. This is exactly what is achieved in proportional control systems, in which the heat input continues at some intermediate level when the temperature is near the desired control point. As the temperature rises somewhat, the controller reduces the heat input *in proportion to the departure from the set control point.* Similarly, the heat input is increased in proportion to a fall in temperature below the set control point. Of course the system becomes more complicated because now the temperature detector must measure the *magnitude of departure from the control point.* (In on-off control, all that the detector has to do is to note that the temperature is above or below the set point.) Also, there must be somewhat more complex interconnection so that the proportionate (or step-by-step) changes in the temperature detector can be translated into proportionate (or step-by-step) action on the part of the valve or motor that controls the fuel or power input.

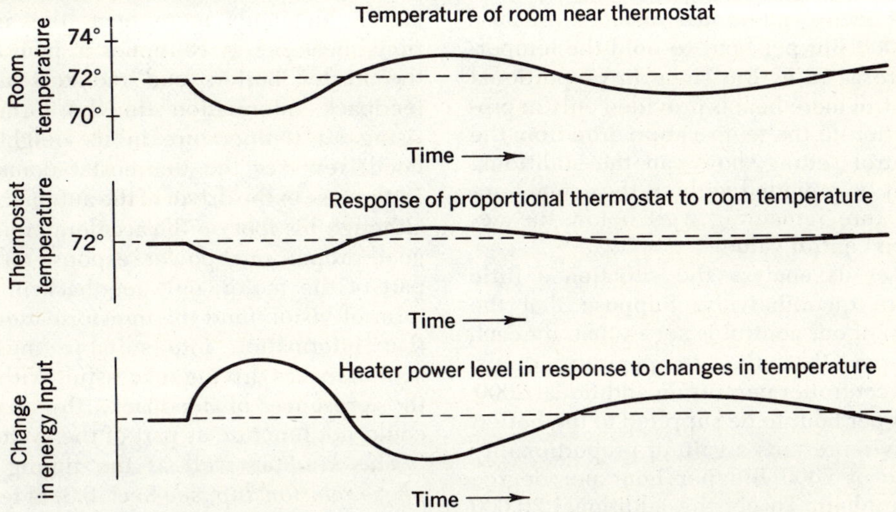

Fig. 6.11. In a proportional control system, the response of the thermostat is proportional to the departure of room temperature from the desired control point and the change in power input to the boiler is proportional to the response of the thermostat.

Let us analyze the action of such a system designed to control the temperature in a room.

When the door of the room opens and lets in a draft of cold air, the thermostat responds as shown by the drop in the upper curve of Fig. 6.11. As shown by the middle curve, at that same time the thermostat control calls for a proportional increase of heat, and the heaters respond as shown by the lower curve. As the draft of cold air becomes warmed somewhat by mixing with the warmer air, the proportional thermostat correspondingly reduces its demand for heat. The net result is that the room temperature is maintained much more closely to the desired 72°F than is possible with on-off control. But the proportional control instruments and equipment tend to be more expensive, and for that reason they are not used where on-off control is adequate.

A serious limitation develops in proportional control systems when the load demand changes so that a different average power level must be applied to hold the variable at the desired control value. To understand this, we note that in proportional control, the output ΔO (Fig. 6.4) has a fixed ratio to the input ΔI. This proportionality ratio, or gain, may be represented by $G = \Delta O/\Delta I$. Assume that the room-temperature control we have been discussing is set to control at 72°F *when the outdoor temperature is around 50°*F. We may assume that this requires an average heat input of 10,000 Btu per hour. Suppose that the outdoor temperature drops to 0°F. Obviously, the heater system must provide a great deal more heat to hold the temperature at

72°F, *say,* 30,000 Btu per hour. We therefore need an additional 30,000 − 10,000 = 20,000 Btu per hour to hold the temperature at 72°F. But since in proportional control more heat is provided only in proportion to the temperature drop from the control setting, how can the additional heat be provided without the actual temperature remaining well below the desired control value?

Let us analyze the situation a little more quantitatively. Suppose that the gain of our control is set so that, for each degree that the temperature drops, the controller permits an additional 2000 Btu per hour to be supplied to the boiler. This represents a gain or proportionality ratio of 2000 Btu per hour per degree fahrenheit. To get the additional 20,000 Btu would require that the temperature of the room go down to 62°F. Or, alternatively, the thermostat setting would have to be moved arbitrarily to about 80°F in order to supply enough heat to hold the room temperature at 72°F as long as the outdoor temperature remained at zero.

This discrepancy could be reduced if the gain were made higher (that is, 1°F could turn on much more than an additional 2000 Btu per hour). But making the gain higher also makes the system more unstable. Other devices can be introduced to change the responsiveness of the controller, such as incorporating into the system an outdoor thermostat that introduces this equivalent of the arbitrary shift of a thermostat setting of 80°F. We need not go into more detail beyond recognizing this severe limitation of proportional control systems.

6.11 Feedback

We must give a little more attention to the important feedback function. When the thermostat of a temperature-control system demands more heat, the additional heat energy continues to pour into the heater boilers and radiators until feedback information (in the form of rising air temperature in its neighborhood) reverses the thermostat demand. In the case of the driver of the automobile, although his foot on the accelerator finds good proportional power response on the part of the motor, only feedback in the form of vision (and the transformation of that information into suitable muscle action) makes driving successful. Without the presence of feedback, the driver could not function as part of the system.

The *kind* as well as the timing (or *phase relationship;* see Sect. 6.3) of feedback are rather important. In the case of the temperature controller, the electrical thermostat reactions must become transformed into heat energy and transfer of this energy to the room if there is to be control of *temperature.* In the case of the driver, the feedback which arrives in the form of sensory information must become interpreted and converted into suitable muscle action on the accelerator pedal to be effective.

In the case of temperature control the feedback must always be *negative.* That is, the rising room temperature causes the thermostat to demand less heat, while a dropping temperature causes it to demand more heat. In the case of driving an automobile, the feedback may be negative (say, when the traffic light turns red and the driver has to let up on the accelerator) or positive (say, when the way is clear for higher speed). When a politician confronts his voting constituents on an important issue, he watches their reactions as he talks, to get some form of feedback, When the response (or feedback) from the audience is "positive," he believes that his statements have been received favorably, whereas a "negative" feedback is likely to make him cautious.

Feedback may take many forms and many types of coupling. Figure 6.12 illustrates a simple modification of an earlier graph. In this illustration some of the output energy is fed back to the input. The box marked "feedback transfer" determines how much of and in what form the output will be fed back. The input is represented by a long arrow with positive increment. The feedback is shown as a small arrow with negative value. In such a setup the net input is *reduced* by the amount of the negative feedback. The effect is to restrain or to limit the output. If the sign of the feedback were positive, the input and the feedback would add and the output would increase continually and build up to destruction, or to the limit of the energy input. A system with feedback is often referred to as a *closed-loop system.* Since such systems incorporate a measure of self-correction, the exact value that the input is permitted to have becomes less critical. This self-correction factor also applies to the automobile driver, who does not have to have a gauge on the foot pedal because the "feedback" of his eyes and ears is enough to guide and restrain his push on the foot pedal.

High values for gain in amplifiers or control circuits tend to make a system unstable, and time lags produce wider oscillations. Negative feedback, on the other hand, tends to stabilize the systems.

6.12 *The elements of control systems*

Now that we have developed some familiarity with control systems, we can identify the functional elements that make up most systems.

THE VARIABLE TO BE CONTROLLED

First there is the variable that the system is expected to cope with or to control within prescribed limits. Actually it is rare that only one variable is present in a system. In the case of room-temperature control the changes in the outdoor temperature constitute an *independent variable,* while the internal temperature represents the controlled variable. Other independent variables may be introduced, such as children running in and

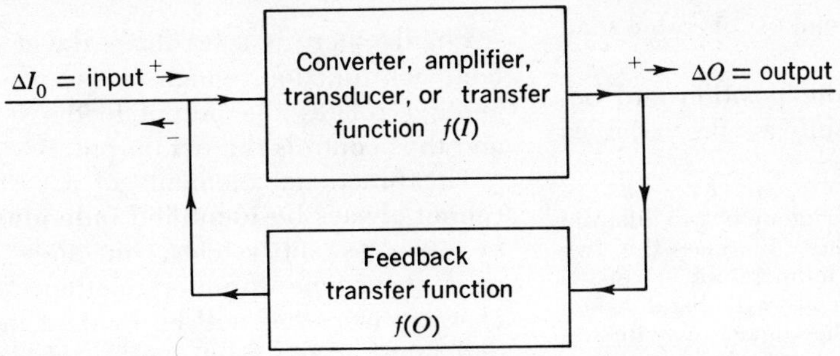

Fig. 6.12. Addition of a feedback transfer function to the transform function of Fig. 6.4.

out of open doors, to cause variable demands for more or less heating.

Similarly, the driver of the automobile has control devices by which he steers and starts and stops the car in relation to the road. But all along the way he is forced to comply with independent demands, such as changing road and traffic conditions, stop signs, and traffic lights, all of which constitute independent variables.

SENSOR DEVICES

Usually there must be some sensor device by which the variable can be measured or gauged. For example, in the case of the temperature measurement we shall learn in Chapter 10 that the temperature of air is actually determined by the velocity of the molecules that make up air. But we cannot gauge the temperature by measuring the velocity of molecules directly. What we can do is to utilize some *effect* that changes with changing molecule velocity. For example: At high air temperature, molecules in a material become more active and bump each other harder and more frequently, causing objects such as fluid in a thermometer or a piece of metal in a thermostat to warm up. A thermostat usually includes a bimetal† that carries an electrical contact; the bimetal changes its position when the air temperature changes and thus makes or breaks an electric circuit.

Similarly, while the position and behavior of the automobile are the variables to be controlled, we gauge these by the use of sensory information (vision, hearing) and the interpretive processes of the brain. The economist also looks for meaningful indices by which to gauge the larger features of national product, industrial trends, and public attitudes. The public utilizes quality and creativity as gauges to evaluate intrinsic or extrinsic return on investment.

ENERGY SOURCE

Whether one deals with a temperature-control system, driving an automobile, or any other situation that involves variables and controls, there must be a *source of energy* by which the job is performed.

MOTOR TRANSFORM DEVICE

In most instances the sensor function must utilize the services of a motor device to restore a variable to its proper value. To do this the motor device, or motor function, utilizes energy from an energy source. In the temperature-control system, the blowers and burners (which are triggered into action by the thermostat) begin to utilize fuel energy to heat the boilers. In the automobile a number of mechanisms come into play to burn the gasoline, to power the steering, and to perform other nondriver functions.

FEEDBACK

Finally there is a feedback device, or feedback function, which in one way or another relates the output to the input and thus controls the net output.

The functional elements of a system cannot always be identified individually or even as subsystems, but they are present in one form or another. One characteristic that will be evident more and more is the wide variety of transformations (transform functions) that are

† A bimetal strip is made up of two different metals bonded together. Because the two metals have different temperature expansion coefficients, the bimetal will bend when heated, thus causing the contact to switch on the system. As it cools, it straightens and contact is terminated.

possible in systems. Molecular speeds are transformed into mechanical bending of a bimetal, which completes an electric circuit and utilizes electric energy. This in turn starts a motor and pump device to feed and burn oil in a boiler, which produces heat that is transported or transferred by various means to another area by other motor devices. Similarly, all the tangible and intangible features of human physical energy and human brain processes become involved with electromechanical and chemical systems in driving an automobile. (For each of these transformations we can apply the more elegant phrase *transform function,* and illustrate the nature of the transformation by means of a mathematical equation, a graph, a listing of data, or a simple picture.)

6.13 Control concepts: cybernetics

The art and science of control theory has had a long and slow history. In the early days it found application in the sailing and steering of ships. With the coming of the steam engine a mechanical governor was needed to keep the speed of the engine constant. In more recent decades a wide variety of instruments, valves, and other equipment have been developed to maintain uniformity in chemical production processes. Servomechanisms were introduced during World War I for control of gunfire. Electric circuitry and electromechanical systems were given intensive study to improve their responsiveness and stability for purposes of controlling high-speed operations. By the 1940's the pace of automation had quickened as the concepts of control theory and of feedback received wider application in the electrical, mechanical, and processing industries. The term *high-fidelity* became a byword in amplifier design as a result of the introduction of negative feedback.

But the concept of feedback seemed to be basic and useful for a much wider range of applications. In 1947 the mathematician Norbert Wiener and Arturo Rosenbleuth compared the phenomena of control and of feedback, as used in technology, to the nervous system and muscle behavior of the human body. They postulated a close coordination of communication relationships between the brain, the sensory organs, and the muscles, and concluded that this resulted from the extensive use of feedback principles. It seemed that a feedback function is responsible for one's ability to reach down and pick up an object and to know how much farther the hand must move to complete the act. Moreover, they found an identity between *feedback* and *information* and the information content of a signal above the noise level. They gave the name *cybernetics* (from the Greek *kybernes* for *steersman*) to the entire field of control and communication theory, whether in the machine or in the animal.†

The concepts of *feedback* and *information* encompassed by cybernetics permit very extensive applications to the biological and social world. Just as the driver of the automobile performs functions in response to the information he derives

† The broad concepts that make up the science of cybernetics as developed by Wiener and his associates were new. The word itself had much older origin, however. It appears that Plato often employed the word "cybernetics" to mean "the steerman's art." His comment in "*Cleitophon,*" "the cybernetics of men, as you, Socrates, often call politics," suggests a wider implication. In 1834 the French physicist Ampère used the word as "means of governing" people.

from seeing and hearing and evaluating the driving situation, so his reactions under other situations are the result of his relationship or interaction with each new environment. Information and feedback are essential to his every move, every decision, almost every thought and learning process.

We shall have many opportunities to refer to the principles that have just been introduced. There will be applications to strictly technical systems, to systems that involve nature's resources, to biological systems, and to social situations.

The importance of the subject suggests that we summarize a few of the ideas that are most pertinent to our purposes.

1. Nature's processes are characterized by continuous dynamic transformations of energy, which may range from the vast magnitudes of astrophysics to the metabolic adaptation of the smallest living organism to its environment.

2. Much of man's own activities also involves the development of processes for conversion and utilization of nature's energy resources for purposes of assuring his survival and comfort. Indeed, the design of systems that integrate physical and chemical variables into cooperative, controlled systems constitutes a main interest of science and industry to bring about modern civilization and the current standard of living of advanced nations.

3. It is now recognized that the elements that make up a controlled system have common characteristics, whether accomplished by machine components, biological elements, thought processes, or social situations.

4. In such systems, the element of *feedback,* or *information,* which interrelates the output (or behavior) of the system and the input variables, constitutes a major factor for the effective operation and stability of systems.

5. The design of every control system requires careful analysis (and usually compromise) to meet the needs of the process. A prime requisite for most control systems is that they be *adequately responsive to changes,* and that they be *stable.* Also needed is an adequate *source of energy* to perform all the functions that are required of the system. The *input* to the system may be some *variable* such as temperature, liquid level, or pressure. Or it may be information that is itself the product of other operations, such as in *computer systems.*

6. The system performs its function by transforming the input to produce an *output* whose energy content is usually amplified, the added energy being derived from the source of energy of the system. The character of the transformation is designed into the system and is identified by its *transform function* to give the change or *amplification gain* to the output.

7. When a feedback (or information) loop permits some of the energy of the output to be fed back to the input, there can be considerable influence on the nature of the net output and on the stability of the system. In general, feedback that opposes changes (negative) in the input will improve the stability of the system, while feedback that arrives at the input in a manner that increases its changes (positive feedback) tends to reduce the stability of a system.

8. The stability of the system suffers and the system "hunts" more violently when the amplification or gain between output and input is too high or when the system responds too quickly to changes in the input variable. The design must include enough damping to reduce excessive overshoot (or violent hunting) of

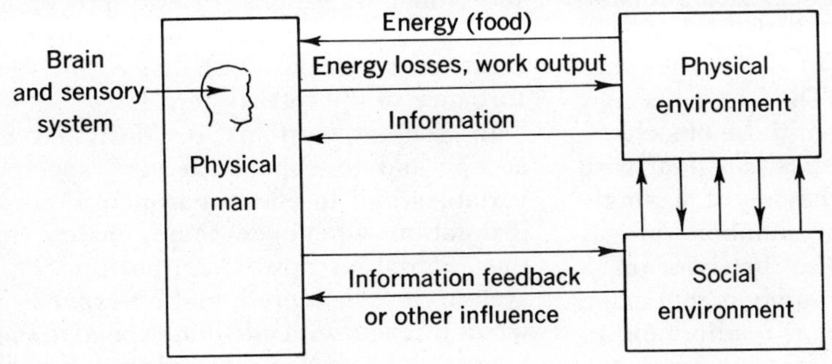

Fig. 6.13. The complex interrelationship of man with his environment.

the system while still providing adequate response. On-off controls offer cost advantages and simplicity, but the need for better control may dictate the use of *proportional* control or of other controls that have more sophisticated design. There can be more than one input to a system, more than one output, and a wide variety of interrelated combinations. In fact, the input may be the statistical output of many interrelated elements or variables.

6.14 Some examples of systems

In Fig. 6.13, which illustrates the relationship of man to his environment, we have identified two aspects of man (his brain and sensory motor system as distinguished from his physical being) and two aspects of his enviornment (the physical and social environments). There is very intimate and extensive interchange between the two aspects of man and between the two aspects of environment, as shown by the proximity and multiple arrows connecting them. Man draws energy and material from the physical environment and returns information and other materials to both.†

In the case of the driving of an automobile, it is difficult to identify all the elements that make up the input to this system. The desire to drive, the sensory activity that provides data to the brain, and the muscle behavior that operates the controls of the car, each is a complex that includes and combines the product of some other part of the system. The energy involved in the seeing, hearing, and judgement operations is negligibly small, but these become greatly magnified by the body's metabolic processes. This transform function of the body is most complex, and is itself made up of innumerable subsystems.

The specific control principles and systems we have discussed thus far are given broader significance by the principles of cybernetics. Cybernetics deals

† It is not easy to distinguish work from information and learning. Physical acts are not readily distinguishable as being separate from sensory response and interpretation that leads to learning, judgment, and decision. Certainly we cannot say that the throwing of a ball, intake of food, reading of newsprint, and a walk around the block are not so much mental processes that lead to future decision or action as they are physical acts.

with elements or variables that are related to each other so intimately that a change in one variable is likely to affect other variables in the system. The elements may be parts of a machine or those of a chemical process. Cybernetics can deal with the very specific behavior of a single molecule among vast numbers of gas molecules or with the behavior of a single cell of the vast numbers that make up an organism. It can as readily (and in general more usefully) consider *the statistical-behavior character of all the gas molecules together, or all the cells of the human body.* It can provide a method for analyzing the economic relationship of a grocer and his customer, or as readily attack questions pertaining to the economics of a whole nation. It establishes functional relationships in the course of changes, emphasizing their coordination, regulation, and control within a systems concept.

From the point of view of cybernetics, the aspect of systems behavior that is of greatest interest is the system's response to a disturbance. This disturbance may be a normal change or a momentary departure (transient) of the input, say as a result of the dropping of temperature of an industrial oven below its control setting when cold material is poured into it. One or more of the input variables or signals may experience changes that sum up to a signal sufficiently large to initiate a major change. For example, many chemical processes go on within the body, such as food intake, digestion, blood cell production, and oxygen utilization. They are not unrelated and all must be considered contributory to whether a person feels well or feels ill. Each process experiences its own daily or hourly variations, which nevertheless may constitute normal operation and good health. There can be occasions, however, when the individual variations in the processes add up to produce sickness of a sort that represents serious imbalance or disturbance of the total system.†

In general, systems are designed to accept and to cope with very specific variables and to effect reasonably quick restoration whenever some change in those variables upsets equilibrium. The system is considered to be *responsive* when it reacts with *adequate* speed to the upset. A system that responds too quickly or introduces corrective steps that are too large is likely to produce instability around the equilibrium point. A system may also be too sensitive to small fluctuations that are of the order of magnitude of background "noise," and for that reason will be unstable.

We might consider the design of an electrical amplifier system such as that used for a quality phonograph system. Figure 6.12 represents a fairly simple circuit for transforming an input through some form of *transducer* to produce an *output.* The *feedback* to the input in this case was designed to counteract or oppose the input, tending to reduce undesirable excursions in the output due to variations other than the sound signals to be amplified. The system constitutes a channel for transmitting and transposing signals, the input signals being *information.* To be effective, the design must usually incorporate suitable capabilities in such terms as capacity, watts, voltage, range, and frequency. These in turn provide the basis for designing suitable *constraints* into the system.

However, a control system is not likely

† As a simple example, the experience of sitting in an awkward position can introduce a combination of neural signals and mental process that suggests the need for a new position and thereby requires a complete readjustment of nerve and muscle systems.

to be designed to control every variable against every change. For example, the body's control of the iris openings of the eyes (to permit only adequate light to reach the retina) has a very specific, limited function and purpose, which excludes sensitivity to other variations of body conditions. The purpose of constraints is to reduce the response of the system to variables that are not considered to be part of the information to be transmitted. There are also natural restraints or constraints on the information and on the variety of information that a channel may transmit. Among these are the limitations and directions imposed by the conservation of energy and the laws of thermodynamics. When a system combines several elements into an integrated organizational and functional interdependence, the interdependence automatically imposes constraints, since the elements are now no longer independent of each other. An amplifier system may have to contend with constraints in the form of costs, against which the designer must balance extra quality or fidelity or amplification.

With only minor modifications the diagram of Fig. 6.12 can represent a quite different system for communication of information. Figure 6.14 illustrates some

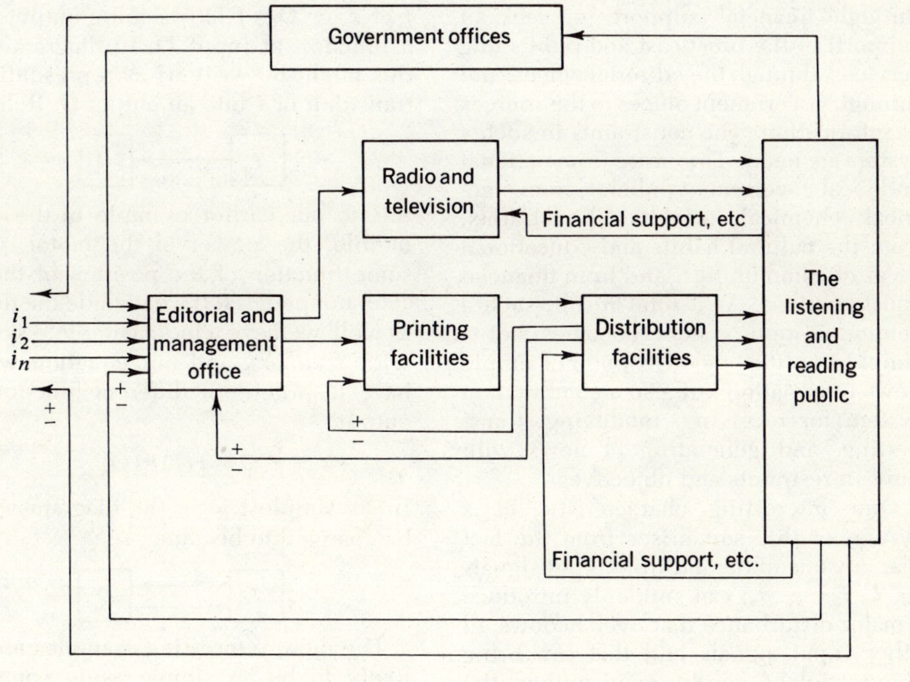

Fig. 6.14. Schematic representation of a system for communicating news to the public.

of the elements that enter into a system for communicating news to the general public. News may be collected from many areas and reported; this news becomes input (i_1, i_2, . . . i_n) to the editorial offices of a news agency. At the editorial offices this information undergoes modification and shaping, and is put into printed form or given electrical broadcast. There will be close liaison among the several blocks that make up the channels for this communication. There will be government influences as well as government sources of information bearing on the editorial and management offices, much of it in the form of feedback reaction to the communication. The listening and reading public applies "feedback" influence through financial support (or lack of support) of the broadcast and publishing services, through the editorial offices, and through government offices to the sources of information. The constraints in such a system are many. They arise from national and local government policies; from electrical, chemical, mechanical restraints; from the cultural habits and educational level of communities; and from financial considerations. As a total result, such a communication system becomes not a simple amplifier and distributor of simple news information but also a combination system for receiving, modifying, transmitting, and generating of news with built-in restraints and objectives.

One interesting characteristic of a system of this sort arises from the fact that any one of the multiple input signals (i_1, i_2, i_3 . . . i_n) can suddenly introduce a major disturbance that overshadows all other input signals and that can bring about violent response in either the forward channels or the feedback channels. Such a disturbance might be an act of war, a strike, a catastrophe, or an event that is especially disliked or especially desirable. There may be quite a few surges of output beyond the desired limits of control before the system settles down again.

One may also conceive that the input (i_1 . . . i_n) can be made up of very many items and elements so that the overall significance of the input is determined by the *statistical character* of the input rather than being overly influenced by any one item.

6.15 Functional relationship: notations

In its simplest form, a cause-and-effect relationship is stated as a simple function such as $y = f(x)$ (meaning y is a function f of x, or $O = f(I)$ (meaning output O is a function of input I). In diagram form this might be written† as representing a transition of I into an output O. Relating

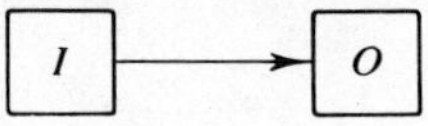

this to our earlier example of the automobile, the power of the motor, O, is some function of the position of the accelerator pedal, I. If we include the driver as well, we have a more complete system *with feedback* and our equation would have to provide a different function for output,

$$O = f(I)F(O)$$

In its simplest form the diagram would be changed to become

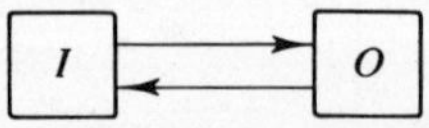

The more interesting examples are not likely to be so simple as to comprise

† The approach in this section is considerably influenced by the treatment given by W. Ross Ashby.

only a single input and single output. Within the total system that includes the auto and driver, there are innumerable smaller systems such as cells, neurons, muscles, organs, machine parts, and electrical controls. A study of such assemblies must set out clear objectives before its approach or results can be made significant. For example, are the functions to be studied primarily those that pertain to keeping the auto on the road, or are they functions that determine the state of the driver's gall bladder, heart beat, or temperature? These subsidiary systems are certainly part of the total system of man and auto, but their details are independent of the specific functions that go with driving the automobile. The situation would be different, of course, if part of a study had to do with the effect of heart or temperature function on the driving, for which purpose a new set of elements would be involved when making up the system to be studied.

The situation is illustrated by Ashby in the following diagram, in which one may trace twenty different† circuits.

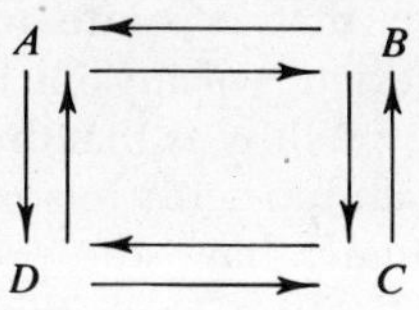

Each subsidiary circuit may have its own mode of feedback and control, and may be either strongly or weakly linked with neighboring circuits. In the case of our driver, vision plays the dominant role in telling him where the auto is going, while his knowledge of the situation is helped by the senses of hearing and by the sensations of his body as the car sways. All the input stimuli have some relation to each other. One may picture a strong relationship (or strong coupling) between vision and hearing and a weaker coupling between heartbeat and vision, as far as driving the car is concerned.

Any study must therefore seek first to identify the functionally significant relationships that are the subject of the study, to identify the elements that bear directly on the functions under study, and to eliminate from consideration those elements that are independent of the selected functions.† This is not easy to do in most cases because there are many varieties of influences and "couplings" that come into play. Often the study must assume a series of situations and obtain results and estimates for a wide variety of combinations of systems.

MODELS

Often it becomes necessary to simplify a system or make it understandable by use of a model or models. This becomes imperative for nearly all biological systems, which are enormously complex. But models can very quickly become detrimental to progress when one loses sight of the simplifications and limitations that are inherent to each model.

6.16 *"Black box" approach*

Ashby considers the interesting case of an experimenter approaching a "black box" that is unknown to him with respect to contents and functions. How should he proceed to investigate and determine the contents and character of the box?

† It is suggested that the reader list the twenty different ways in which a signal may travel through the system, starting at *A* and returning ultimately to *A* in each case. For example, *ABCDA* or *ABCBCDA*.

† The importance of such an approach in the study of human behavior is seen in the Gestalt psychology view of phenomena.

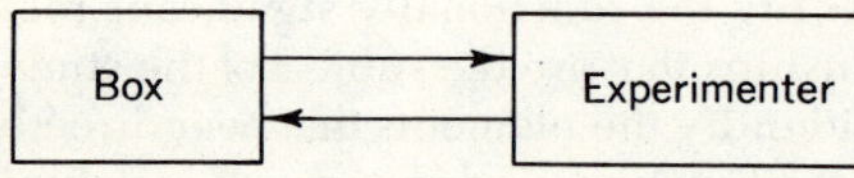

Fig. 6.15. Black-box relationship.

(The procedure can be especially important if one imagines that the "black box" could be an explosive bomb.) There is, immediately, a relationship between the experimenter and the box, of the type shown in Fig. 6.15. The diagram illustrates the nature of exchanges and feedback that take place as the experimenter explores the problem by various means. The "means" presumably might include such things as pushing and pulling of the box and levers. For a systematic search, each move would be recorded along with the "state" of the box that accompanied each move. In time the experimenter would presumably be able to identify the "state" of the box for each type of input, and possibly also the function for each type of input. Many of the systems with which we deal are actually made up of "black boxes," and the functional characteristics of the total assembly may be determined by the characteristics of each box and by the nature of their coupling together.† But we may fail to characterize the system because a combination of black boxes may produce an unexpected function that is quite unrelated to the characteristics of any one box. An example given by Ashby is that the approximately twenty amino acids in a bacterium do not individually have the property of being self-replicating, but their combination does introduce this property.

Real-life problems tend to have many "black boxes," often interconnected in such manner as to obscure the specific role of each box, each subsystem. One may make progress in the analysis of the total system and its parts by systematic analysis of "responses" or "states" to questions and input stimuli. One may seek to discover factors that produce

† The reader may picture the similar situation that exists when he first meets a person who is to become his associate on some project.

certain extreme "responses" of "states." The use of computers helps handle large quantities of data and identify common elements or contrasts. But progress in attacking complex problems depends more often on good use of judgment, experience, intuition or insight, persistence, and some luck. It is not always easy to identify and isolate the specific functions that are of importance for the system's functioning. There may be multitudinous other elements within the total system that do not bear on the specific functions under study.

6.17 *The closed-loop amplifier system*†

It will be helpful to look a little more closely at the quantitative aspects of a system that has feedback characteristics. Figure 6.16 illustrates a system in which an input signal E_i (which may be in volts and related to temperature, pressure, blood count, or other variable) constitutes the control variable. The system may be designed to do something that is proportional to or determined by this control variable *i*. If the system is a servomechanism, input E_i may represent the angle of rotation of a small motor and output E_o the angle of rotation of a larger motor, the objective being to keep the two motors in step with each other. Or E_i may be the input voltage from a measuring circuit that has high resistance and low power and which is to be converted to an identical voltage in a low-resistance circuit to operate a loudspeaker or solenoid or some other device that requires more power than is available at the input end of the circuit. (Throughout this discussion keep in mind that there must be a source of energy to make this conversion possible, as is illustrated in Fig. 6.16.)

The signal E_i may have a fixed value or may vary with time. It feeds into a comparator element, where E_i is compared (added) to the signal coming as feedback. It both the input signal and

† The sections printed in color represent optional reading material.

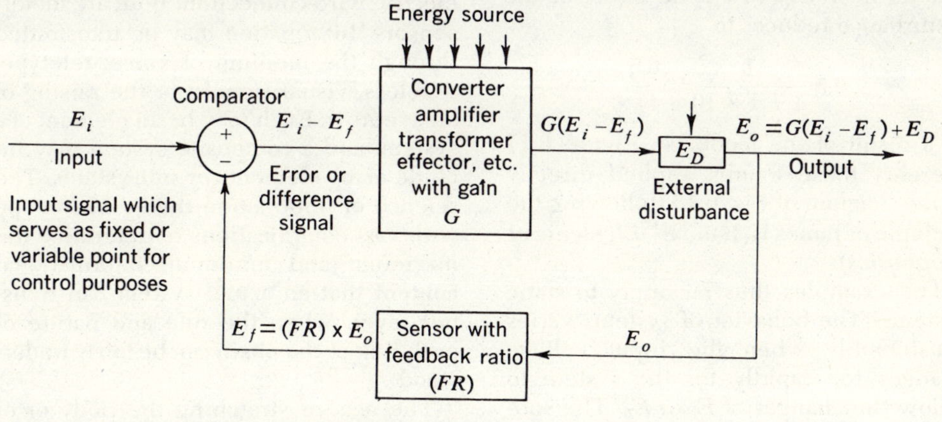

Fig. 6.16. *A closed-loop amplifying system.*

or

$$E_O = E_i\left(\frac{G}{1+(FR)G}\right) + E_D\left(\frac{1}{1+(FR)G}\right) \tag{6-2}$$

For the above example,

$$E_O = 1\left(\frac{10}{1+(1\times 10)}\right) + 0\left(\frac{1}{1+1\times 10}\right)$$

$$= \frac{10}{11}\text{ volt}$$

The error or difference voltage therefore becomes $E_i - E_f = 1 - (10/11) = 1/11$ volt. The amplifier must be capable of acting on this voltage if the system is designed to work on such magnitudes of error. One way to improve this is to increase the gain G of the amplifier. If, in this case, the gain G is increased from 10 to 100 while keeping the feedback ratio and E_D the same, the error voltage reduces to 1/101 volt from 1/11 volt. It can be seen that the gain can change markedly without introducing serious error in such a system.

Finally suppose there develops a disturbance E_D amounting to 0.5 volt, with the gain $G = 10$ and $FR = 1$. From the last term of Eq. (6-2), the effect of the disturbance reduces to

$$0.5\,\frac{1}{1+1\times 10} = \frac{0.5}{11}$$

as a result of the feedback. Figure 6.17 presents these figures applied directly to the diagram of Fig. 6.16 (following the example of James E. Randall, *Elements of Biophysics*).

The examples thus far apply to static systems. The behavior of systems varies considerably when the input voltage changes too rapidly for the system to follow the changes of E_i or E_D. The subject of controller stability has received a great deal of attention in connection with servomechanism design (for remote control of airplane movements, and similar applications) and electric circuit design for communication, but we cannot delve into that aspect of controller theory. However, one related consideration is "noise," mentioned briefly in an earlier discussion. This also has had considerable study because of the important effect on the capacity of circuitry to convey "information."

6.18 The nature of "living" systems

In the discussions of control systems thus far we have not distinguished between systems involving machine components and those involving living systems. Nor is it our intention to do so now. The fact is that, except for varying complexity, the very same concepts may be applied to living as well as nonliving or machine systems. Each of the sensory organs through which we communicate with the much vaster system of nature is itself designed, oriented, and functionally controlled to achieve certain specific goals or "purposes." It does not matter whether we discuss a nerve cell or an electric wire connection. Both are motor-sensors. Information may be transmitted through the medium of voice, teletype, wireless, visual signals, or the raising of an eyebrow. Each may be an element of a system, and a composite system may include many elements or subsystems. The science of information theory must cope with vast complications to determine the maximum and minimum informational content that an actual system can transmit, even when the role and nature of each link of the chain can be fairly understood.

The acts of stretching the body or of reaching to pick up an object entail the function of a fairly complex system of

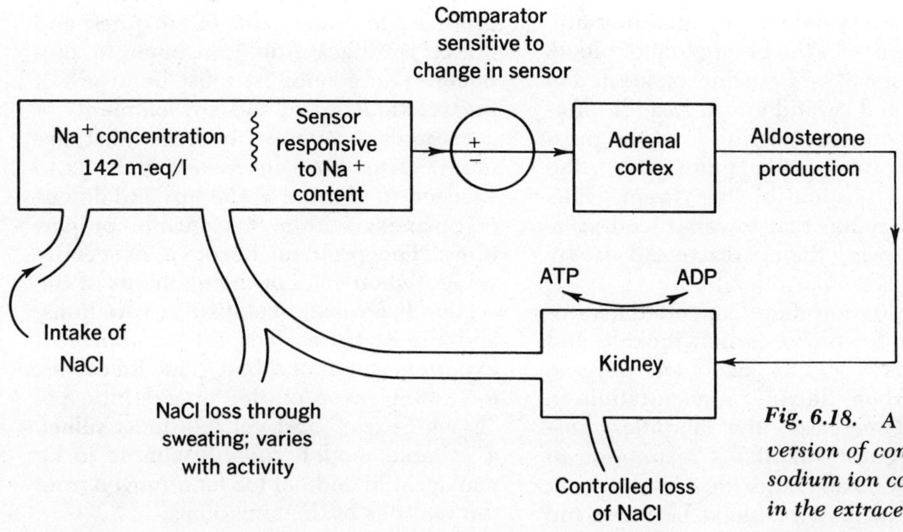

Fig. 6.18. A simplified version of control of sodium ion concentration in the extracellular fluid.

regulation and control. This has been demonstrated by Karl Smith's† experiments with delayed visual feedback in visual motor behavior, which showed that what a person sees is delayed in reaching him when he is performing various other tasks such as writing.

The *intention* to stretch or to pick up an object is itself a complex function, developing in the mind as a product of other activities and influences. The *command signal,* in the form of nerve impulses, originate in the motor cortex of the brain and initiates action in the muscle contractile proteins. There is an amplification, G, which may be expressed as change in muscle length for a unit change in the motor neurons that initiate the discharges. The muscle spindle acts as a *sensor-transducer* to produce nerve impulses, in proportion to muscle length extension, to send back to the brain as *feedback* on the extension. The original impulses and the feedback impulses are integrated in the spinal cord and give indication of the *error* or *difference* from completion of the intended act. The spindle proprioceptors serve to provide constant information on the state and tone of the muscle system, and assure smooth action of the body. When an individual is deprived of their help, muscle activity tends to be abrupt rather than smooth, requiring dependence on visual sense of position to the extent that he cannot stand when blindfolded.

A person suffering from Parkinson's disease retains some benefit from proprioceptive information, but tends to overshoot when reaching for an object—a motion that recalls damped oscillation.

In later studies of biological systems we shall have occasion to study in some detail a few of the regulatory systems on which life depends. Figure 6.18 illustrates how the sodium ion concentration

† See K. V. Smith.

is maintained constant in the fluid that surrounds the individual cells. There is an elaborate system for maintaining uniform pressure in the circulation of blood. Pressure-sensitive transducers, located in the aorta and carotid arteries, send information about the magnitude of the pressure to an integrating center within the medullary portion of the brain. This results in action that lowers blood pressure by slowing the heart rate and also by producing vascular dilatation.

For respiration there is needed a minimal value for blood carbon dioxide and an adequate supply of blood oxygen. When carbon dioxide concentration in the blood increases, the medullary respiratory center stimulates respiration to eliminate carbon dioxide. The transit time for the flow of blood between the lungs and the respiratory center is only a few seconds under normal conditions (see Randall, p. 108). Body temperature is maintained by a delicate balance between heat loss (from warm-blooded or homeothermic animals) and heat production within the animal through metabolism. The "thermostat" that controls this balance is located in the hypothalmus of the brain and receives information from various temperature transducers of the body to guide its own function.

The regulatory system can extend beyond the body to include the interactions involving climate, geography, geology, agriculture, theology, government, disease, or any other influences. The elements of determinate function, disturbances, control variables, amplification, feedback, informational content, are all three, but they may take the forms of imposed law, self-imposed law, self-imposed restraints, religious restraints, moral obligations, and many other forms that are even less tangible.

The regulatory principles apply to commercial production plants where orders for goods become converted to products for sale, with often quick and direct feedback from consumer to producer. The economist must be aware of the relationship of the key elements of a nation's economy in terms that are identical to those discussed, if he is to succeed in regulating the ups and downs of business within manageable proportions. The problem becomes especially severe when each of the elements of the system is a result of statistical variations, and the statistics lack the assurance of experience or of numbers. The difficulties too often savor of the uncertainties of "black boxes," and yet one must select a suitable model, suitably simple to be manageable and not too far removed from the realities of the situation.

The student is urged to study carefully all the details that have been included in this section on controls. In time he will find that many of the topics that are to come in later chapters will fall more easily into place. For nature and man exist and continue as a result of a balance of forces and utilization of energy, the whole constituting a system that is in a state of reasonable balance and regulation and yet continually changing toward wholly new forms.

In conclusion, we hope that this brief introduction to systems and cybernetics will encourage each reader to view the events of his life with keener appreciation for the *interrelationship* of the factors that bear on the events, and especially for *feedback* influences. A word of caution is in order, however, with respect to overextended use of the term *cybernetics* to situations wherein the relationships are too complex or too obscure, and wherein there are not present the control systems elements which we have discussed.

Questions/Discussions

The assignments for this chapter are intended to give the reader opportunity to discover for himself how broadly the concepts and techniques involving *systems, feedback, control, stability-instability,* and *cybernetics* apply to phenomena in nature and to all aspects of human social relations. It is suggested that from two to four weeks be allowed for completion of this work.

1. For purposes of review, tabulate the five elements of control systems (described in Sec. 6.12) that apply in the following personal situations. Explain also whether the feedback is positive or negative in each case.

(a) The control of temperature of your home.
(b) The factors that control your waking up on a weekday morning.
(c) The factors that control your breakfasting.
(d) One situation or experience of your day that includes strong positive feedback.
(e) A situation or experience of your day that includes strong negative feedback.

2. Select three phenomena or situations, taken from any three of the following categories, and analyze their "systems" aspects in the following terms:

(a) The dependent and independent variables that are involved in each, either as "input" to the system or as disturbances.
(b) The sensor devices or transform functions required at the input end for each variable.
(c) The energy sources.
(d) The motor devices or processes, and the related transform functions.
(e) The gain or amplification between output and input.
(f) The nature of feedback influences (distinguishing between positive and negative feedback and phase relationships) related to each input and each output.
(g) The nature of subsystems that are included.
(h) The factors that make for stability and instability in the total system or subsystems.
(i) The graphical representation of the above elements and processes, with indication of polarity (direction) of feedback between each output and input.

The phenomena or situations are to be drawn from any three of the following seven categories:

I Electromechanical, pneumatic systems, chemical or production processes
II Geophysical or meteorological processes
III Biological processes (plants, animals), ecological relationships
IV Medical, pathological experiences
V Economics (international, national, or personal), business operations
VI Behavioral, cultural, ethical, moral, theological, and psychological aspects of social experiences

Note: It is suggested that each "case" be given adequate discussion and one to two pages of graphical representation. Because of the importance of the subject of "systems," it is suggested that these analyses be given time for class discussion. Group effort on the part of the students is encouraged, although each must present his own final case study.

The author, the first American Nobel Prize physicist, traces the determinations of the velocity of light, one of the handful of constants of nature.

4 Velocity of Light

A. A. Michelson

A chapter from his book, *Studies in Optics,* published in 1927.

The velocity of light is one of the most important of the fundamental constants of Nature. Its measurement by Foucault and Fizeau gave as the result a speed greater in air than in water, thus deciding in favor of the undulatory and against the corpuscular theory. Again, the comparison of the electrostatic and the electromagnetic units gives as an experimental result a value remarkably close to the velocity of light—a result which justified Maxwell in concluding that light is the propagation of an electromagnetic disturbance. Finally, the principle of relativity gives the velocity of light a still greater importance, since one of its fundamental postulates is the constancy of this velocity under all possible conditions.

The first attempt at measurement was due to Galileo. Two observers, placed at a distance of several kilometers, are provided with lanterns which can be covered or uncovered by a movable screen. The first observer uncovers his light, and the second observer answers by uncovering his at the instant of perceiving the light from the first. If there is an interval between the uncovering of the lantern by the first observer and his perception of the return signal from the second (due allowance being made for the delay between perception and motion), the distance divided by the time interval should give the velocity of propagation.

Needless to say, the time interval was far too small to be appreciated by such imperfect appliances. It is nevertheless worthy of note that the principle of the method is sound, and, with improvements that are almost intuitive, leads to the well-known method of Fizeau. The first improvement would clearly be the substitution of a mirror instead of the second observer. The second would consist in the substitution of a series of equidistant apertures in a rapidly revolving screen instead of the single screen which covers and uncovers the light.

The first actual determination of the velocity of light was made in 1675 by Römer as a result of his observation of the eclipses of the first satellite of Jupiter. These eclipses, recurring at very nearly equal intervals, could be calculated, and Römer found that the observed and the calculated values showed an annual discrepancy. The eclipses were later by an interval of sixteen minutes and twenty-six seconds[1] when the earth is farthest from Jupiter than when nearest to it. Römer correctly attributed this difference to the time required by light to traverse the earth's orbit. If this be taken as 300,000,000 kilometers and the time interval as one thousand seconds, the resulting value for the velocity of light is 300,000 kilometers per second.

Another method for the determination of the velocity of light is due to Bradley, who in 1728 announced an apparent annual deviation in the direction of the fixed stars from their mean position, to which he gave the name "aberration." A star whose direction is at right angles to the earth's orbital motion appears displaced in the direction of motion by an angle of $20''.445$. This displacement Bradley attributed to the finite velocity of light.

With a telescope pointing in the true direction of such a star, during the time of passage of the light from ob-

[1] The value originally given by Römer, twenty-two minutes, is clearly too great.

jective to focus the telescope will have been displaced in consequence of the orbital motion of the earth so that the image of the star falls behind the crosshairs. In order to produce coincidence, the telescope must be inclined forward at such an angle a that the tangent is equal to the ratio of the velocity v of the earth to the velocity of light,

$$\tan a = \frac{v}{V},$$

or, since $v = \pi D/T$, where D is the diameter of the earth's orbit and T the number of seconds in the year,

$$\tan a = \frac{\pi D}{VT},$$

from which the velocity of light may be found; but, as is also the case with the method of Römer, only to the degree of accuracy with which the sun's distance, $\frac{1}{2}D$, is known; that is, with an order of accuracy of about 1 per cent.[1]

In 1849 Fizeau announced the result of the first experimental measurement of the velocity of light. Two astronomical telescope objectives L_1 and L_2 (Fig. 73) are placed facing each other at the two stations. At the focus of the first is an intense but minute image a of the source of light (arc) by reflection from a plane-parallel plate N. The light from this image is rendered approximately parallel by the first objective. These parallel rays, falling on the distant objective, are brought to a focus at the surface of a mirror, whence the path is retraced and an image formed which coincides with the original image a, where it is observed by the ocular E. An accurately divided toothed wheel W is given a uniform rotation,

[1] The value of the velocity of light has been obtained, by experimental methods immediately to be described, with an order of accuracy of one in one hundred thousand, so that now the process is inverted, and this result is employed to find the sun's distance.

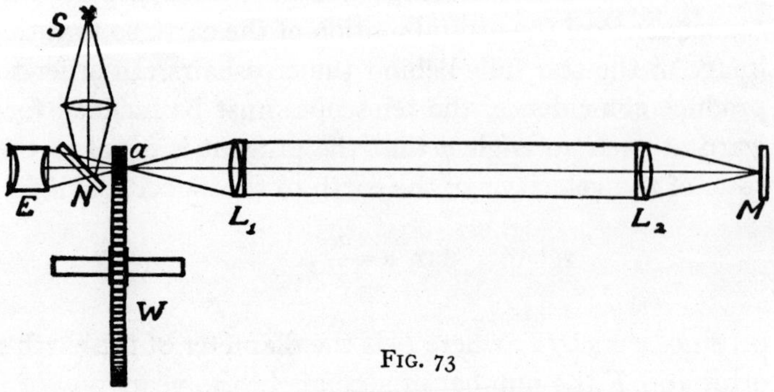

FIG. 73

thus interrupting the passage of the light at a. If, on returning, the light is blocked by a tooth, it is eclipsed, to reappear at a velocity such that the next succeeding interval occupies the place of the former, and so on.

If n is the number of teeth and N the number of turns per second, K the number of teeth which pass during the double journey of the light over the distance D,

$$V=\frac{2NnD}{K}.$$

It is easier to mark the minima than the maxima of intensity, and accordingly

$$K=\frac{2p-1}{2}$$

if p is the order of the eclipse. Let δK be the error committed in the estimate of K (practically the error in estimation of equality of intensities on the descending and the ascending branches of the intensity curve). Then

$$\frac{\partial V}{V}=\frac{\partial K}{K}.$$

Hence it is desirable to make K as great as possible. In Fizeau's experiments this number was 5 to 7, and should have given a result correct to about one three-hundredth. It was, in fact, about 5 per cent too large.

A much more accurate determination was undertaken by Cornu in 1872 in which K varied from 3 to 21, the result as given by Cornu being 300,400, with a probable error of one-tenth of 1 per cent. In discussing Cornu's results, however, Listing showed that these tended toward a smaller value as the speed increased, and he assigns this limit as the correct value, namely, 299,950. Perrotin, with the same apparatus, found 299,900.

Before Fizeau had concluded his experiments, another project was proposed by Arago, namely, the utilization of the revolving mirror by means of which Wheatstone had measured the speed of propagation of an electric current. Arago's chief interest in the problem lay in the possibility of deciding the question of the relative velocities in air and water as a crucial test between the undulatory and the corpuscular theories. He pointed out, however, the possibility of measuring the absolute velocity.

The plan was to compare the deviations of the light from an electric spark reflected directly from the revolving mirror with that which was reflected after traversing a considerable distance in air (or in water). The difficulty in executing such an experiment lay in the uncertainty in the direction in which the two reflected images of the spark were to appear (which might be anywhere in 360°). This difficulty was solved by Foucault in 1862 by the following ingenious device whereby the return light is always reflected in the same direction (apart from the deviation due to the retardation which it is required to measure), notwithstanding the rotation of the mirror.

Following is the actual arrangement of apparatus by which this is effected. Light from a source S falls upon an objective L, whence it proceeds to the revolving mirror R, and is thence reflected to the concave mirror C

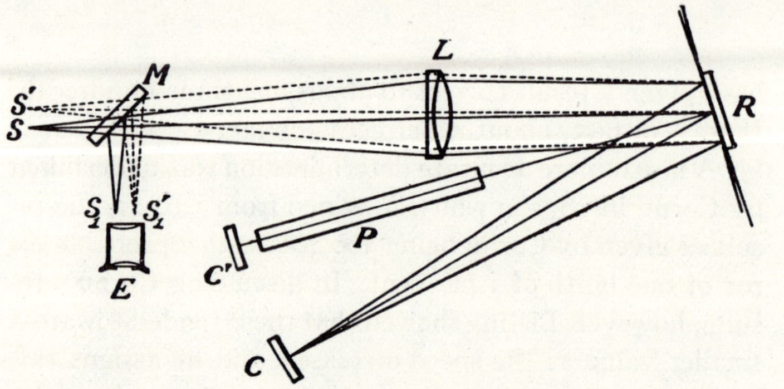

FIG. 74

(whose center is at R), where it forms a real image of the source. It then retraces its path, forming a real image which coincides with the source even when the revolving mirror is in slow motion. Part of the light is reflected from the plane-parallel glass M, forming an image at a where it is observed by the micrometer eyepiece E.

If now the revolving mirror is turning rapidly, the return image, instead of coinciding with its original position, will be deviated in the direction of rotation through an angle double that through which the mirror turns while the light makes its double transit. If this angle is α and the distance between mirrors is D, and the revolving mirror makes N turns per second,

$$\alpha = 2\pi N \frac{2D}{V},$$

or

$$V = \frac{4\pi ND}{\alpha}.$$

In principle there is no essential difference between the two methods. In the method of the toothed wheel the angle α corresponds to the passage of K teeth, and is therefore $\alpha = 2\pi K/n$, so that the formula previously found, $V = \frac{2NnD}{K}$, now becomes $V = \frac{4\pi ND}{\alpha}$, the same as for the revolving mirror. The latter method has, however, the same advantage over the former that the method of mirror and scale has over the direct reading of the needle of a galvanometer.

On the other hand, an important advantage for the method of the toothed wheel lies in the circumstance that the intensity of the return image is one-half of that which would appear if there were no toothed wheel, whereas with the revolving mirror this fraction is $\frac{n\beta}{rD}$ if the mirror has n facets), where β is the angular aperture of the concave mirror, and f is the focal length of the mirror, r is the distance from slit to revolving mirror, and D is the distance between stations.

In the actual experiments of Foucault, the greatest distance D was only 20 m (obtained by five reflections from concave mirrors), which, with a speed of five hundred turns per second, gives only 160″ for the angle 2α which is to be measured. The limit of accuracy of the method is about one second, so that under these conditions the results of Foucault's measurements can hardly be expected to be accurate to one part in one hundred and sixty. Foucault's result, 298,000, is in fact too small by this amount.[1]

In order to obtain a deflection 2α sufficiently large to measure with precision it is necessary to work with a much larger distance. The following plan renders this possible, and in a series of experiments (1878) the distance D was about 700 m and could have been made much greater.

[1] Apart from the mere matter of convenience in limiting the distance D to the insignificant 20 m (on account of the dimensions of the laboratory), it may be that this was in fact limited by the relative intensity of the return image as compared with that of the streak of light caused by the direct reflection from the revolving mirror, which in Foucault's experiments was doubtless superposed on the former. The intensity of the return image varies inversely as the cube of the distance, while that of the streak remains constant.

The image-forming lens in the new arrangement is placed between the two mirrors, and (for maximum intensity of the return image) at a distance from the revolving mirror equal to the focal length of the lens. This necessitates a lens of long focus; for the radius of measurement r (from which α is determined by the relation $\delta = r \tan \alpha$, in which δ is the measured displacement of the image) is given by $r = \frac{f^2}{D}$, if f is the focal length of the lens; whence r is proportional to f^2. In the actual experiment, a non-achromatic lens of 25-m focus and 20-cm diameter was employed, and with this it was found that the intensity of the return light was quite sufficient even when the revolving mirror was far removed from the principal focus.

With so large a displacement, the inclined plane-parallel plate in the Foucault arrangement may be suppressed, the direct (real) image being observed. With 250 to 300 turns per second, a displacement of 100 to 150 mm was obtained which could be measured with an error of less than one ten-thousandth.

The measurement of D presents no serious difficulty. This was accomplished by means of a steel tape whose coefficient of stretch and of dilatation was carefully determined, and whose length under standard conditions was compared with a copy of the standard meter. The estimated probable error was of the order of 1 : 200,000.

The measurement of the speed of rotation presents some points of interest. The optical "beats" between the revolving mirror and an electrically maintained tuning fork were observed at the same time that the coincidence of the deflected image with the crosshairs of the eyepiece was maintained by hand regulation of an air blast which actuated the turbine attached to the revolving mirror. The number of vibrations of the fork *plus* the number of beats per second gives the number of revolutions per second in terms of the rate of the fork. This, however, cannot be relied upon except for a short interval, and it was compared before and after every measurement with a standard fork. This fork, whose temperature coefficient is well determined, is then compared, as follows, directly with a free pendulum.

For this purpose the pendulum is connected in series with a battery and the primary of an induction coil whose circuit is interrupted by means of a platinum knife edge attached to the pendulum passing through a globule of mercury. The secondary of the induction coil sends a flash through a vacuum tube, thus illuminating the edge of the fork and the crosshair of the observing microscope. If the fork makes an exact whole number (256) of vibrations during one swing of the pendulum, it appears at rest; but if there is a slight excess, the edge of the fork appears to execute a cycle of displacement at the rate of n per second. The rate of the fork is then $N \pm n$ per second of the free pendulum. This last is finally compared with a standard astronomical clock.[1] The order of accuracy is estimated as 1 : 200,000.

The final result of the mean of two such determinations of the velocity of light made under somewhat similar conditions but at a different time and locality is 299,895.

A determination of the velocity of light by a modification of the Foucault arrangement was completed by Newcomb in 1882. One of the essential improvements consisted in the use of a revolving steel prism with square section twice as long as wide. This permits the sending and receiving of the light on different parts of the mirror, thus eliminating the effect of direct reflection. It should also be mentioned that very accurate means were provided for measuring the deflection, and finally that the

[1] The average beat of such a clock may be extremely constant although the individual "seconds" vary considerably.

speed of the mirror was registered on a chronograph through a system of gears connected with the revolving mirror. Newcomb's result is 299,860.

The original purpose of the Foucault arrangement was the testing of the question of the relative velocities of light in air and in water. For this purpose a tube filled with water and closed with plane-parallel glasses is interposed. There are then two return images of the source which would be superposed if the velocities were the same. By appropriately placed diaphragms these two images may be separated, and if there is any difference in velocities this is revealed by a relative displacement in the direction of rotation. This was found greater for the beam which had passed through the water column, and in which, therefore, the velocity must have been less. This result is in accordance with the undulatory theory and opposed to the corpuscular theory of light.

The experiments of Foucault do not appear to have shown more than qualitative results, and it should be of interest, not only to show that the velocity of light is less in water than in air, but that the ratio of the velocities is equal to the index of refraction of the liquid. Experiments were accordingly undertaken with water, the result obtained agreeing very nearly with the index of refraction. But on replacing the water by carbon disulphide, the ratio of velocities obtained was 1.75 instead of 1.64, the index of refraction. The difference is much too great to be attributed to errors of experiment.

Lord Rayleigh found the following explanation of the discrepancy. In the method of the toothed wheel the disturbances are propagated in the form of isolated groups of wave-trains. Rayleigh finds that the velocity of a group is not the same as that of the separate waves except in a medium without dispersion. The simplest form of group analytically considered is that produced by two simple harmonic wave-trains of slightly different frequencies and wave-lengths. Thus, let

$$y=\cos(nt-mx)+\cos(n_1t-m_1x)\ ,$$

in which $n=2\pi/T$, and $m=2\pi/\lambda$, T being the period and λ the wave-length. Let $n-n_1=\partial n$, and $m-m_1=\partial m$. Then

$$y=2\cos\tfrac{1}{2}(\partial nt-\partial mx)\cos(nt-mx)\ .$$

This represents a series of groups of waves such as illustrated in Figure 75.

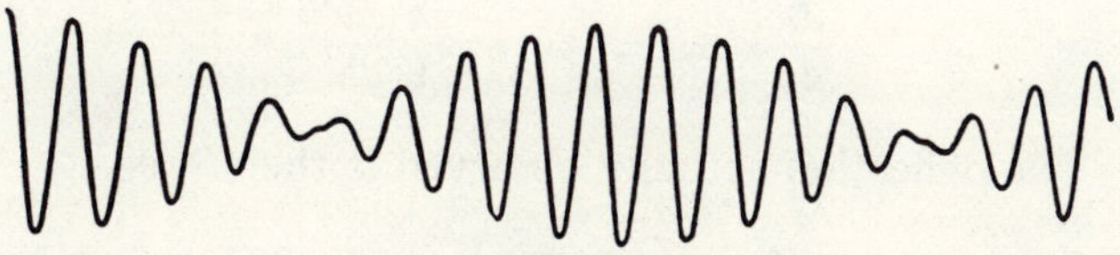

FIG. 75

The velocity of the waves is the ratio $V=n/m$, but the velocity of the group (e.g., the velocity of propagation of the maximum or the minimum) will be

$$V'=\partial n/\partial m\ ,$$

or, since $n=mV$,

$$V'=\frac{\partial(mV)}{\partial m}=V+m\frac{\partial V}{\partial m}=V\left(1+\frac{m\partial V}{V\partial m}\right)\ ,$$

or, since $m=2\pi/\lambda$,

$$V'=V\left(1-\frac{\lambda}{V}\frac{\partial V}{\partial\lambda}\right)\ .$$

The demonstration is true, not only of this particular form of group, but (by the Fourier theorem) can be applied to a group of any form.

It is not quite so clear that this expression applies to the measurements made with the revolving mirror. Lord Rayleigh shows that in consequence of the Doppler effect there is a shortening of the waves at one edge of the beam of light reflected from the revolving mirror and a lengthening at the opposite edge, and since the velocity of propagation depends on the wave-length in a dispersive medium, there will be a rotation of the individual wave-fronts.

If ω is the angular velocity of the mirror, and ω_1 that of the dispersional rotation,

$$\omega_1 = \frac{dV}{dy} = \frac{dV}{d\lambda}\frac{d\lambda}{dy},$$

where y is the distance from the axis of rotation. But

$$\frac{d\lambda}{dy} = 2\omega\frac{\lambda}{V} \therefore \omega_1 = 2\omega\frac{\lambda}{V}\frac{dV}{d\lambda}.$$

The deflection actually observed is therefore

$$T(2\omega + \omega_1),$$

where T is the time required to travel distance $2D$; or

$$\frac{4D}{V}\omega\left(1+\frac{\lambda}{V}\frac{dV}{d\lambda}\right),$$

hence the velocity measured is

$$V'' = V \div \left(1+\frac{\lambda}{V}\frac{dV}{d\lambda}\right),$$

or, to small quantities of the second order,

$$V'' = V' = \text{group velocity}.$$[1]

The value of $\left(1+\frac{\lambda}{\mu}\frac{d\mu}{d\lambda}\right)$ for carbon disulphide for the mean wave-length of the visible spectrum is 0.93. Accordingly,

$$\frac{V_0}{V'} = \frac{V_0}{V}\frac{1}{0.93} = \frac{1.64}{0.93} = 1.76,$$

which agrees with the value found by experiment.

[1] J. W. Gibbs (*Nature*, 1886) shows that the measurement is in reality exactly that of groups and not merely an approximation.

RECENT MEASUREMENTS OF THE VELOCITY OF LIGHT

In the expression for V, the velocity of light as determined by the revolving mirror, $V=4\pi ND/\alpha$, there are three quantities to be measured, namely, N, the speed of the mirror; D, the distance between stations; and α, the angular displacement of the mirror. As has already been mentioned, the values of N and D may be obtained to one part in one hundred thousand or less. But α cannot be measured to this order of accuracy. It has been pointed out by Newcomb[1] that this difficulty may be avoided by giving the revolving mirror a prismatic form and making the distance between the two stations so great that the return light is reflected at the same angle by the next following face of the prism.

The following is an outline of a proposed attempt to realize such a project between Mount Wilson and Mount San Antonio near Pasadena, the distance being about 35 km. For this, given a speed of rotation of 1,060 turns per second, the angular displacement of the mirror during the double journey would be 90°; or, if the speed were half as great, an angle of 45° would suffice.[2] Accordingly, the revolving mirror may have the form of an octagon. It is, of course, very important that the angles should be equal, at least to the order of accuracy desired.

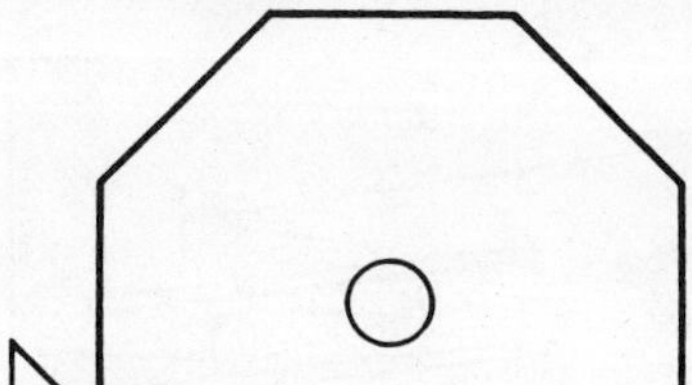

FIG. 76

This has already been attained as follows. The octagon, with faces polished and angles approximately correct, is applied to the test angle $a'b'$ made up of a 45° prism cemented to a true plane. The faces b_1b are made parallel by the interference fringes observed in

[1] *Measures of the Velocity of Light.* Nautical Almanac Office, 1882.

[2] It may be noted that with eight surfaces the resulting intensity will be four times as great as with the revolving plane-parallel disk.

monochromatic light. In general, the faces a_1a will not be parallel, and the angle between them is measured by the distance and inclination of the interference bands. The same process is repeated for each of the eight angles, and these are corrected by repolishing until the distance and inclination are the same for all, when the corresponding angles will also be equal. It has been found possible in this way to produce an octagon in which the average error was of the order of one-millionth, that is, about one-tenth to one-twentieth of a second.[1]

Another difficulty arises from the direct reflection and the scattered light from the revolving mirror. The former may be eliminated, as already mentioned, by slightly inclining the revolving mirror, but to avoid the scattered light it is essential that the return ray be received on a different surface from the outgoing.

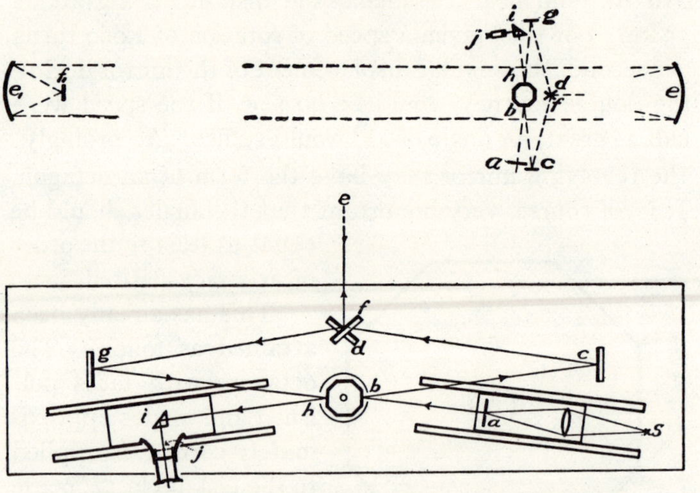

FIG. 77.—Light path a, b, c, d, e, e_1, f_1, l_1, e, f, g, h, i, j

[1] It may be noted that while a distortion may be expected when the mirror is in such rapid rotation, if the substance of the mirror (glass, in the present instance) is uniform, such distortion could only produce a very slight curvature and hence merely a minute change of focus.

Again, in order to avoid the difficulty in maintaining the distant mirror perpendicular to the incident light, the return of the ray to the home station may be accomplished exactly as in the Fizeau experiment, the only precaution required being the very accurate focusing of the beam on the small plane (better, concave) mirror at the focus of the distant collimator.

Finally, it is far less expensive to make both sending and receiving collimators silvered mirrors instead of lenses.

In Figure 77 is shown the arrangement of apparatus which fulfilled all these requirements.

Three determinations were undertaken between the home station at the Mount Wilson Observatory and Mount San Antonio 22 miles distant. The rate of the electric tuning fork was 132.25 vibrations per second, giving four stationary images of the revolving mirror when this was rotating at the rate of 529 turns per second. The fork was compared before and after every set of the observations with a free pendulum whose rate was found by comparison with an invar pendulum furnished and rated by the Coast and Geodetic Survey.

The result of eight measurements in 1924 gave

$$V_a = 299{,}735 \,.$$

Another series of observations with a direct comparison of the same electric fork with the Coast and Geodetic Survey pendulum[1] was completed in the summer of 1925 with a resulting value

$$V_a = 299{,}690 \,.$$

A third series of measurements was made in which the electric fork was replaced by a free fork making 528 vibra-

[1] This comparison was made by allowing the light from a very narrow slit to fall on a mirror attached to the pendulum. An image of the slit was formed by means of a good achromatic lens, in the plane of one edge of the fork, where it was observed by an ordinary eyepiece.

tions per second maintained by an "audion circuit," thus insuring a much more nearly constant rate. The result of this measurement gave

$$V_a = 299{,}704\ .$$

Giving these determinations the weights 1, 2, and 4, respectively, the result for the velocity in air is

$$V_a = 299{,}704\ .$$

Applying the correction of 67 km for the reduction to *vacuo* gives finally $V = 299{,}771$.

This result should be considered as provisional, and depends on the value of D, the distance between the two stations which was furnished by the Coast and Geodetic Survey, and which it is hoped may be verified by a repetition of the work.

It was also found that a trial with a much larger revolving mirror gave better definition, more light, and steadier speed of rotation; so that it seems probable that results of much greater accuracy may be obtained in a future investigation.

FINAL MEASUREMENTS

Observations with the same layout were resumed in the summer of 1926, but with an assortment of revolving mirrors.

The first of these was the same small octagonal glass mirror used in the preceding work. The result obtained this year was $V = 299{,}813$. Giving this a weight 2 and the result of preceding work weight 1 gives 299,799 for the weighted mean.

The other mirrors were a steel octagon, a glass 12-sider, a steel 12-sider, and a glass 16-sider.

The final results are summarized in Table VII.

TABLE VII

Turns per Second	Mirror	Number of Observations	Vel. of Light *in Vacuo*
528..........	Glass oct.	576	299,797
528..........	Steel oct.	195	299,795
352..........	Glass 12	270	299,796
352..........	Steel 12	218	299,796
264..........	Glass 16	504	299,796

Weighted mean, 299,796 ± 1

Table VIII shows the more reliable results of measurements of V with distance between stations, method used, and the weight assigned to each.

TABLE VIII

Author	D	Method	Wt.	V
Cornu.........	23 km	Toothed wheel	1	299,990
Perrotin.......	12	Toothed wheel	1	299,900
M_1 and M_2.....	0.6	Rev. mirror	1	299,880
Newcomb*.....	6.5	Rev. mirror	3	299,810
M_3...........	35	Rev. mirror	5	299,800

* Newcomb's value omitting all discordant observations was 298,860.

Bees, water fleas, and horseshoe crabs navigate by polarized light. Sunglasses, camera filters, and glare-free auto headlights are among other applications.

5 Popular Applications of Polarized Light

William A. Shurcliff and Stanley S. Ballard

A chapter from the book *Polarized Light* published in 1964.

If there is a logical order in which the various applications of polarizers and polarized light should be considered, the authors have never discovered it. The policy adopted here is to consider the most popular and "humanistic" applications first, and the more scientific and esoteric applications last.

POLARIZATION AND THE HUMAN EYE

The most humanistic fact about polarization of light is that it can be detected directly by the naked eye. Nearly anyone, if told carefully what to look for, can succeed in this. Sometimes he can even determine the form and azimuth of polarization.

What the observer actually "sees" is a certain faint pattern known as Haidinger's brush and illustrated in Fig. 10-1. The brush is so faint and ill-defined that it will escape notice unless the field of view is highly uniform: a clear blue sky makes an ideal background, and a brightly illuminated sheet of white paper is nearly as good. The best procedure for a beginner is to hold a linear polarizer in front of his eye, stare fixedly through it toward a clear blue sky, and, after five or ten seconds, suddenly turn the polarizer through 90°. Immediately the brush is seen. It fades away in two or three seconds, but reappears if the polarizer is again turned through 90°. The brush itself is sym-

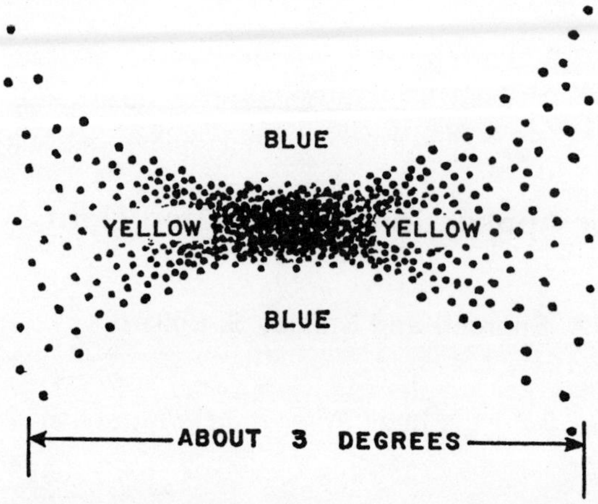

FIG. 10-1 Approximate appearance of Haidinger's brush when the vibration direction of the beam is vertical.

metric, double-ended, and yellow in color; it is small, subtending an angle of only about 2° or 3°. The adjacent areas appear blue, perhaps merely by contrast. The long axis of the brush is approximately perpendicular to the direction of electric vibration in the linearly polarized beam, i.e., perpendicular to the transmission axis of the polarizer used.

Circular polarization, too, can be detected directly by eye, and even the handedness can be determined. When an observer facing a clear blue sky places a right circular polarizer in front of his eye, he sees the yellow brush and finds that its long axis has an upward-to-the-right, downward-to-the-left direction, i.e., an azimuth of about +45°. This is true, of course, irrespective of the orientation of the polarizer, since a circle has no top or bottom. If he employs a *left* circular polarizer, he finds the brush to have a −45° orientation. In each case the pattern fades away rapidly, but can be restored to full vigor by switching to a polarizer of opposite handedness. Instead of using a circular polarizer the observer can use a single linear polarizer in series with a 90° retarder, the latter being held nearer to the eye. Turning the retarder through 90° reverses the handedness of the circular polarization.

Some people see the brush easily; others have difficulty. A few see the brush when looking innocently at the partially polarized blue sky, i.e., without using any polarizer at all, and even without meaning to see the brush. Some people see the brush more distinctly by linearly polarized light than by circularly polarized light, and for others the reverse is true. An observer may find the brush to have a slightly different orientation depending on which eye is used.

The spectral energy distribution of the light is important. If the light is rich in short-wavelength (blue) radiation, the brush is very noticeable, but if the short-wavelength radiation is eliminated by means of a yellow filter, the brush fails to appear. Use of a blue filter tends to accentuate the brush.

Although the phenomenon was discovered in 1844, by the Austrian mineralogist Haidinger, the cause is not yet fully understood. Presumably the thousands of tiny blue-light-absorbing bodies in the central (foveal) portion of the retina are dichroic and are oriented in a radial pattern, for example, a pattern such that the absorption axis of each body lies approximately along a radius from the center of the fovea. Incident linearly polarized light will then be absorbed more strongly in some parts of the pattern than in other parts and consequently some parts will fatigue more than others. When the vibration direction of the light is suddenly changed, the varying degrees of fatigue are revealed as a subjective radial pattern. Presumably no such dichroism or orientation pattern applies to longer wavelength (yellow and red) light; consequently a yellow sensation dominates in those regions where fatigue-to-blue has occurred.

The fact that circular polarization, also, may be detected perhaps implies that some transparent portion of the eye is weakly birefringent and acts like a retarder, converting circularly polarized light to linearly or elliptically polarized light. The direction of the major axis of the ellipse depends only on the direction of the fast axis of the retarding layer and hence remains fixed—unless the observer tips his head.

Perhaps physicists will some day write matrices to describe the retarding layers and dichroic areas of the eye. Poets were the first to see magic fire and jewels in the human eye; physicists will be the first to see matrices!

Bees, too, can detect the vibration direction of linearly polarized light. The experiments of the biologist K. von Frisch during World War II showed that bees "navigate" back and forth between hive and source of honey by using the sun as a guide. More interesting, when the sun is obscured by a large area of clouds the bees can still navigate successfully if they can see a bit of blue sky: they can detect the azimuth of linear polarization of the blue light and navigate with respect to it. One way of demonstrating the bee's ability to detect the azimuth of polarization is to place the bee in a large box the top of which consists of a huge sheet of linear polarizer, such as H-sheet. Each time the experimenter turns the polarizer to a different azimuth, the bee changes his direction of attempted travel correspondingly.

Certain other animals also can detect the polarization of skylight and navigate by it. This includes ants, beetles, and the fruit fly *Drosophila*. Probably many other examples will be discovered.

POLARIZATION OF SKY LIGHT

Blue-sky light traveling in a direction roughly at right angles to the sun's rays is partially polarized. When an observer holds a linear polarizer in front of his eye and gazes in a direction perpendicular to the direction of the sun, he finds that rotating the polarizer slowly causes the sky to change from bright to dark successively. The degree of polarization of sky light may reach 70 or 80 percent when the air is clear and dust-free, the sun is moderately low in the sky, and the observation direction is near the zenith.

The polarization is a result of the scattering of the sun's rays by the molecules in the air. Rayleigh's well-known inverse-fourth-power law relating scattering intensity to wavelength accounts for the blue color of the scattered light, and the asymmetry associated with the 90° viewing angle accounts for the polarization, as explained in Chapter 5. Some multiple scattering occurs, and this reduces the degree of polarization somewhat; when the observer ascends to a higher altitude, the amount of air involved is reduced, multiple scattering is reduced, and the degree of polarization is increased. A further increase results when a yellow or red filter is used to block the short-wavelength component of the light and transmit the long-wavelength component —the latter component is less subject to multiple scattering. (The situation is very different for infrared radiation of wavelength exceeding 2 microns: much of this radiation is produced by emission from the air itself, rather than by scattering, and this exhibits little or no polarization.)

Some persons are capable of detecting the polarization of sky light directly by eye, by virtue of the Haidinger brush phenomenon discussed in a preceding section; a few individuals find the brush noticeable enough to be a nuisance. Ordinarily, of course, it escapes notice and plays little part in the affairs of man. Its practical use by bees, ants, etc., has been indicated, and the importance to photographers is discussed in a later section.

POLARIZATION OF LIGHT UNDER WATER

A surprising fact about the polarization found in light present beneath the surface of the ocean (or of a pond) is that the predominant direction of electric vibration is horizontal. The opposite might be expected, since most of the light that enters the water enters *obliquely* from above, and the most strongly reflected component of obliquely incident light is the horizontally vibrating component. But oceanographers and biologists, working at depths of 5 to 30 feet in waters off Bermuda and in the Mediterranean Sea, have found the main cause of submarine polarization to be the scattering of the light by microscopic particles suspended in the water. Sunlight and sky light enter the water from above, and the average direction of illumination is roughly vertical; consequently the polarization form of the scattered light that travels horizontally toward an underwater observer is partially polarized with the electric vibration direction horizontal. The situation is much the same as that discussed in Chapter 5, except that the incident light has a more steeply downward direction and the asymmetric scattering is by microscopic particles instead of molecules.

Typically, the degree of polarization is 5 to 30 percent, an amount found to be important to a variety of underwater life. The water flea *Daphnia* tends to swim in a direction perpendicular to the electric vibration direction, for reasons not yet known. When tests are conducted in a tank filled with water that is free of suspended particles, so that the submarine illumination is practically unpolarized, *Daphnia* ceases to favor any one direction. But if suspended matter is added, thus restoring the polarization, *Daphnia* resumes the custom of traveling perpendicular to the vibration direction.

The arthropod *Limulus* (horse shoe crab) easily detects the polarization of the underwater light and is presumed to navigate with respect to the electric vibration direction. The same is true of the crustacean *Mysidium gracile* and various other forms of marine life. Most tend to swim perpendicularly to the vibration direction; some swim parallel to it; a few swim at different relative orientations depending on the time of day. For all of these animals, polarization is a compass that works even under water!

POLARIZING SUNGLASSES

The lenses of ordinary sunglasses employ absorbing materials that are isotropic, and accordingly the incident light is attenuated by a fixed factor irrespective of polarization form. This is unfortunate. The fact is that "glare" consists predominantly of light having a horizontal vibration direction. Why? For these reasons:

(a) The main source of light (sun and sky) is overhead, and consequently the main flux of light is downward.

(b) The surfaces that are most strongly illuminated by the downward flux are horizontal surfaces.

(c) Such surfaces are usually viewed obliquely, since a person seldom looks straight down.

(d) Most outdoor objects are of dielectric material.

(e) Light reflected obliquely from a horizontal dielectric surface is partially linearly polarized with the dominant vibration direction *horizontal,* as explained in Chapter 4.

Polarizing sunglasses take full advantage of this fact. The lenses are made of dichroic material (H-sheet, usually) oriented with the transmission axis vertical, as indicated in Fig. 10-2a, so

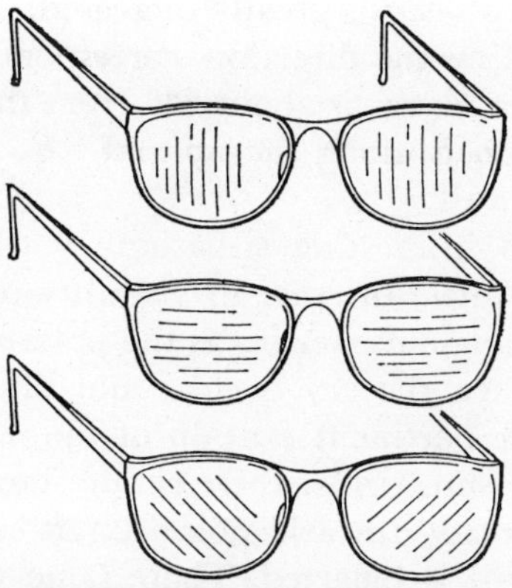

FIG. 10-2 Three types of polarizing spectacles. In (a) the transmission axis is vertical, for eliminating glare reflected from horizontal surfaces. In (b) the axis is horizontal, for eliminating reflections from vertical windows of trains, store-fronts (show-windows), etc. In (c) the axis directions are 45° and −45°, a standard arrangement used in viewing polarization-coded stereoscopic pictures.

that almost all of the horizontal vibrations are absorbed. The component having vertical vibration direction is transmitted. Usually some isotropic absorber is included in the lenses to absorb ultraviolet light strongly and blue and red light to a moderate extent; the sunglasses then have a greenish hue which has nothing to do with the polarization.

Motorists and vacationists find that polarizing sunglasses are helpful not only in reducing the brightness of the field of view as a whole, but also in enhancing the beauty of the scene. Because specularly reflected light is absorbed preferentially, roads, trees, grassy fields, etc., appear softer and more deeply colored through polarizers. Specularly reflected light tends to veil nature's inherent beauty; polarizing sunglasses remove the veil.

Fishermen and boatsmen enjoy another benefit from wearing polarizing sunglasses. They want to be able to see fish, rocks, etc., beneath the surface of the water, yet the light from such objects is dim and is usually lost in the "noise" of the sky light reflected obliquely from the surface. Since the reflected light is highly polarized with horizontal vibration direction, the polarizing sunglasses absorb this component strongly, and the visibility of the underwater objects is greatly increased. The increase is greatest when the viewing direction corresponds to the polarizing angle, which, for water, is about 53° from the normal. When the viewing direction is along the normal, i.e., straight down, there is no increase at all.

There is one interesting situation in which polarizing sunglasses produce little increase in visibility of underwater objects even when the angle of viewing is the polarizing angle. This situation occurs when the sky is clear and blue, the sun is low in the sky, and the pertinent portion of the sky is at 90° from the direction of the sun. Under these circumstances the light striking the water is already linearly polarized at such an azimuth that almost none of it is reflected. There is no task left for the sunglasses to perform—there is no reflected glare to suppress. The underwater objects are seen with great clarity. Persons unfamiliar with the polarization of sky light and with the dependence of oblique reflection on polarization form are likely to ascribe the remarkable clarity to "especially clear water" rather than to absence of reflection.

CAMERA FILTERS

Photographers often wish to enhance the contrast between blue sky and white clouds. Thirty years ago they did this by employing a yellow filter, which absorbed most of the blue light from the clear sky but transmitted most of the white light from the clouds. Using ordinary black-and-white film, they obtained excellent contrast by this method. Today, photographers are using color film increasingly, and the use of yellow filters is no longer permissible since it would eliminate all blue colors from the finished photograph.

The only known solution is to exploit the difference in polarization between blue sky and white clouds. Light from most portions of the blue sky is partially linearly polarized, as explained in a preceding section, and light from clouds is unpolarized. Therefore a neutral-color, linear polarizer mounted at the optimum azimuth in front of the lens will absorb a large fraction (e.g., 80 percent) of the sky light while transmitting a large fraction (nearly half) of the light from the clouds; thus the contrast is increased by a factor of two or three. The factor is less if the air is hazy, and more if the air is extremely clear (as in Arizona) and if the camera is aimed about 90° from the direction of the sun.

The usual way of choosing the azimuth of the polarizer is crude, but perhaps adequate. The photographer holds the polarizer in front of his eye, finds by trial and error which azimuth maximizes the contrast of the clouds in question, and then attempts to mount the polarizer on the camera without changing the azimuth of the polarizer. One type of polarizing filter for cameras is equipped with a small "satellite" polarizer mounted at the end of a short arm and aligned permanently with the main polarizer. The photographer installs the main polarizer in front of the lens, looks through the small polarizer and turns the arm to whatever azimuth maximizes the contrast. Both polarizers then have this optimum orientation. The satisfactoriness of the azimuth can be checked visually at any time. Instead of using these empirical methods, a scientifically minded photographer can proceed by dead reckoning, i.e., by following this well-known rule: Mount the polarizer so that its transmission axis lies in the plane determined by camera, sun, and object photographed. (So oriented, the polarizer performs a valuable additional service: it eliminates most of the specularly reflected light from trees, roads, etc., and enhances the softness and depth of color of the scene.)

When a photographer standing on a sidewalk tries to photograph objects situated behind a store window, the reflection of the street scene from the window may threaten to spoil the photograph. An excellent solution is to place the camera off to one side so that the window is seen obliquely at about the polarizing angle, and mount a linear polarizer in front of the lens; the polarizer is turned so that its transmission axis is horizontal, and the polarized light reflected from the window is absorbed. The authors have a friend who has applied this same principle to a pair of special spectacles he wears while touring the country by railroad. The lenses consist of polarizers oriented with the transmission axis horizontal, as indicated in Fig. 10-2b; thus when he gazes out of the train window in oblique forward direction, the reflected images of passengers and newspapers are wiped out, and the scenery appears in its pristine glory.

USE OF CIRCULAR POLARIZERS IN ELIMINATING PERPENDICULARLY REFLECTED LIGHT

Eliminating perpendicularly reflected lights is a different problem from that of eliminating obliquely reflected light. The process of oblique reflection at Brewster's angle causes the reflected beam to be linearly polarized, and accordingly a linear polarizer can eliminate the reflected beam entirely. But the process of *normal* reflection, i.e., with incident and reflected beams *perpendicular* to the smooth glossy surface in question, produces no polarization at all. How, then, can the specularly reflected light be eliminated while light originating behind the surface is transmitted freely?

The question is an important one to radar operators scanning the cathode-ray-oscilloscope screens on which dim greenish spots representing airborne objects appear. The screen proper is situated in a large evacuated tube, and the greenish light emerges through a curved glass window at the front end of the tube. (Sometimes the window is flat; sometimes a safety plate of glass or plastic is mounted close in front of it.) Often the operator has difficulty in seeing the greenish spots, not only because they are faint, but also because they may be masked by various extraneous images reflected by the front surface of the window, e.g., reflections of room lights and of people, clothing, papers, etc., situated near the operator. Extinguishing the room lights would eliminate these reflections, but would make it impossible for the operator to read instructions or make notes. What he needs is some kind of filter that will transmit the light originating behind the window and absorb the light reflected approximately perpendicularly from it.

This need is filled by the circular polarizer. Such a device, if mounted close in front of the window, will transmit nearly half of the light that originates behind the window, yet will eliminate about 99 percent of the room light that is reflected perpendicularly from it. The circular polarizer acts on the room light *twice:* it circularly polarizes room light that is approaching the window, then absorbs the reflected component. The logic behind this requires explanation. Two key facts must be kept in mind:

(1) A beam that is reflected perpendicularly and specularly by a smooth glossy surface has the same degree of polarization as the incident beam, since the reflection process does not introduce randomness of any kind.

(2) The reflection process reverses the handedness of polarization, because handedness is defined with respect to the propagation direction and the reflection process reverses the propagation direction.

If the polarizer is of right-circular type, as in the arrangement shown in Fig. 10-3, room light that passes through and ap-

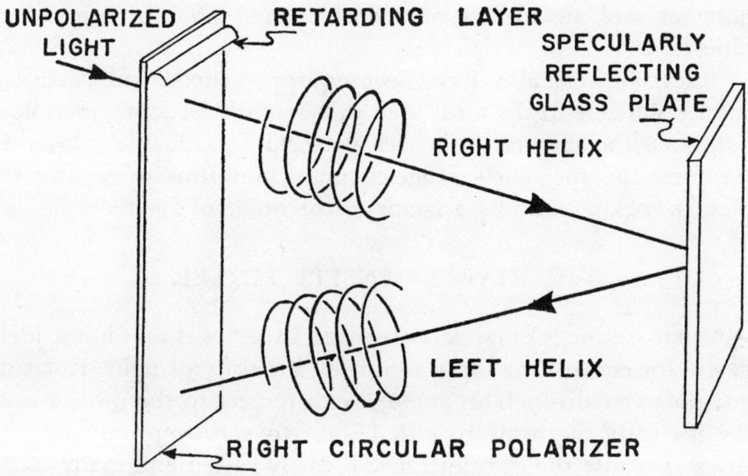

FIG. 10-3 Use of a circular polarizer in absorbing light reflected by a surface approximately perpendicular to the incident beam. Note that the reflection process reverses the handedness of circular polarization.

proaches the window is right-circularly polarized; the reflected light is *left*-circularly polarized and hence is *totally absorbed by the polarizer*. In effect, the circular polarizer "codes" the light, the window reverses the coding, and the polarizer then annihilates the reverse-coded beam. If both faces of the window are ideally flat and smooth, if the light is incident exactly along the normal, and if the polarizer is truly of circular type, the reflected light is totally absorbed. Usually the conditions are less ideal: the rear surface of the window usually serves as support for the luminescent screen and has a matte appearance; the window is usually curved and much of the troublesome room light incident on the window makes an angle of 10° or 20° or more with the normal; and the polarizer, although circular with respect to some wavelengths, is elliptical with respect to others. Nevertheless, the improvement provided by the polarizer is large, and the amount of faint detail that the operator can see on the screen is greatly increased.

One precaution must be mentioned: reflections from the polarizer itself must be avoided. This is usually accomplished by tilting the polarizer forward so that the only reflected images the observer sees are images of a dark-colored floor or other dark objects.

Television sets, also, have been equipped successfully with circular polarizers. If the set is used in a brightly lit room, or is used outdoors, the circular polarizer performs a valuable service in trapping the specularly reflected glare and thus increasing the picture-vs-glare ratio by a factor of the order of 10.

VARIABLE-DENSITY FILTER

A pair of linear polarizers arranged in series is an almost ideal device for controlling the transmitted intensity of light. Rotating one polarizer through an angle θ with respect to the other causes the intensity of the transmitted light to vary approximately as $\cos^2 \theta$. Because the transmittance is easily varied and easily calculated, the pair of polarizers has found much favor in the eyes of designers of spectrophotometers and other devices for controlling and measuring light intensity.

Specially designed sunglasses employing pairs of linear polarizers in place of lenses have been used successfully by aviators and others. One polarizer of each pair can be rotated through an angle as large as 90°, and a linkage connecting the two pairs insures that the attenuation is the same for both eyes. By moving one small lever, the wearer can vary the transmittance throughout a range of about 10,000 to 1.

Controllable pairs of very-large-diameter polarizers have been used as windows of railroad cars and ocean liners. A person sitting near such a window turns a small knob to rotate one polarizer with respect to the other and thus reduce the intensity of the transmitted light to any extent desired.

One of the authors has experimented with a variable-density filter employing *three* linear polarizers in series, in order that a transmittance range of 10^8 to 1 could be achieved. The device worked well and, as expected, obeyed a cosine-fourth, rather than a cosine-square law.

THREE-DIMENSIONAL PHOTOGRAPHY AND THE USE OF POLARIZERS FOR CODING

Millions of polarizers found their way into the motion picture theaters of North America in 1952 and 1953 when stereoscopic (three-dimensional, or 3-D) movies achieved brief prominence. Each spectator wore a pair of polarizing spectacles called viewers, and polarizers were mounted in front of the projectors.

A photographer who enjoys looking at 3-D still pictures in his living room needs no polarizers. Usually he employs a small viewing box containing a light source and two lenses, one for each eye; a black partition, or septum, divides the box into right and left halves. The picture, consisting of two small transparencies mounted about two inches apart in a side-by-side arrangement on a cardboard frame, is inserted in the box so that the right-eye transparency lines up with the right lens and the left-eye transparency lines up with the left lens. (The two transparencies are, of course, slightly different because they were taken by cameras situated about two or three inches apart; the spacing used approximates the spacing of the two eyes.) The side-by-side arrangement of the two transparencies and the presence of the septum insure that the observer's right eye sees only the right transparency and the left eye sees only the left transparency. No cross-communication, or "cross-talk," can occur. Consequently the observer enjoys an impressively realistic stereoscopic illusion.

When 3-D motion-picture films are projected in a theater, many complications arise. Separate projectors must be provided for the right-eye and left-eye movie films, and the two projectors must be synchronized within about 0.01 second. Since there is just one large screen and this is to be viewed by hundreds of spectators, there can be no septum. Indeed, no practical geometrical method of preventing cross-talk is known.

Before the advent of mass-produced polarizers in the 1930's, an *analglyph* system of preventing cross-talk was invented. It applied wavelength coding to the two projected beams. The right-eye picture was projected through a long-wavelength (red) filter, and the left-eye picture was projected through a shorter-wavelength (green) filter. The spectator's viewers contained right and left lenses of red and green plastic, respectively, and accordingly each lens transmitted light from the appropriate projector and absorbed light from the other. Thus each eye received just the light intended for it. The system succeeded as a short-term novelty: stereoscopic illusions were created. But the system had two major defects: chromatic "retinal rivalry" between the two eyes, and incompatibility with the showing of colored motion pictures.

In the 1930's the problem was solved with éclat by a polarization-coding system, demonstrated with great impact at the New York World's Fair of 1939 and improved in later years. As indicated in Fig. 10-4, a linear polarizer oriented with its transmission axis at $-45°$ is placed in front of the projector used for the right-eye pictures, and a polarizer at $+45°$ is placed in front of the projector used for the left-eye pictures. Thus the two beams striking the movie screen are orthogonally coded. The lenses of the spectator's viewers consist of correspondingly oriented linear polarizers, and so each eye receives only light that originates in the appropriate projector. Superb stereoscopic illusions result. Since the polarizers perform well at all wavelengths in the visual range, color movies can be presented as easily and faithfully as can black-and-white movies.

The polarizers placed in front of the projectors consist, ordinarily, of K-sheet; as explained in Chapter 3, K-sheet is highly resistant to heat, and any polarizing filter placed close in front of a powerful projector is bound to heat up considerably since it necessarily absorbs about half the light. The lenses of the 3-D viewers are usually of HN-38 sheet; it has high major transmittance k_1 and small minor transmittance k_2, and it is inex-

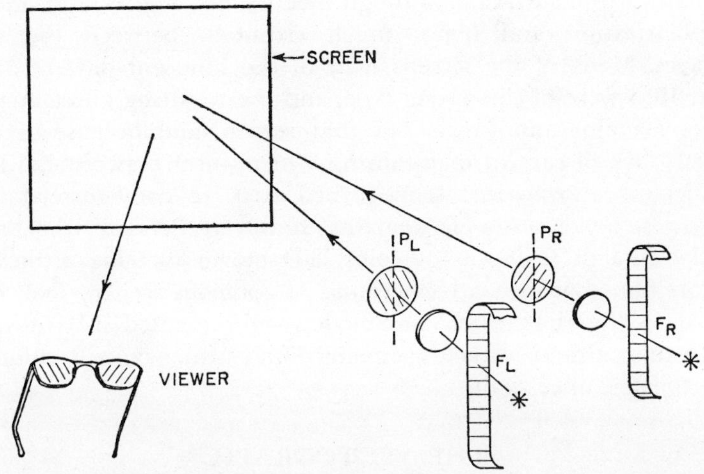

FIG. 10-4 Arrangement for projecting polarization-coded stereoscopic motion-picture films by means of two side-by-side projectors. Films F_R and F_L containing the "right-eye pictures" and "left-eye pictures" are mounted in the right and left projectors, which are equipped with linear polarizers P_R and P_L oriented at $-45°$ and $+45°$ respectively. The viewer contains correspondingly oriented polarizers, and accordingly each eye sees only the images intended for it.

pensive. The viewers are cheap enough (about 10¢ each) that they can be discarded after a single use.

The polarization-coding scheme has one limitation: if the spectator tilts his head to one side, the polarizers in his viewers no longer line up accurately with the respective polarizers on the projectors. Thus cross-talk occurs: the right eye sees faintly the image meant for the left eye, and vice versa: each eye sees a faint ghost image in addition to the main image. The spectator does not enjoy this. The difficulty could be avoided if the linear polarizers were replaced by high-quality, achromatic circular polarizers, but unfortunately no method is known for producing achromatic circular polarizers economically.

The effectiveness of any polarization-coding projection system is destroyed if the screen depolarizes the light appreciably. Screens that have a smooth aluminum coating usually conserve polarization to the extent of about 99 percent, but those having a matte white surface or a rough metallic coating produce much depolarization and hence much cross-talk between the two images. Many of the screens used in the innocent days of 1952 and 1953 were of the wrong type, and the resulting ghost images were a major annoyance. For that reason, and because of frequent lack of care in maintaining synchronism between the two projectors, movie-goers soon turned back to conventional 2-D pictures. Some nostalgia remains, however. Persons who were lucky enough to see a full-color, 3-D movie showing attractive actors filmed against a background of gorgeous scenery look forward to the time when well-made, well-presented 3-D movies, with their almost miraculous realism and intimacy, will animate the theaters once again.

THE VECTOGRAPH

The type of three-dimensional photography discussed in the preceding section is parallel-projected 3-D photography. The two motion-picture films are situated side-by-side, and two projectors are operated in parallel. During the late 1930's a radically new approach, called *vectography,* was developed by E. H. Land, J. Mahler, and others. In this system, the two films are arranged in series, bonded together. Because of the permanent series arrangement, many problems disappear. Only one projector is needed, and perfect synchronism is "guaranteed at the factory." Each pair of pictures (each vectograph) is projected as a single unit, in the same projector aperture and at the same time, and onto the same area of the same screen. If the film breaks, it can be spliced with no concern as to preservation of synchronism.

The method can succeed only if means are provided for preserving the identity of the two coincident projected beams. Again, polarization-coding is the answer. However, because the two images are bonded together in series, the coding must occur within the images themselves. In the system used by Land and Mahler each image consists of varying quantities of linearly dichroic molecules aligned in a common direction, and the directions employed in the two images are mutually at right angles. Dark areas in any one image contain a high concentration of dichroic molecules; light areas contain little or no dichroic material; but irrespective of concentration, the alignment direction is always the same. For the other image, the alignment direction is always orthogonal to the first. It is to be noted that the images contain no silver and no other isotropic absorber. Only aligned absorbers having high dichroic ratio are used.

A communications engineer would describe the vectograph by saying that it provides two distinct channels. Each is assigned to one image. Each is independent of the other. Since the vectograph images themselves perform the polarization coding, no polarizer is used in front of the projector; indeed, the interposition of such a filter would play havoc with the system. As before, the screen must preserve the polarization and the spectator's viewers must perform the appropriate decoding, or discriminating, act. Excellent stereoscopic effects are achieved. However, the production of vectograph film is a costly undertaking involving very specialized equipment, and constant attention is needed to maintain high enough dichroic ratio so that the channels are truly independent and ghost images are avoided.

Vectograph pictures of the "still" type are easier and cheaper to make than vectograph movies. Stereo pairs of aerial photographs of mountainous country, if presented in vectograph form, give a navigator (wearing an appropriate viewer) a very realistic impression of the terrain, and a map maker can prepare an accurate contour map from the vectograph with ease.

POLARIZING HEADLIGHTS

It is ironic that the main goal of Land and others in developing high-quality, large-area, low-cost polarizers has never been achieved. The polarizers are used with great success in dozens of applications, but not the application that was uppermost in the minds of the inventors.

Their goal was to eliminate glare from automobile headlights. In an era when dual-lane highways, circumferential bypasses, and other safety engineering advances were virtually unknown and the aim and focus of automobile headlights were highly erratic, the glare that confronted motorists at night was almost unbearable, and was an important cause of accidents. As early as 1920 several illumination engineers recognized that the glare could be eliminated by means of polarizers—if large-area polarizers could somehow be produced. If every headlight lens were covered by a linear polarizer oriented with the transmission axis horizontal and every windshield were covered with a linear polarizer oriented with its axis vertical, no direct light from the headlights of Car A could pass through the windshield of oncoming Car B. Drivers in both cars could see road-markings, pedestrians, and so forth, but neither would experience any glare from the other's headlights. Moreover, it would be permissible for each driver to use his *high* beam continuously, and accordingly his ability to see pedestrians, etc., would be greater than before, despite the fact that each polarizer would transmit only about half of the light incident on it.

It was soon recognized that the analyzing polarizer should not be made a permanent part of the windshield, but should be incorporated in a small visor situated just in front of the driver's eyes. During the day, when headlights were not in use, the visor could be swung out of the way. It was also recognized that care should be taken to make sure the headlight polarizers had sufficient light-leak, i.e., sufficiently large k_2 value, that the headlights of oncoming cars would not disappear entirely!

Land and his colleagues moved rapidly. They invented a whole series of polarizers, each superior to its predecessor. The first successful type, J-sheet, employed aligned, microscopic crystals of the dichroic mineral herapathite; the method of manufacture is described in Chapter 3. Then came H-sheet, which was better in nearly every respect and in addition was easier to make. Finally, K-sheet appeared; it had most of the superb qualities of the earlier materials and the added virtue of being unaffected by fairly high temperature, such as 215°F. To persons seeking polarizers for use in headlights, K-sheet appeared to be the pot of gold at the end of a polarized rainbow.

Concurrently, several better ways of orienting the polarizers were proposed. One attractive scheme was to orient the headlight polarizers and the visor polarizer at the identical azimuth, namely −45°, as indicated in Fig. 10-5. Then, even a polarization-conserving object in the path of the headlights would appear

FIG. 10-5 Automobile equipped with headlight polarizers and a visor polarizer oriented at −45°. When two such cars approach one another, each driver is protected from the glare from the headlights of the other.

to the driver to be brightly illuminated. (This would not be the case if his visor polarizer were crossed with his headlight polarizers.) The −45° system disposed of the headlight glare problem adequately: if two cars A and B both equipped in this manner approached one another at night, each driver's visor would be crossed with the other car's headlight polarizers, and neither driver would experience any glare.

Using the Mueller calculus, Billings and Land compared a wide variety of polarizer orientation schemes, and found several to be particularly attractive. Perhaps the best system was one called "−55°, −35°." The transmission axes of the headlight polarizers and visor polarizer are at 55° and 35° from the vertical, respectively, an arrangement that minimizes complications stemming from the obliquity of the portion of the windshield situated just in front of the driver.

Despite the successes on all technical fronts, the project bogged down. To this day no one knows just why. Probably many little reasons were responsible. Among these were the following:

(1) The polarizers absorbed slightly more than half of the light incident on them, and accordingly the automobile manufacturers felt that they would have to increase the power of the lamps themselves and perhaps use larger generators and batteries also.

(2) Some windshields were moderately birefringent; therefore

they would act like retarders, alter the polarization form of the incident light, and allow some glare to leak through.

(3) Nearly every year the automobile manufacturers increased the backward tilt of the windshields; such tilt tends to alter the polarization form of light having an oblique vibration direction, and hence leads to glare-leak.

(4) Passengers, as well as drivers, would require visors, since passengers also dislike glare.

(5) Pedestrians might find that the glare was worse than ever, unless they too employed polarizing visors or spectacles.

(6) The system would succeed only if adopted by *all* car manufacturers, and therefore no one manufacturer would gain any promotional advantage from it.

(7) The first few drivers to put the system to use would get little benefit from it for at least a year or two, i.e., until millions of other cars were similarly equipped.

(8) It was difficult to decide when and how to force the owners of old cars to install the necessary polarizers on their cars.

(9) The patents on the only fully satisfactory polarizers were held by a single company.

(10) To introduce the system would require formal, coordinated action by all States.

(11) Improvements in headlight design and aiming, the increasing numbers of dual-lane highways, and the brighter street lamps used in citics and suburbs led some people to believe that the need for a polarization-type of glare control was no longer acute.

However, persons who have actually experienced the polarization method of glare removal are convinced that the drawbacks are trivial compared to the benefits.

Perhaps some day the system will be tried out on a pilot scale in a small, isolated community, where all the cars could be equipped with polarizers in a few weeks. Perhaps an island of moderate size would make a good test ground. If the system is found to be highly successful there, it will presumably spread throughout every country that teems with automobiles.

An explanation of how the eye works, by the biologist who won a Nobel Prize for contributions to this field.

6 Eye and Camera

George Wald

A *Scientific American* article, 1950

OF all the instruments made by man, none resembles a part of his body more than a camera does the eye. Yet this is not by design. A camera is no more a copy of an eye than the wing of a bird is a copy of that of an insect. Each is the product of an independent evolution; and if this has brought the camera and the eye together, it is not because one has mimicked the other, but because both have had to meet the same problems, and frequently have done so in much the same way. This is the type of phenomenon that biologists call convergent evolution, yet peculiar in that the one evolution is organic, the other technological.

Over the centuries much has been learned about vision from the camera, but little about photography from the eye. The camera made its first appearance not as an instrument for making pictures but as the *camera obscura* or dark chamber, a device that attempted no more than to project an inverted image upon a screen. Long after the optics of the camera obscura was well understood, the workings of the eye remained mysterious.

In part this was because men found it difficult to think in simple terms about the eye. It is possible for contempt to breed familiarity, but awe does not help one to understand anything. Men have often approached light and the eye in a spirit close to awe, probably because they were always aware that vision provides their closest link with the external world. Stubborn misconceptions held back their understanding of the eye for many centuries. Two notions were particularly troublesome. One was that radiation shines out of the eye; the other, that an inverted image on the retina is somehow incompatible with seeing right side up.

I am sure that many people are still not clear on either matter. I note, for example, that the X-ray vision of the comic-strip hero Superman, while regarded with skepticism by many adults, is not rejected on the ground that there are no X-rays about us with which to see. Clearly Superman's eyes supply the X-rays, and by directing them here and there he not only can see through opaque objects, but can on occasion shatter a brick wall or melt gold. As for the inverted image on the retina, most people who learn of it concede that it presents a problem, but comfort themselves with the thought that the brain somehow compensates for it. But of course there is no problem, and hence no compensation. We learn early in infancy to associate certain spatial relations in the outside world with certain patterns of nervous activity stimulated through the eyes. The spatial arrangements of the nervous activity itself are altogether irrelevant.

It was not until the 17th century that the gross optics of image formation in the eye was clearly expressed. This was accomplished by Johannes Kepler in 1611, and again by René Descartes in 1664. By the end of the century the first treatise on optics in English, written by William Molyneux of Dublin, contained several clear and simple diagrams comparing the projection of a real inverted image in a "pinhole" camera, in a camera obscura equipped with a lens and in an eye.

Today every schoolboy knows that the eye is like a camera. In both instruments a lens projects an inverted image of the surroundings upon a light-sensitive surface: the film in the camera and the retina in the eye. In both the opening of the lens is regulated by an iris. In both the inside of the chamber is lined with a coating of black material which absorbs stray light that would otherwise be reflected back and forth and obscure the image. Almost every schoolboy also knows a difference between the camera and the eye. A camera is focused by moving the lens toward or away from the film; in the eye the distance between the lens and the retina is fixed, and focusing is accomplished by changing the thickness of the lens.

The usual fate of such comparisons is that on closer examination they are exposed as trivial. In this case, however, just the opposite has occurred. The more we have come to know about the mechanism of vision, the more pointed and fruitful has become its comparison with photography. By now it is clear that the relationship between the eye and the camera goes far beyond simple optics, and has come to involve much of the

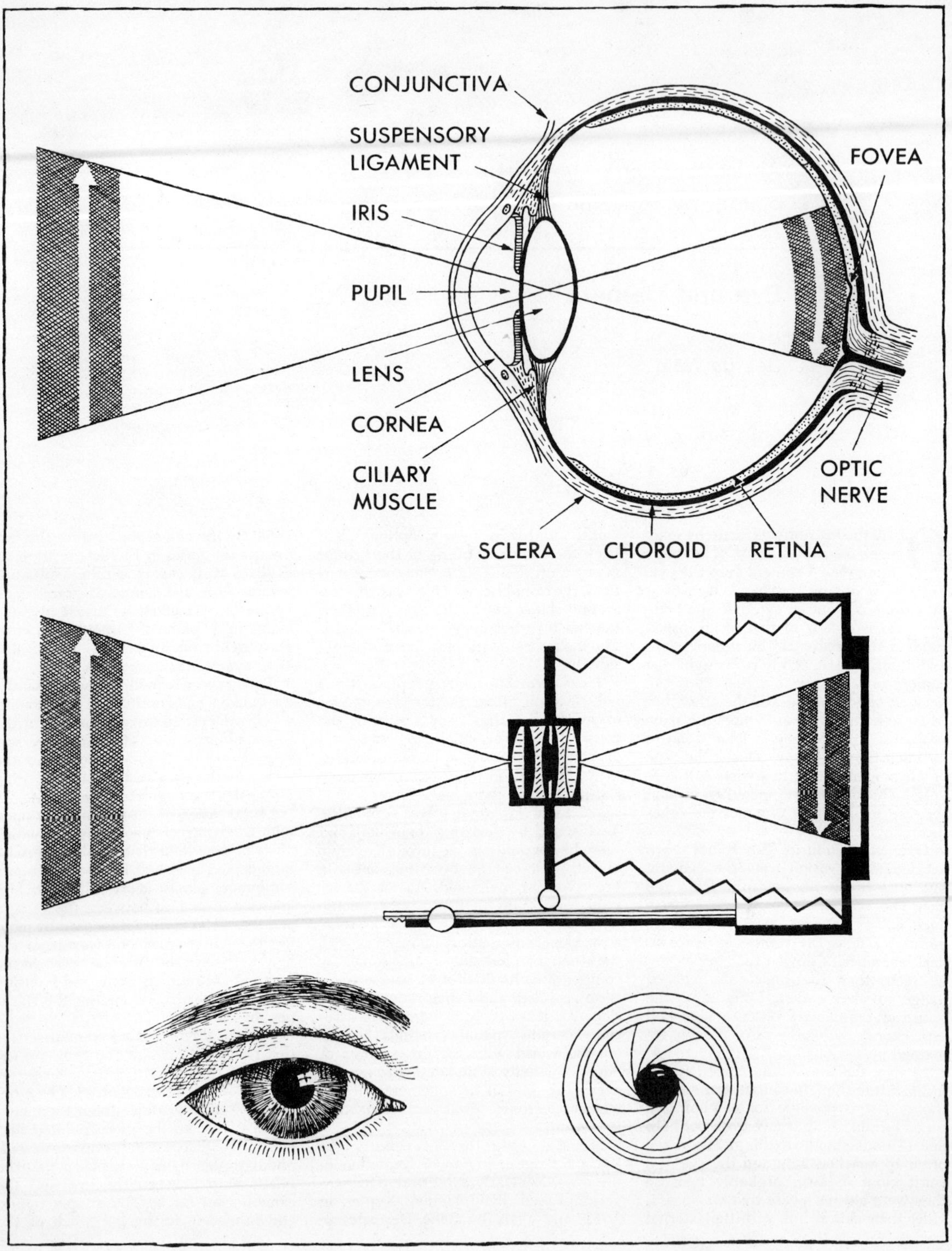

OPTICAL SIMILARITIES of eye and camera are apparent in their cross sections. Both utilize a lens to focus an inverted image on a light-sensitive surface. Both possess an iris to adjust to various intensities of light. The single lens of the eye, however, cannot bring light of all colors to a focus at the same point. The compound lens of the camera is better corrected for color because it is composed of two kinds of glass.

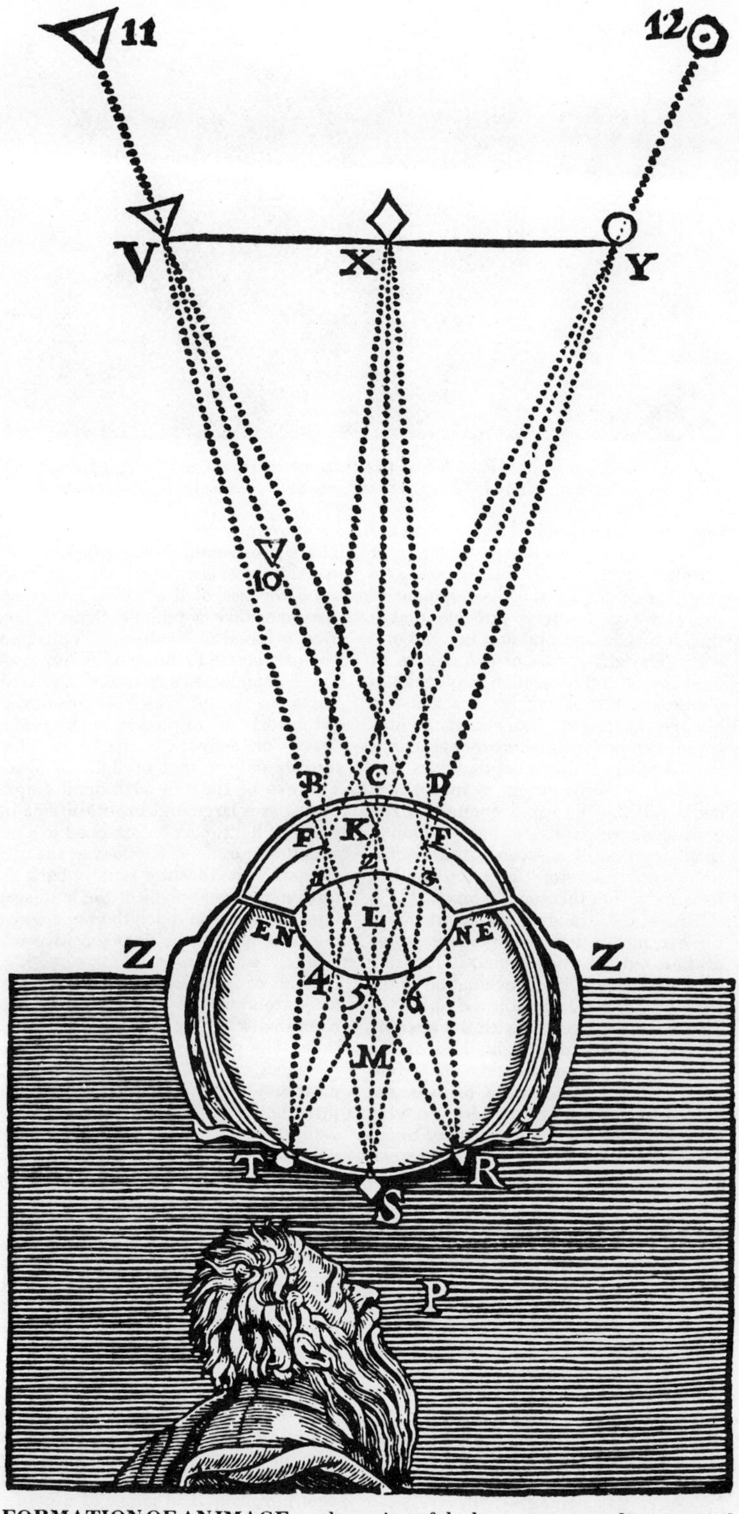

FORMATION OF AN IMAGE on the retina of the human eye was diagrammed by Rene Descartes in 1664. This diagram is from Descartes' *Dioptrics.*

essential physics and chemistry of both devices.

Bright and Dim Light

A photographer making an exposure in dim light opens the iris of his camera. The pupil of the eye also opens in dim light, to an extent governed by the activity of the retina. Both adjustments have the obvious effect of admitting more light through the lens. This is accomplished at some cost to the quality of the image, for the open lens usually defines the image less sharply, and has less depth of focus.

When further pressed for light, the photographer changes to a more sensitive film. This ordinarily involves a further loss in the sharpness of the picture. With any single type of emulsion the more sensitive film is coarser in grain, and thus the image cast upon it is resolved less accurately.

The retina of the eye is grainy just as is photographic film. In film the grain is composed of crystals of silver bromide embedded in gelatin. In the retina it is made up of the receptor cells, lying side by side to form a mosaic of light-sensitive elements.

There are two kinds of receptors in the retinas of man and most vertebrates: rods and cones. Each is composed of an inner segment much like an ordinary nerve cell, and a rod- or cone-shaped outer segment, the special portion of the cell that is sensitive to light. The cones are the organs of vision in bright light, and also of color vision. The rods provide a special apparatus for vision in dim light, and their excitation yields only neutral gray sensations. This is why at night all cats are gray.

The change from cone to rod vision, like that from slow to fast film, involves a change from a fine- to a coarse-grained mosaic. It is not that the cones are smaller than the rods, but that the cones act individually while the rods act in large clumps. Each cone is usually connected with the brain by a single fiber of the optic nerve. In contrast large clusters of rods are connected by single optic nerve fibers. The capacity of rods for image vision is correspondingly coarse. It is not only true that at night all cats are gray, but it is difficult to be sure that they are cats.

Vision in very dim light, such as starlight or most moonlight, involves only the rods. The relatively insensitive cones are not stimulated at all. At moderately low intensities of light, about 1,000 times greater than the lowest intensity to which the eye responds, the cones begin to function. Their entrance is marked by dilute sensations of color. Over an intermediate range of intensities rods and cones function together, but as the brightness increases, the cones come to dominate vision. We do not know that

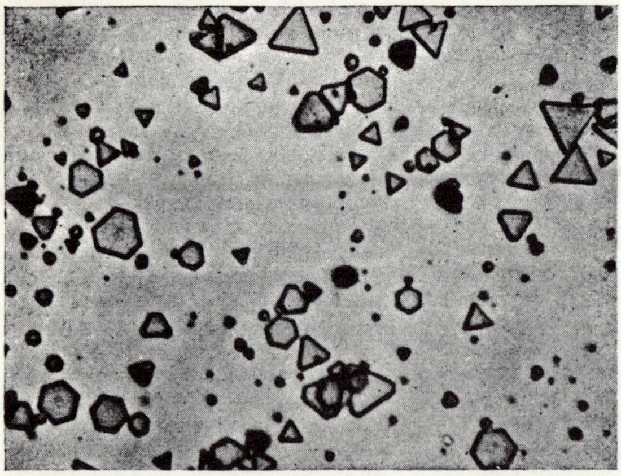

GRAIN of the photographic emulsion, magnified 2,500 times, is made up of silver-bromide crystals in gelatin.

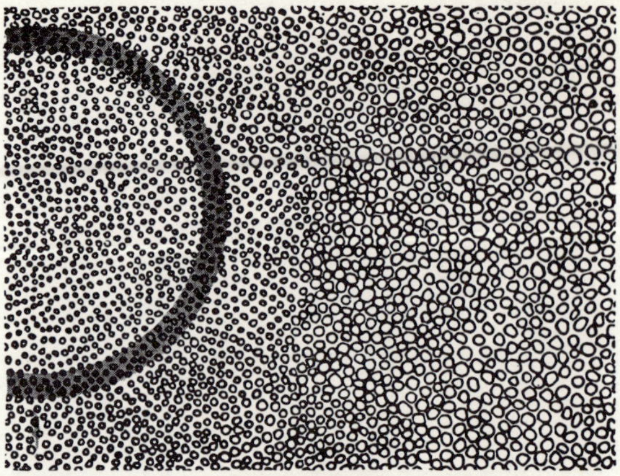

"GRAIN" of the human retina is made up of cones and rods (*dots at far right*). Semicircle indicates fovea.

the rods actually stop functioning at even the highest intensities, but in bright light their relative contribution to vision falls to so low a level as to be almost negligible.

To this general transfer of vision from rods to cones certain cold-blooded animals add a special anatomical device. The light-sensitive outer segments of the rods and cones are carried at the ends of fine stalks called myoids, which can shorten and lengthen. In dim light the rod myoids contract while the cone myoids relax. The entire field of rods is thus pulled forward toward the light, while the cones are pushed into the background. In bright light the reverse occurs: the cones are pulled forward and the rods pushed back. One could scarcely imagine a closer approach to the change from fast to slow film in a camera.

The rods and cones share with the grains of the photographic plate another deeply significant property. It has long been known that in a film exposed to light each grain of silver bromide given enough developer blackens either completely or not at all, and that a grain is made susceptible to development by the absorption of one or at most a few quanta of light. It appears to be equally true that a cone or rod is excited by light to yield either its maximal response or none at all. This is certainly true of the nerve fibers to which the rods and cones are connected, and we now know that to produce this effect in a rod—and possibly also in a cone—only one quantum of light need be absorbed.

It is a basic tenet of photochemistry that one quantum of light is absorbed by, and in general can activate, only one molecule or atom. We must attempt to understand how such a small beginning can bring about such a large result as the development of a photographic grain or the discharge of a retinal receptor. In the photographic process the answer to this question seems to be that the absorption of a quantum of light causes the oxidation of a silver ion to an atom of metallic silver, which then serves as a catalytic center for the development of the entire grain. It is possible that a similar mechanism operates in a rod or a cone. The absorption of a quantum of light by a light-sensitive molecule in either structure might convert it into a biological catalyst, or an enzyme, which could then promote the further reactions that discharge the receptor cell. One wonders whether such a mechanism could possibly be rapid enough. A rod or a cone responds to light within a small fraction of a second; the mechanism would therefore have to complete its work within this small interval.

One of the strangest characteristics of the eye in dim light follows from some of these various phenomena. In focusing the eye is guided by its evaluation of the sharpness of the image on the retina. As the image deteriorates with the opening of the pupil in dim light, and as the retinal capacity to resolve the image falls with the shift from cones to rods, the ability to focus declines also. In very dim light the eye virtually ceases to adjust its focus at all. It has come to resemble a very cheap camera, a fixed-focus instrument.

In all that concerns its function, therefore, the eye is one device in bright light and another in dim. At low intensities all its resources are concentrated upon sensitivity, at whatever sacrifice of form; it is predominantly an instrument for seeing light, not pattern. In bright light all this changes. By narrowing the pupil, shifting from rods to cones, and other stratagems still to be described, the eye sacrifices light in order to achieve the utmost in pattern vision.

Images

In the course of evolution animals have used almost every known device for forming or evaluating an image. There is one notable exception: no animal has yet developed an eye based upon the use of a concave mirror. An eye made like a pinhole camera, however, is found in Nautilus, a cephalopod mollusk related to the octopus and squid. The compound eye of insects and crabs forms an image which is an upright patchwork of responses of individual "eyes" or ommatidia, each of which records only a spot of light or shade. The eye of the tiny arthropod Copilia possesses a large and beautiful lens but only one light receptor attached to a thin strand of muscle. It is said that the muscle moves the receptor rapidly back and forth in the focal plane of the lens, scanning the image in much the same way as it is scanned by the light-sensitive tube of a television camera.

Each of these eyes, like the lens eye of vertebrates, represents some close compromise of advantages and limitations. The pinhole eye is in focus at all distances, yet to form clear images it must use a small hole admitting very little light. The compound eye works well at distances of a few millimeters, yet it is relatively coarse in pattern resolution. The vertebrate eye is a long-range, high-acuity instrument useless in the short distances at which the insect eye resolves the greatest detail.

These properties of the vertebrate eye are of course shared by the camera. The use of a lens to project an image, however, has created for both devices a special group of problems. All simple lenses are subject to serious errors in image formation: the lens aberrations.

Spherical aberration is found in all lenses bounded by spherical surfaces. The marginal portions of the lens bring rays of light to a shorter focus than the central region. The image of a point in space is therefore not a point, but a little "blur circle." The cost of a camera is largely determined by the extent to

CONES of the catfish *Ameiurus* are pulled toward the surface of the retina (*top*) in bright light. The rods remain in a layer below the surface.

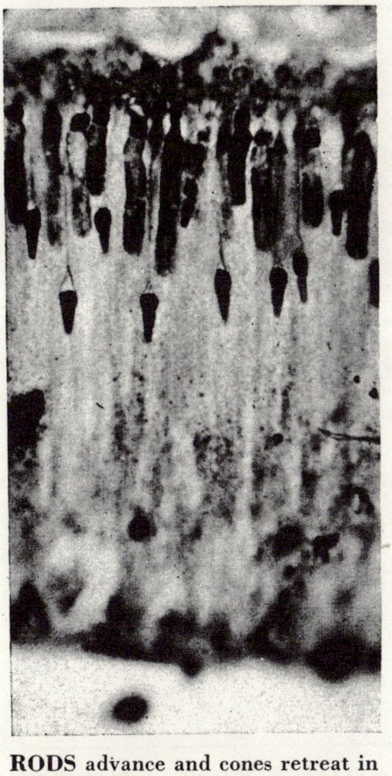

RODS advance and cones retreat in dim light. This retinal feature is not possessed by mammals. It is peculiar to some of the cold-blooded animals.

which this aberration is corrected by modifying the lens.

The human eye is astonishingly well corrected—often slightly overcorrected—for spherical aberration. This is accomplished in two ways. The cornea, which is the principal refracting surface of the eye, has a flatter curvature at its margin than at its center. This compensates in part for the tendency of a spherical surface to refract light more strongly at its margin. More important still, the lens is denser and hence refracts light more strongly at its core than in its outer layers.

A second major lens error, however, remains almost uncorrected in the human eye. This is chromatic aberration, or color error. All single lenses made of one material refract rays of short wavelength more strongly than those of longer wavelength, and so bring blue light to a shorter focus than red. The result is that the image of a point of white light is not a white point, but a blur circle fringed with color. Since this seriously disturbs the image, even the lenses of inexpensive cameras are corrected for chromatic aberration.

It has been known since the time of Isaac Newton, however, that the human eye has a large chromatic aberration. Its lens system seems to be entirely uncorrected for this defect. Indeed, living organisms are probably unable to manufacture two transparent materials of such widely different refraction and dispersion as the crown and flint glasses from which color-corrected lenses are constructed.

The large color error of the human eye could make serious difficulties for image vision. Actually the error is moderate between the red end of the spectrum and the blue-green, but it increases rapidly at shorter wavelengths: the blue, violet and ultraviolet. These latter parts of the spectrum present the most serious problem. It is a problem for both the eye and the camera, but one for which the eye must find a special solution.

The first device that opposes the color error of the human eye is the yellow lens. The human lens is not only a lens but a color filter. It passes what we ordinarily consider to be the visible spectrum, but sharply cuts off the far edge of the violet, in the region of wavelength 400 millimicrons. It is this action of the lens, and not any intrinsic lack of sensitivity of the rods and cones, that keeps us from seeing in the near ultraviolet. Indeed, persons who have lost their lenses in the operation for cataract and have had them replaced by clear glass lenses, have excellent vision in the ultraviolet. They are able to read an optician's chart from top to bottom in ultraviolet light which leaves ordinary people in complete darkness.

The lens therefore solves the problem of the near ultraviolet, the region of the spectrum in which the color error is greatest, simply by eliminating the region from human vision. This boon is distributed over one's lifetime, for the lens becomes a deeper yellow and makes more of the ordinary violet and blue invisible as one grows older. I have heard it said that for this reason aging artists tend to use less blue and violet in their paintings.

The lens filters out the ultraviolet for the eye as a whole. The remaining devices which counteract chromatic aberration are concentrated upon vision in bright light, upon cone vision. This is good economy, for the rods provide such a coarse-grained receptive surface that they would be unable in any case to evaluate a sharp image on the retina.

As one goes from dim to bright light, from rod to cone vision, the sensitivity of the eye shifts toward the red end of the spectrum. This phenomenon was described in 1825 by the Czech physiologist Johannes Purkinje. He had noticed that with the first light of dawn blue objects tend to look relatively bright compared with red, but that they come to look relatively dim as the morning advances. The basis of this change is a large difference in spectral sensitivity between rods and cones. Rods have their maximal sensitivity in the blue-green at about 500 millimicrons; the entire spectral sensitivity of the cones is transposed toward the red, the maximum lying in the yellow-green at about 562 millimicrons. The point of this difference for our present argument is that as one goes from dim light, in which pattern vision is poor in any case, to bright light, in which it becomes acute, the sensitivity of the eye moves away from the region of the spectrum in which the chromatic aberration is large toward the part of the spectrum in which it is least.

The color correction of the eye is completed by a third dispensation. Toward the center of the human retina there is a small, shallow depression called the fovea, which contains only cones. While the retina as a whole sweeps through a visual angle of some 240 degrees, the fovea subtends an angle of only about 1.7 degrees. The fovea is considerably smaller than the head of a pin, yet with this tiny patch of retina the eye accomplishes all its most detailed vision.

The fovea also includes the fixation point of the eye. To look directly at something is to turn one's eye so that its image falls upon the fovea. Beyond the boundary of the fovea rods appear, and they become more and more numerous as the distance from the fovea increases. The apparatus for vision in bright light is thus concentrated toward the center of the retina, that for dim light toward its periphery. In very dim light, too dim to excite the cones, the fovea is blind. One can see objects then only by looking at them slightly askance to catch their images on areas rich in rods.

In man, apes and monkeys, alone of all known mammals, the fovea and the region of retina just around it is colored yellow. This area is called the yellow patch, or *macula lutea*. Its pigmentation lies as a yellow screen over the light receptors of the central retina, subtending a visual angle some five to 10 degrees in diameter.

Several years ago in our laboratory at Harvard University we measured the color transmission of this pigment in the living human eye by comparing the spectral sensitivities of cones in the yellow patch with those in a colorless peripheral area. The yellow pigment was also extracted from a small number of human maculae, and was found to be xanthophyll, a carotenoid pigment that occurs also in all green leaves. This pigment in the yellow patch takes up the absorption of light in the violet and blue regions of the spectrum just where absorption by the lens falls to very low values. In this way the yellow patch removes for the central retina the remaining regions of the spectrum for which the color error is high.

So the human eye, unable to correct its color error otherwise, throws away those portions of the spectrum that would make the most trouble. The yellow lens removes the near ultraviolet for the eye as a whole, the macular pigment eliminates most of the violet and blue for the central retina, and the shift from rods to cones displaces vision in bright light bodily toward the red. By these three devices the apparatus of most acute vision avoids the entire range of the spectrum in which the chromatic aberration is large.

Photography with Living Eyes

In 1876 Franz Boll of the University of Rome discovered in the rods of the frog retina a brilliant red pigment. This bleached in the light and was resynthesized in the dark, and so fulfilled the elementary requirements of a visual pigment. He called this substance visual red; later it was renamed visual purple or rhodopsin. This pigment marks the point of attack by light on the rods: the absorption of light by rhodopsin initiates the train of reactions that end in rod vision.

Boll had scarcely announced his discovery when Willy Kühne, professor of physiology at Heidelberg, took up the study of rhodopsin, and in one extraordinary year learned almost everything about it that was known until recently. In his first paper on retinal chemistry Kühne said: "Bound together with the pigment epithelium, the retina behaves not merely like a photographic plate, but like an entire photographic workshop, in which the workman continually renews

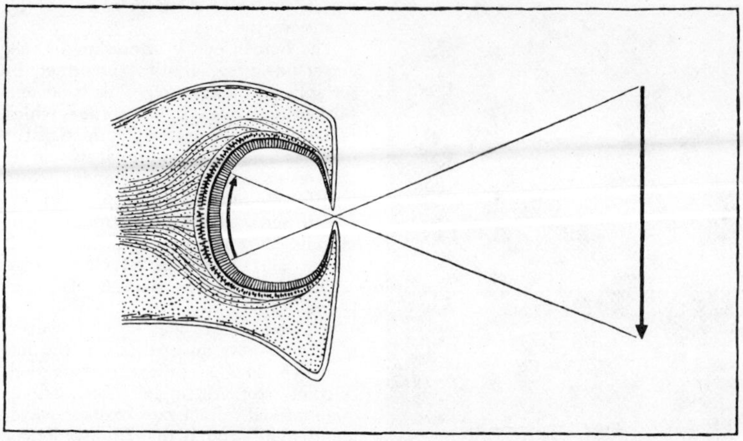

PINHOLE-CAMERA EYE is found in Nautilus, the spiral-shelled mollusk which is related to the octopus and the squid. This eye has the advantage of being in focus at all distances from the object that is viewed. It has the serious disadvantage, however, of admitting very little light to the retina.

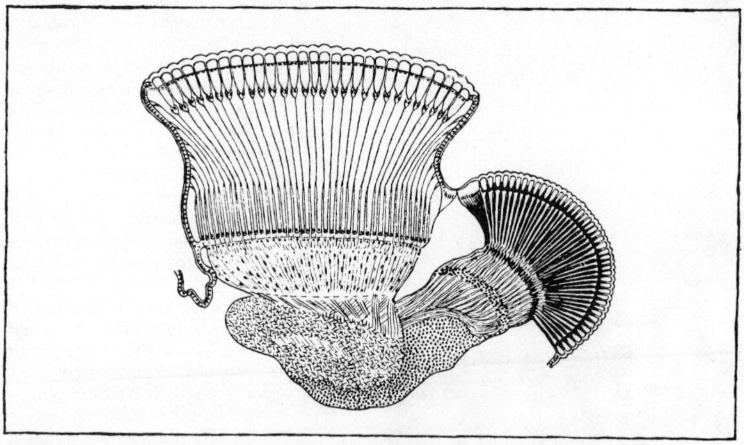

COMPOUND EYE is found in insects. Each element contributes only a small patch of light or shade to make up the whole mosaic image. This double compound eye is found in the mayfly *Chloeon*. The segment at the top provides detailed vision; the segment at the right, coarse, wide-angled vision.

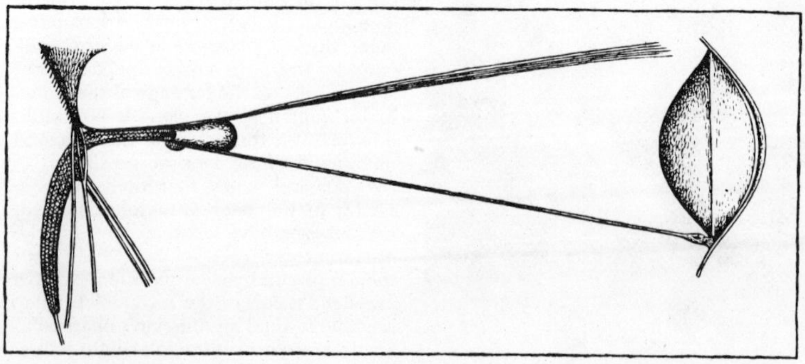

SCANNING EYE is found in the arthropod Copilia. It possesses a large lens (*right*) but only one receptor element (*left*). Attached to the receptor are the optic nerve and a strand of muscle. The latter is reported to move the receptor back and forth so that it scans the image formed by the lens.

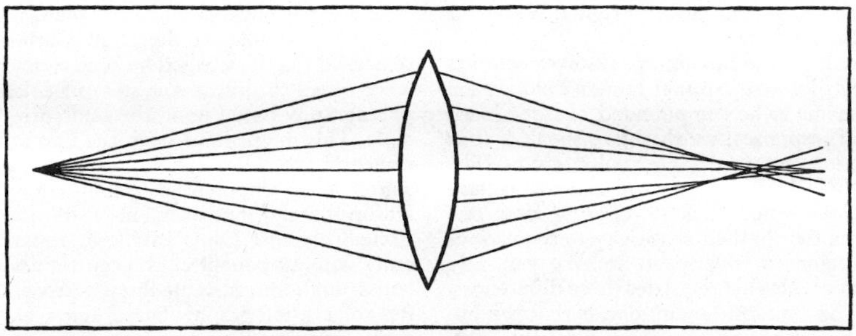

SPHERICAL ABERRATION occurs when light is refracted by a lens with spherical surfaces. The light which passes through the edge of the lens is brought to a shorter focus than that which passes through the center. The result of this is that the image of a point is not a point but a "blur circle."

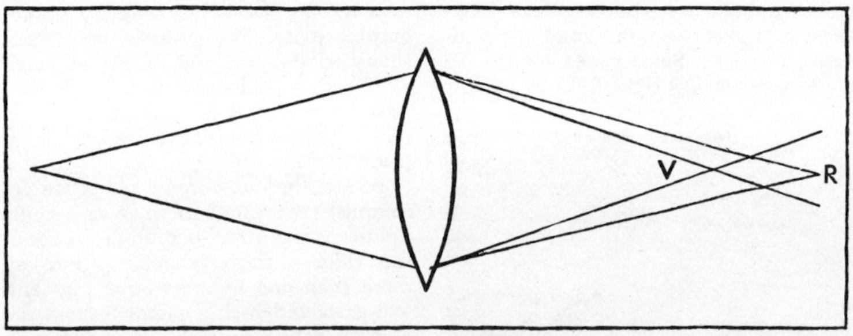

CHROMATIC ABERRATION occurs when light of various colors is refracted by a lens made of one material. The light of shorter wavelength is refracted more than that of longer wavelength, *i.e.*, violet is brought to a shorter focus than red. The image of a white point is a colored blur circle.

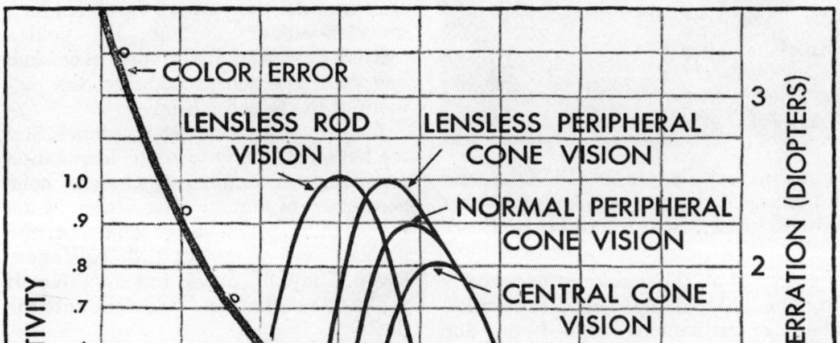

CHROMATIC ABERRATION of the human eye is corrected by various stratagems which withdraw the cones from the region of maximum aberration, *i.e.*, the shorter wavelengths. The horizontal coordinate of this diagram is wavelength in millimicrons; the colors are indicated by initial letters.

the plate by laying on new light-sensitive material, while simultaneously erasing the old image."

Kühne saw at once that with this pigment which was bleached by light it might be possible to take a picture with the living eye. He set about devising methods for carrying out such a process, and succeeded after many discouraging failures. He called the process optography and its products optograms.

One of Kühne's early optograms was made as follows. An albino rabbit was fastened with its head facing a barred window. From this position the rabbit could see only a gray and clouded sky. The animal's head was covered for several minutes with a cloth to adapt its eyes to the dark, that is to let rhodopsin accumulate in its rods. Then the animal was exposed for three minutes to the light. It was immediately decapitated, the eye removed and cut open along the equator, and the rear half of the eyeball containing the retina laid in a solution of alum for fixation. The next day Kühne saw, printed upon the retina in bleached and unaltered rhodopsin, a picture of the window with the clear pattern of its bars.

I remember reading as a boy a detective story in which at one point the detective enters a dimly lighted room, on the floor of which a corpse is lying. Working carefully in the semidarkness, the detective raises one eyelid of the victim and snaps a picture of the open eye. Upon developing this in his darkroom he finds that he has an optogram of the last scene viewed by the victim, including of course an excellent likeness of the murderer. So far as I know Kühne's optograms mark the closest approach to fulfilling this legend.

The legend itself has nonetheless flourished for more than 60 years, and all of my readers have probably seen or heard some version of it. It began with Kühne's first intimation that the eye resembles a photographic workshop, even before he had succeeded in producing his first primitive optogram, and it spread rapidly over the entire world. In the paper that announces his first success in optography, Kühne refers to this story with some bitterness. He says: "I disregard all the journalistic potentialities of this subject, and willingly surrender it in advance to all the claims of fancy-free coroners on both sides of the ocean, for it certainly is not pleasant to deal with a serious problem in such company. Much that I could say about this had better be suppressed, and turned rather to the hope that no one will expect from me any corroboration of announcements that have not been authorized with my name."

Despite these admirable sentiments we find Kühne shortly afterward engaged in a curious adventure. In the nearby town of Bruchsal on November 16, 1880, a young man was beheaded by guillotine. Kühne had made arrangements to receive the corpse. He had prepared a dimly lighted room screened with red and yellow glass to keep any rhodopsin left in the eyes from bleaching further. Ten minutes after the knife had fallen he obtained the whole retina from the left eye, and had the satisfaction of seeing and showing to several colleagues a sharply demarcated optogram printed upon its surface. Kühne's drawing of it is reproduced at the bottom of the next page. To my knowledge it is the only human optogram on record.

Kühne went to great pains to determine what this optogram represented. He says: "A search for the object which served as source for this optogram remained fruitless, in spite of a thorough inventory of all the surroundings and reports from many witnesses. The delinquent had spent the night awake by the light of a tallow candle; he had slept from four to five o'clock in the morning; and had read and written, first by candlelight until dawn, then by feeble daylight until eight o'clock. When he emerged in the open, the sun came out for an instant, according to a reliable observer, and the sky became somewhat brighter during the seven minutes prior to the bandaging of his eyes and his execution, which followed immediately. The delinquent, however, raised his eyes only rarely."

Color

One of the triumphs of modern photography is its success in recording color. For this it is necessary not only to graft some system of color differentiation and rendition upon the photographic process; the finished product must then fulfill the very exacting requirement that it excite the same sensations of color in the human eye as did the original subject of the picture.

How the human eye resolves colors is not known. Normal human color vision seems to be compounded of three kinds of responses; we therefore speak of it as trichromatic or three-color vision. The three kinds of response call for at least three kinds of cone differing from one another in their sensitivity to the various regions of the spectrum. We can only guess at what regulates these differences. The simplest assumption is that the human cones contain three different light-sensitive pigments, but this is still a matter of surmise.

There exist retinas, however, in which one can approach the problem of color vision more directly. The eyes of certain turtles and of certain birds such as chickens and pigeons contain a great predominance of cones. Since cones are the organs of vision in bright light as well as of color vision, these animals necessarily function only at high light intensities. They are permanently night-blind, due to a poverty or complete absence of rods. It is for this reason that chickens must roost at sundown.

In the cones of these animals we find a system of brilliantly colored oil globules, one in each cone. The globule is situated at the joint between the inner and outer segments of the cone, so that light must pass through it just before entering the light-sensitive element. The globules therefore lie in the cones in the position of little individual color filters.

One has only to remove the retina from a chicken or a turtle and spread it on the stage of a microscope to see that the globules are of three colors: red, orange and greenish yellow. It was suggested many years ago that they provide the basis of color differentiation in the animals that possess them.

RETINAL PHOTOGRAPH, or an optogram, was drawn in 1878 by the German investigator Willy Kühne. He had exposed the eye of a living rabbit to a barred window, killed the rabbit, removed its retina and fixed it in alum.

In a paper published in 1907 the German ophthalmologist Siegfried Garten remarked that he was led by such retinal color filters to invent a system of color photography based upon the same principle. This might have been the first instance in which an eye had directly inspired a development in photography. Unfortunately, however, in 1906 the French chemist Louis Lumière, apparently without benefit of chicken retinas, had brought out his autochrome process for color photography based upon exactly this principle.

To make his autochrome plates Lumiere used suspensions of starch grains from rice, which he dyed red, green and blue. These were mixed in roughly equal proportions, and the mixture was strewn over the surface of an ordinary photographic plate. The granules were then squashed flat and the interstices were filled with particles of carbon. Each dyed granule served as a color filter for the patch of silver-bromide emulsion that lay just under it.

Just as the autochrome plate can accomplish color photography with a single light-sensitive substance, so the cones of the chicken retina should require no more than one light-sensitive pigment. We extracted such a pigment from the chicken retina in 1937. It is violet in color, and has therefore been named iodopsin from *ion*, the Greek word for violet. All three pigments of the colored oil globules have also been isolated and crystallized. Like the pigment of the human macula, they are all carotenoids: a greenish-yellow carotene; the golden mixture of xanthophylls found in chicken egg yolk; and red astaxanthin, the pigment of the boiled lobster.

Controversy thrives on ignorance, and we have had many years of disputation regarding the number of kinds of cone concerned in human color vision. Many investigators prefer three, some four, and at least one of my English colleagues seven. I myself incline toward three. It is a good number, and sufficient unto the day.

The appearance of three colors of oil globule in the cones of birds and turtles might be thought to provide strong support for trichromatic theories of color vision. The trouble is that these retinas do in fact contain a fourth class of globule which is colorless. Colorless globules have all the effect of a fourth color; there is no doubt that if we include them, bird and turtle retinas possess the basis for four-color vision.

Latent Images

Recent experiments have exposed a wholly unexpected parallel between vision and photography. Many years ago Kühne showed that rhodopsin can be extracted from the retinal rods into clear water solution. When such solutions are

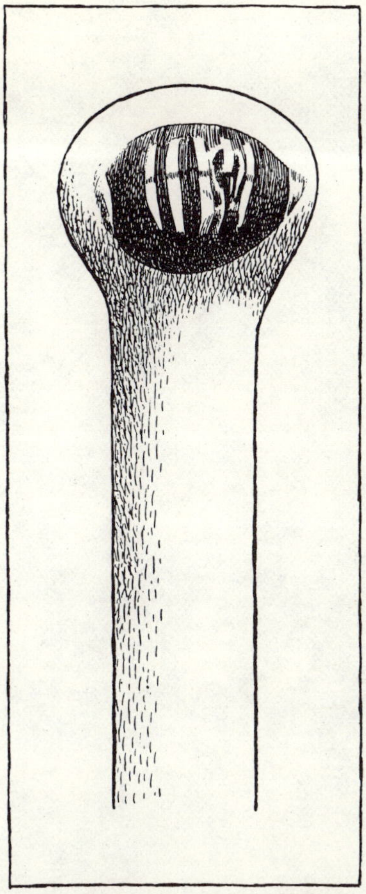

FROG OPTOGRAM showing a barred pattern was made by the German ophthalmologist Siegfried Garten. The retina is mounted on a rod.

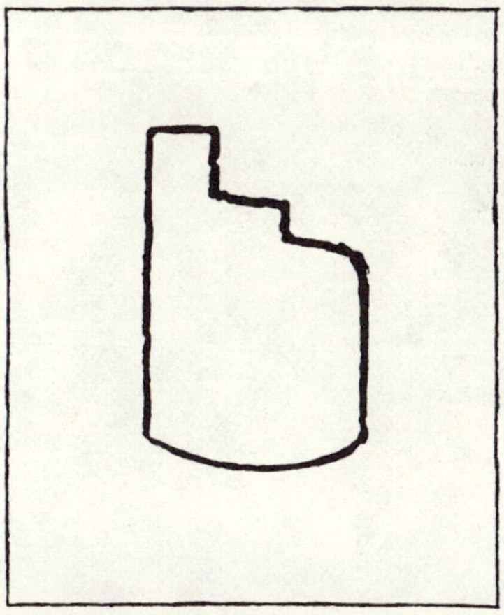

HUMAN OPTOGRAM was drawn by Kühne after he had removed the retina of a beheaded criminal. Kühne could not determine what it showed.

exposed to light, the rhodopsin bleaches just as it does in the retina.

It has been known for some time that the bleaching of rhodopsin in solution is not entirely accomplished by light. It is started by light, but then goes on in the dark for as long as an hour at room temperature. Bleaching is therefore a composite process. It is ushered in by a light reaction that converts rhodopsin to a highly unstable product; this then decomposes by ordinary chemical reactions—"dark" reactions in the sense that they do not require light.

Since great interest attaches to the initial unstable product of the light reaction, many attempts were made in our laboratory and at other laboratories to seize upon this substance and learn its properties. It has such a fleeting existence, however, that for some time nothing satisfactory was achieved.

In 1941, however, two English workers, E. E. Broda and C. F. Goodeve, succeeded in isolating the light reaction by irradiating rhodopsin solutions at about −73 degrees Celsius, roughly the temperature of dry ice. In such extreme cold, light reactions are unhindered, but ordinary dark processes cannot occur. Broda and Goodeve found that an exhaustive exposure of rhodopsin to light under these conditions produced only a very small change in its color, so small that though it could be measured one might not have been certain merely by looking at these solutions that any change had occurred at all. Yet the light reaction had been completed, and when such solutions were allowed to warm up to room temperature they bleached *in the dark*. We have recently repeated such experiments in our laboratory. With some differences which need not be discussed, the results were qualitatively as the English workers had described them.

These observations led us to re-examine certain early experiments of Kühne's. Kühne had found that if the retina of a frog or rabbit was thoroughly dried over sulfuric acid, it could be exposed even to brilliant sunlight for long periods without bleaching. Kühne concluded that dry rhodopsin is not affected by light, and this has been the common understanding of workers in the field of vision ever since.

It occurred to us, however, that dry rhodopsin, like extremely cold rhodopsin, might undergo the light reaction, though with such small change in color as to have escaped notice. To test this possibility we prepared films of rhodopsin in gelatin, which could be dried thoroughly and were of a quality that permitted making accurate measurements of their color transmission throughout the spectrum.

We found that when dry gelatin films of rhodopsin are exposed to light, the same change occurs as in very cold rhodopsin. The color is altered, but so slightly as easily to escape visual observation. In any case the change cannot be described as bleaching; if anything the color is a little intensified. Yet the light reaction is complete; if such exposed films are merely wetted with water, they bleach in the dark.

We have therefore two procedures—cooling to very low temperatures and removal of water—that clearly separate the light from the dark reactions in the bleaching of rhodopsin. Which of these reactions is responsible for stimulating rod vision? One cannot yet be certain, yet the response of the rods to light occurs so rapidly that only the light reaction seems fast enough to account for it.

What has been said, however, has a further consequence that brings it into direct relation with photography. Everyone knows that the photographic process also is divided into light and dark components. The result of exposing a film to light is usually invisible, a so-called "latent image." It is what later occurs in the darkroom, the dark reaction of development, that brings out the picture.

This now appears to be exactly what happens in vision. Here as in photography light produces an almost invisible result, a latent image, and this indeed is probably the process upon which retinal excitation depends. The visible loss of rhodopsin's color, its bleaching, is the result of subsequent dark reactions, of "development."

One can scarcely have notions like this without wanting to make a picture with a rhodopsin film; and we have been tempted into making one very crude rhodopsin photograph. Its subject is not exciting—only a row of black and white stripes—but we show it at the right for what interest it may have as the first such photograph. What is important is that it was made in typically photographic stages. The dry rhodopsin film was first exposed to light, producing a latent image. It was then developed in the dark by wetting. It then had to be fixed; and, though better ways are known, we fixed this photograph simply by redrying it. Since irradiated rhodopsin bleaches rather than blackens on development, the immediate result is a positive.

Photography with rhodopsin is only in its first crude stages, perhaps at the level that photography with silver bromide reached almost a century ago. I doubt that it has a future as a practical process. For us its primary interest is to pose certain problems in visual chemistry in a provocative form. It does, however, also add another chapter to the mingled histories of eye and camera.

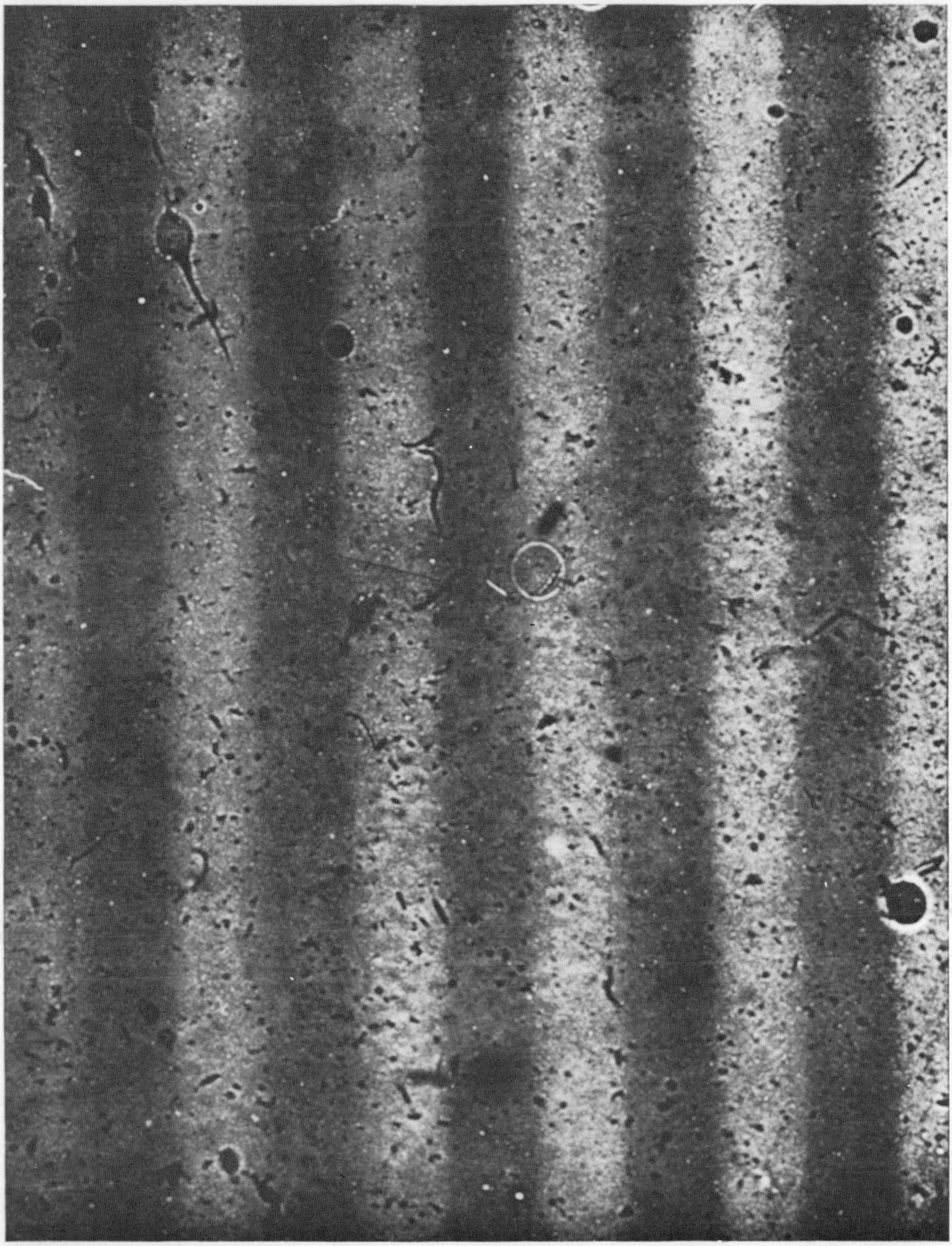

RHODOPSIN PHOTOGRAPH was made by the author and his associates Paul K. Brown and Oscar Starobin. Rhodopsin, the light-sensitive red pigment of rod vision, had been extracted from cattle retinas, mixed with gelatin and spread on celluloid. This was then dried and exposed to a pattern made up of black and white stripes. When the film was wetted in the dark with hydroxylamine, the rhodopsin bleached in the same pattern.

A device that operates on the principles of optics and molecular physics—and that has an astonishing range of applications.

7 The Laser — What it is and Does

J. M. Carroll

A chapter from a popular book, 1964.

Introduction

In 1960, electronics scientists and engineers began to see things in a different light.

It was a rich ruby light: not "kindled in the vine," as the Persian poet Omar Khayyám said, but emitted by the atoms of a synthetic gem stone.

The light came from the laser, a new device with wide potential application in science, medicine, industry, and national defense.

WHAT'S IN A NAME?

The word laser is an acronym, or a word made up of the first letters of several other words. Laser stands for *L*ight *A*mplification by *S*timulated *E*mission of *R*adiation. It was coined by analogy with another acronym: *maser*. Maser stands for *M*icrowave *A*mplification by *S*timulated *E*mission of *R*adiation.

The maser works on the same basic principle as the laser but, of course, emits microwave energy rather than light. Masers are used as input amplifiers (preamplifiers) of radio telescopes and space-tracking receivers that magnify feeble signals gleaned from outer space.

No one is completely satisfied with the name "laser" because lasers do not really amplify light in a strict sense;

instead they generate light with particular characteristics that engineers and scientists find useful. In electronics terminology a device that generates radiation is called an oscillator, not an amplifier.

Furthermore, most lasers do not emit visible light at all but rather infrared, or invisible, light. It is conceivable that devices working on the same principle as the laser and maser may someday emit ultraviolet or so-called black light, X rays, or even gamma rays.

Scientists who moved from maser research into laser research insist on calling the laser an optical maser. But it can be argued that it is ridiculous to talk of "optical microwave amplification by stimulated emission of radiation" since "optical" means one thing and "microwave" quite another.

Proponents of the term "optical maser" counter by saying that "maser" doesn't stand for microwave amplification by stimulated emission of radiation at all, but rather for *molecular* amplification by stimulated emission of radiation.

To the comment that masers do not amplify molecules comes the answer that they depend for their action on the behavior of the molecules of the substance.

Well, *some* masers and lasers do depend on molecular effects. But more depend on the behavior of submolecular particles: atoms, ions (atoms that have lost one or more electrons), perhaps even electrons themselves.

Recently the term *quantum device* has been applied to both masers and lasers, and this seems to make sense, since the action of both the laser and the maser can be explained by the science of *quantum mechanics*. In fact, some scientists and engineers interested in lasers and masers are attempting to form within the Institute of Electrical and Electronics Engineers a professional group on quantum electronics. And though for the present the term "laser" seems deeply ingrained in the scientific vocabulary, let's remember that the science we call electronics was once known as thermionic engineering!

WHAT'S SPECIAL ABOUT A LASER?

The important thing about laser light is that it is coherent. The individual light rays are all of the same wavelength or color, and are all in step. A laser beam differs from a beam of ordinary light in both character and effectiveness in the same way that a platoon of well-drilled soldiers differs from a ragtag, disorganized mob.

When light waves from a laser march in step, they can perform amazing feats. The reason is that their energy is not dissipated as the beam spreads out. This makes for an intense concentration of energy at a very sharply defined point. It also greatly extends the range of a light source.

Three of the many spectacular achievements of the laser demonstrate how the properties of coherent light can be put to work:

• Because its light does not spread out even at great distances, a laser can illuminate the surface of the moon with a two-mile-wide circle of light.

Laser beam on moon (black dot) compared with area of radar beam (shaded area) (Raytheon)

• Because its energy is concentrated at a fine point, it can send a short, searing pinpoint of light into the human eyeball to weld a detached retina back into place and restore sight.

• And since its radiation is so intense, it can burn holes in a steel plate ⅛ inch thick at a distance of several feet.

These abilities have given rise to a whole range of applications. Laser range finders are used both by artillery officers to sight their guns and by surveyors. In outer space, where there is no atmosphere to absorb the light, the laser will supplement conventional radar and radio for space-vehicle navigation and communications.

Lasers can cut metal and other materials. But it is highly unlikely that a laser will ever replace an engine lathe or an oxyacetelene torch in most machining and metal-cutting operations. Lasers are being used in the precision machining of metals and in machining brittle materials such as diamonds.

A laser can weld metals as well as retinas. But here, too, its use is for precise work, as in making microelectronic circuits. Nevertheless, large lasers mounted atop high mountain peaks are being developed to provide a defense against intercontinental-ballistic-missile warheads.

To the scientist, the laser is already a valuable tool in absorption spectroscopy or the identification of compounds by the particular wavelengths of light that they absorb.

Radiant Energy

How can a beam of light burn a hole in a steel plate? It can do so because light is a form of radiant energy, and a laser concentrates much radiant energy in a very tiny spot. Radiant energy exists in many forms besides visible light. It exists as radio waves, ultraviolet and infrared light, X rays, gamma rays, and even cosmic rays.

WAVELENGTH AND FREQUENCY

It is sometimes convenient to think of radiant energy as waves, that is, *electromagnetic waves.* Then the different forms of radiant energy can be classified by their *wavelengths* and arranged according to wavelength in a spectrum. We have all seen the waves made by a pebble thrown into a quiet pond. They are a series of alternating crests and troughs. The wavelength is defined as the distance between two adjacent crests or two adjacent troughs.

Now, when a wave goes from crest to trough and back to crest again, it is said to have gone through one cycle, or alternation. The number of cycles that a wave executes in one second is known as the *frequency* of the wave.

Light waves and all other electromagnetic waves travel at the same speed, which is 186,000 miles, or 300,000,000 meters, a second. All scientific measurements are made in metric system. In the metric system the basic unit of length is the meter—a little over three feet.

RADIO SPECTRUM

The alternating current supplied by the power company is an electromagnetic wave that executes 60 cycles a second; thus, in 1/60 of a second, or the time of one alternation, the wave will travel 300,000,000/60, or 5,000,000 meters—roughly the distance from New York to Los Angeles. The *electromagnetic spectrum* arranges the different kinds of electromagnetic energy according to decreasing wavelength.

Everyone is familiar with the red, orange, yellow, green, blue, and violet spectrum of the rainbow after a spring shower. The same separation of white light into its color components occurs when we pass light through a glass prism. A spectrum arranges the frequency components of white light according to decreasing wavelengths. Similar

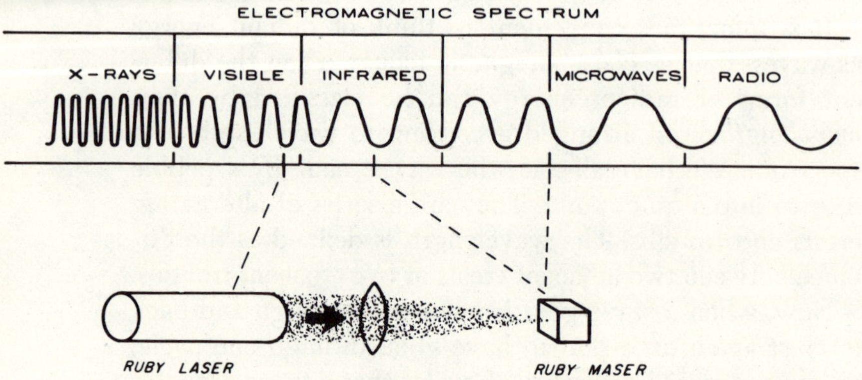

Electromagnetic spectrum from radio frequencies to X rays (Hughes)

spectra exist in the infrared and ultraviolet regions, but we can't see them. They can, however, be photographed by using special film. Radio waves also form part of the electromagnetic spectrum.

A radio broadcasting station with a frequency of 1,000 kilocycles per second (or cycles per second times 1,000) has a wave 300 meters long. A radar set used for navigation at sea has a wavelength of about 10 centimeters (one centimeter equals 1/100 meter), or approximately 4 inches.

VISIBLE SPECTRUM AND INFRARED

Radiant energy is invisible to the human eye only until we get to a wavelength of 0.00000075 meter, which we see as red light. Since the meter is an ungainly unit for measuring wavelengths of light, physicists use what is called the angstrom unit, abbreviated Å. One angstrom equals 1/10,000,000,000 meter. Therefore we can say the visible spectrum extends from 7,500 Å (deep red) to 4,000 Å, or blue. In between are regions of orange (about 6,000 Å), yellow (about 5,900 Å), and green (about 5,300 Å).

The visible spectrum is bounded by longer waves of

infrared that we sense as heat. For example, a jet engine exhaust has a wavelength of 40,000 Å, while the heat of the human body has a wavelength of about 99,000 Å.

FROM SUN TANS TO COSMIC RAYS

The short wavelength, or blue end of the spectrum, is bounded by the ultraviolet region. Sun-tanning ultraviolet rays have a wavelength of about 3,000 Å. Still shorter are X rays (150 to 10 Å) and gamma rays (1.4 to 0.1 Å). Gamma rays are associated with nuclear reactions, and account for some of the deadly effects of atomic and hydrogen bombs and of radioactive waste materials. At the high end of the spectrum are cosmic rays (0.01 to 0.001 Å), those weird visitors from outer space whose effects (they can cause biological mutations) are awesome indeed but about which very little is understood.

Scientists have known for a long time that the energy of radiation is proportional to its frequency. We cannot sense the presence of radio waves even though we stand close by the antenna of a powerful broadcasting station. Yet if we put a hand in front of a radar antenna, we may feel a slight sensation of warmth. The energy of ultraviolet waves will become painfully evident to some who sun bathe not wisely but too well. The penetrating power of X rays and gamma rays makes them useful for making shadowgraphs of the human skeleton and internal organs for medical diagnosis and for inspecting manufactured parts for hidden flaws. Indeed, hard, or short, X rays and gamma rays are used to destroy malignant tissue in the treatment of cancer and related diseases.

The energy of each wavelet of radiation is called a "quantum." It is measured by the frequency of the radiation multiplied by Planck's constant (this is equal to 6.625×10^{-27} erg seconds—26 zeros in front of the first

6). The intensity of a source of radiation depends upon the number of quanta emitted from it that pass a designated boundary at a given time.

FLUORESCENCE

The action of the laser is allied to another, more familiar, phenomenon, that of fluorescence. Fluorescence is said to occur when radiant energy hits the atoms or molecules of some particular material and in turn causes that substance to emit further radiant energy. Fluorescence has this important property: the emitted radiation is always at a lower frequency (longer wavelength) than the initial radiation.

Here's how scientists explain fluorescence: Every atom and molecule has certain energy states that it can occupy. When the atoms absorb energy, they move to higher energy states. Conversely, when they return to lower energy states, they give up energy, or emit radiation.

Imagine an atom to be a coil spring. When there is no compression on the spring, it is in its ground, or rest, state. When you compress the spring, you add potential energy to the system. When you release the spring, it bounces back and vibrates, giving up what is called its kinetic energy.

In the picture tube of your television set, electrons bombard a phosphor screen on the back of the faceplate. The kinetic energy, or energy of motion of the rapidly moving electrons, excites the atoms of the phosphor. As these atoms relax, the faceplate of the picture tube glows, and you see the television program because of fluorescence.

When a radiologist examines you with a fluoroscope, X rays penetrate your body and excite the atoms of a phosphor screen. As the atoms of the phosphor coating relax, the fluoroscope screen glows green, producing a shadowgraph of the part of the body being visualized.

In a neon sign, an alternating current creates an electromagnetic field that agitates the molecules of neon gas filling the tube. Because collisions of rapidly moving neon molecules raise these molecules to higher energy levels, they relax, emitting the orange-red glow characteristic of a neon sign.

Of course, the common fluorescent lamp works on the same principle of energy exchange. The inner walls of the lamp tube are coated with beryllium oxide. Inside the tube, there is an intense arc discharge between electrodes at either end of the lamp tube. This arc discharge is rich in ultraviolet light that energizes the phosphor molecules. As these molecules relax, the lamp emits a blue-white light similar to natural daylight.

We now have seen several examples of quantum energy exchanges, but no one ever burned a hole in a steel plate or illuminated the moon with a neon sign or with a fluorescent lamp. What, then, does the laser have that its less powerful cousins lack?

Frequency Coherence

The answer is: the laser's coherence. In all the previous examples of the phenomenon of fluorescence, the emitted radiation had a broad spectrum. Because it was emitted in random fashion, some wavelets added together while others opposed each other.

Frequency coherence makes a big difference. It means that all the emitted energy has the same wavelength. When this happens, you can have a useful output indeed. Take the babble of voices at a cocktail party as an example of incoherent sound. The sound doesn't carry very far and it is not especially meaningful. But if you were to concentrate all that sound energy into the blast of a police whistle or siren, you could awaken half a city.

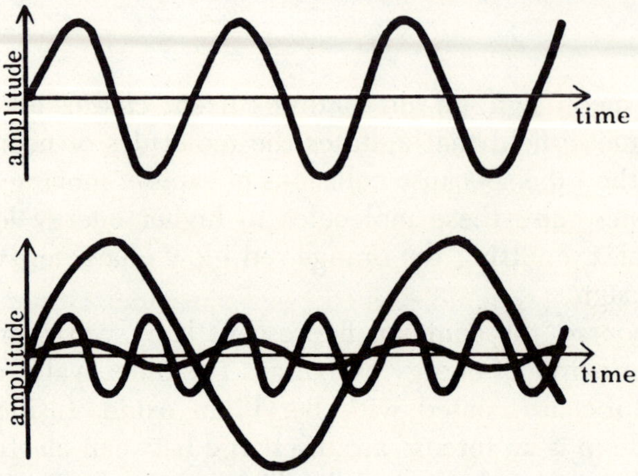

Frequency coherent radiation, top, and frequency incoherent radiation, bottom (Raytheon)

Engineers learned many years ago that they could communicate more efficiently and more meaningfully when they concentrated all the output of a radio transmitter at a single frequency. But frequency coherence has other advantages besides efficiency. A beam of coherent light can be modulated much as a radio signal can be. Modulation is a process by which intelligence such as music or speech is impressed upon a so-called carrier signal such as a radio wave.

An incoherent light beam can be modulated in only the most elementary manner—such as by switching it on and off, as with the visual blinker lights used to send Morse code between ships. But the frequency-coherent laser beam can be modulated by such complex signals as speech, music, or even a television picture.

Frequency-coherent light also lends itself to frequency multiplication, the technique whereby a closely controlled but relatively low radio frequency can be raised to a higher output frequency. The output of a ruby laser at

6,943 Å has been doubled to 3,472 Å. The input was deep red and the output blue-violet, almost ultraviolet. The reason the wavelengths of laser light are given so precisely is that the emission of laser light depends on the shifting of electrons between atomic orbits, and each wavelength is characteristic of one particular orbital shift, or so-called quantum jump.

Laser beams can also be mixed. For example, a ruby laser operates in two slightly different modes. These modes can be mixed in a microwave phototube. The frequency difference between the modes yields a microwave signal that can be handled by conventional radio or television techniques. This property has permitted some engineers to modulate laser beams with television pictures and to recover the television signal after transmission for several feet.

Scientists find the frequency coherence of the laser especially gratifying. Before the discovery of the laser, only signals in the lower, or radio, end of the spectrum could be produced coherently. Radio techniques were limited to producing signals whose wavelength was on the order of a millimeter or so.

If monochromatic (or single-frequency) signals were desired anywhere else in the spectrum, they had to be produced by placing an appropriate filter in front of an incoherent source. This method was unsatisfactory for two reasons: it was very inefficient, since the source had to produce many times the energy that could be usefully employed; and, second, since no filtered output is ever truly coherent, modulation, frequency multiplication, and mixing were always unsatisfactory. But now a whole new section of the spectrum, ranging from the "near" (to visible light, that is) infrared to near ultraviolet, is open to investigation, and there is evidence that the existing gaps at the high and low ends of this laser operating range can be filled by using related techniques.

SPATIAL COHERENCE

Frequency coherence is only part of the picture. The output of a laser is also spatially coherent. This means that all wavelets start in step with each other. Spatial coherence also adds to the efficiency of a device. The difference

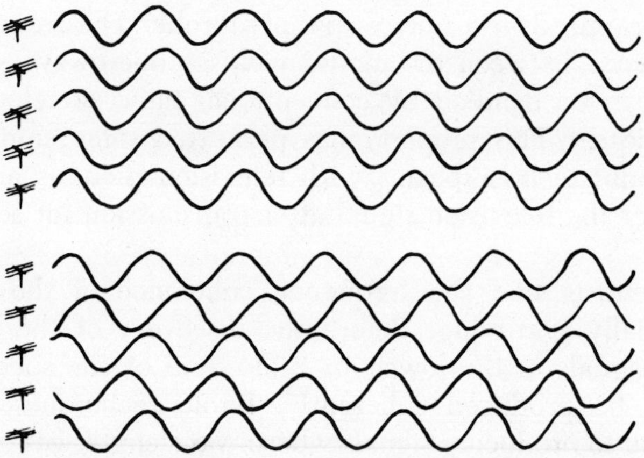

Spatially coherent radiation, top, and spatially incoherent radiation, bottom (Raytheon)

between spatial incoherence and spatial coherence is like the difference between a disorganized group of castaways of a raft each paddling in his own way and the smooth, efficient performance of a well-trained crew rowing an eight-oared racing shell.

RUBY LASERS

The ruby laser was the first device to generate coherent light successfully. The rubies used in lasers are synthetic gem stones. They are made by fusing aluminum and chromium oxides to produce large crystals. The amount of chromium in a synthetic ruby is small—about five hundredths of 1 percent. But it is that chromium upon which laser action depends.

The ruby crystal is cylindrical, about ¼ inch in diameter and 1½ to 2 inches long. It appears pink to the eye. That is because there are two absorption bands in a ruby—one at 5,600 Å and the other at 4,100 Å—which means that when you hold a ruby up to the light, yellow-green light and blue light are absorbed. This subtraction of yellow, green, and blue from white light (which is a mixture of all colors) gives the remaining light transmitted to the eye its distinctive pink hue. Actually, there is also some natural fluorescence in a ruby, but it is all but imperceptible to the eye.

A laser crystal must be polished to optical flatness on both ends. Both ends are also silvered, one with a heavy coat while the other, or output end, is lightly silvered with a coat that permits it to reflect only about 92 percent of light incident on it.

Exploded view of ruby laser showing ruby, mirrors, and helical flashtube (Hughes)

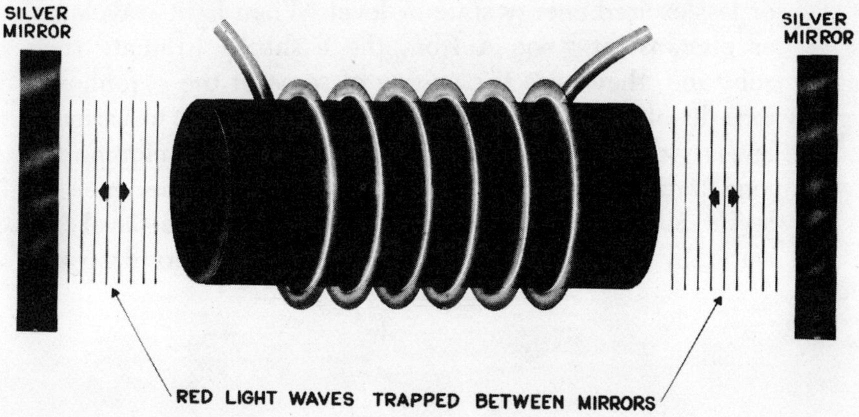

The ruby rod is now placed within a helical-shaped xenon flashtube, the kind of tube widely used in electronic flash attachments for cameras. The process of irradiating the ruby rod with a xenon flashtube is called optical pumping. The output of the flash lamp is rich in the yellow-green region.

The energy level of an atom (an ion is just an atom that has lost one or more electrons) depends upon the condition of its electrons. Now, an atom is like a miniature solar system. It has a positive nucleus at its center in place of the sun, and a specific number of planet-like electrons. These electrons revolve around the nucleus and spin on their own axes. Unlike the planets of the solar system, however, each electron can occupy not just one but several orbits. Moreover, the electrons can revolve around the nucleus with different azimuthal momenta (speed) and even change their direction of spin. Each change in orbit, momentum, or spin corresponds to a discrete energy level.

For example, when energy is imparted to an atom, an electron may move to an orbit more remote from the nucleus. The atom is said to absorb energy and to have been raised to a higher or more excited energy state or level. If the electron then returns to its original orbit, the atom gives up energy; it may now emit light of a certain precise wavelength. The atom is said to relax to a lower or less excited energy state or level. When light wavelets, or *photons*, at 5,600 Å from the flashtube irradiate the ruby rod, they raise the energy of some of the chromium ions dissolved in the ruby from ground state ① to various levels lying within the absorption band. Then the chromium ions immediately begin to drop from these higher energy levels. Some drop right back to the ground state—level ① —as they do in natural fluorescence. But others drop to

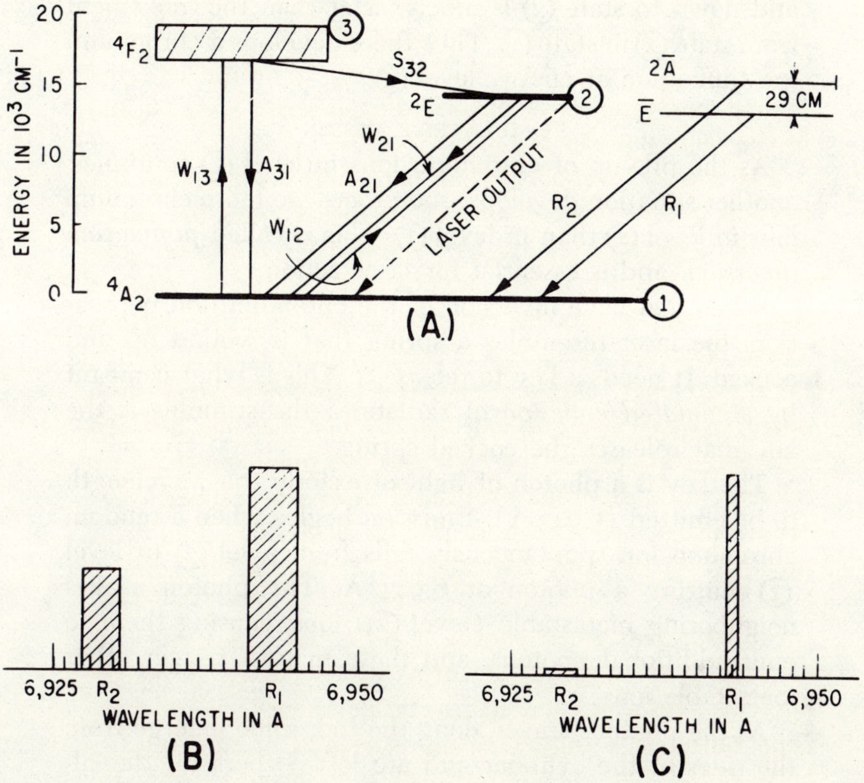

Energy level transitions in a ruby laser as described in text (A); low-level pumping (B) and high-level pumping (C) showing how latter mode concentrates energy at one wavelength *(Electronics)*

an intermediate or so-called *metastable state* ②. If left alone, the latter chromium ions would continue their drop to level ①, and the result would just be natural fluorescence. But these ions dally for a short but measurable time in level ②, and this is what makes laser action possible.

While the chromium ions are trying to get back to level ①, the flashtube keeps on irradiating more chromium ions.

In fact, the two-step movement from state ① to state ③ and down to state ② is much faster than the movement from state ② to state ①. Thus there develops a chromium-ion traffic jam at energy level ②.

STIMULATED EMISSION

As the pile-up of chromium ions in level ② continues, another situation develops: soon there are more chromium ions in level ② than in level ①. This is called *population inversion,* and is essential for laser action.

When you have inversion of the chromium ion population, the laser resembles a spring that is wound up and cocked. It needs a key to release it. This is what is meant by *stimulated emission* of radiation: the stimulus is the key that releases the cocked spring.

The key is a photon of light of exactly the wavelength to be emitted (6,943 Å). Emission begins when a random chromium ion spontaneously falls from level ② to level ① emitting a photon at 6,943 Å. The photon strikes neighboring metastable (level ②) ions, causing them to emit additional photons, and these in turn trigger other metastable ions.

As the photons travel along the rod, some emerge from the sides of the cylinder and are lost. Others hit the silvered ends of the cylinder and are reflected back into the rod. The reflections tend to favor those photons that are traveling parallel to the long axis of the cylinder. And so, there is now a stream of photons bouncing back and forth between the silvered ends of the cylinder. The pho-

Two following pages: How a ruby laser works. Pumping light irradiates ruby rod (A) raising some atoms to their metastable state (B). One atom spontaneously emits coherent radiation (C) triggering other nearby atoms (D). Photons emitted parallel to sides bounce back and forth between mirrors triggering other atoms (E) until light pulse (F) bursts from slightly transparent end *(Electronics)*

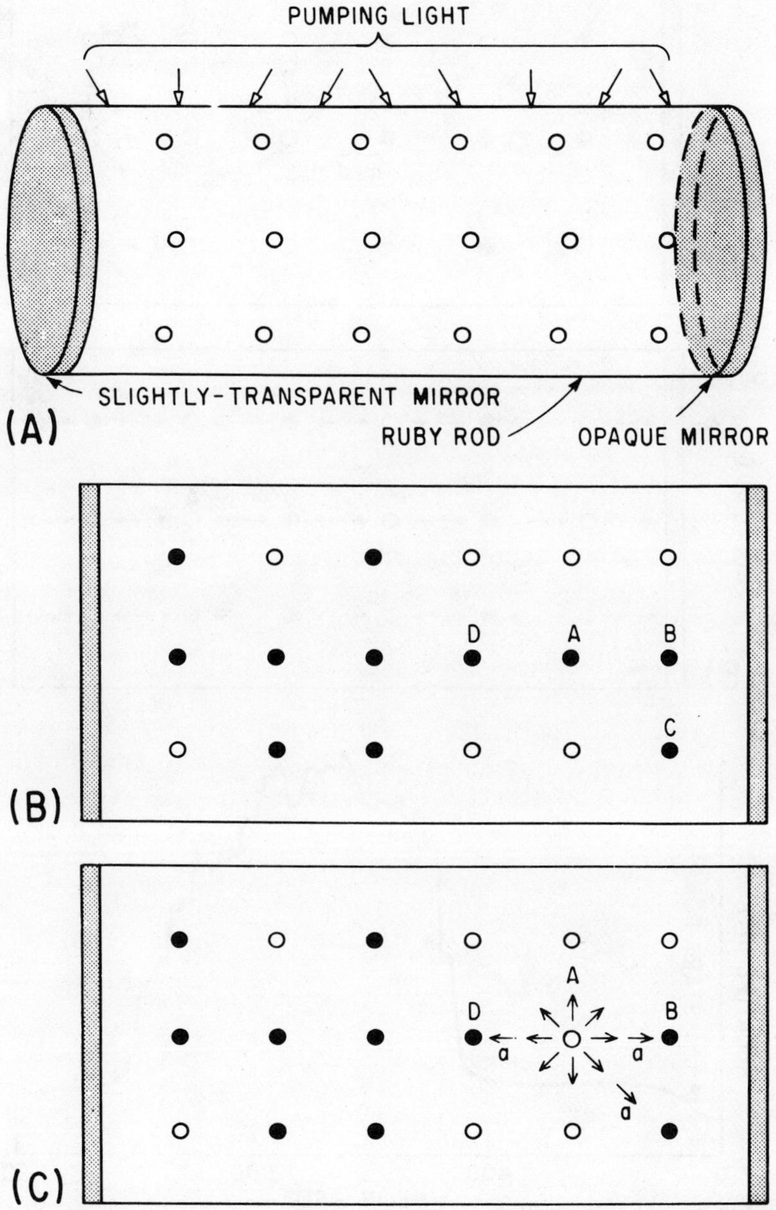
PUMPING LIGHT
SLIGHTLY-TRANSPARENT MIRROR
(A)
RUBY ROD
OPAQUE MIRROR
D
A
B
C
(B)
A
D
B
a
a
a
(C)

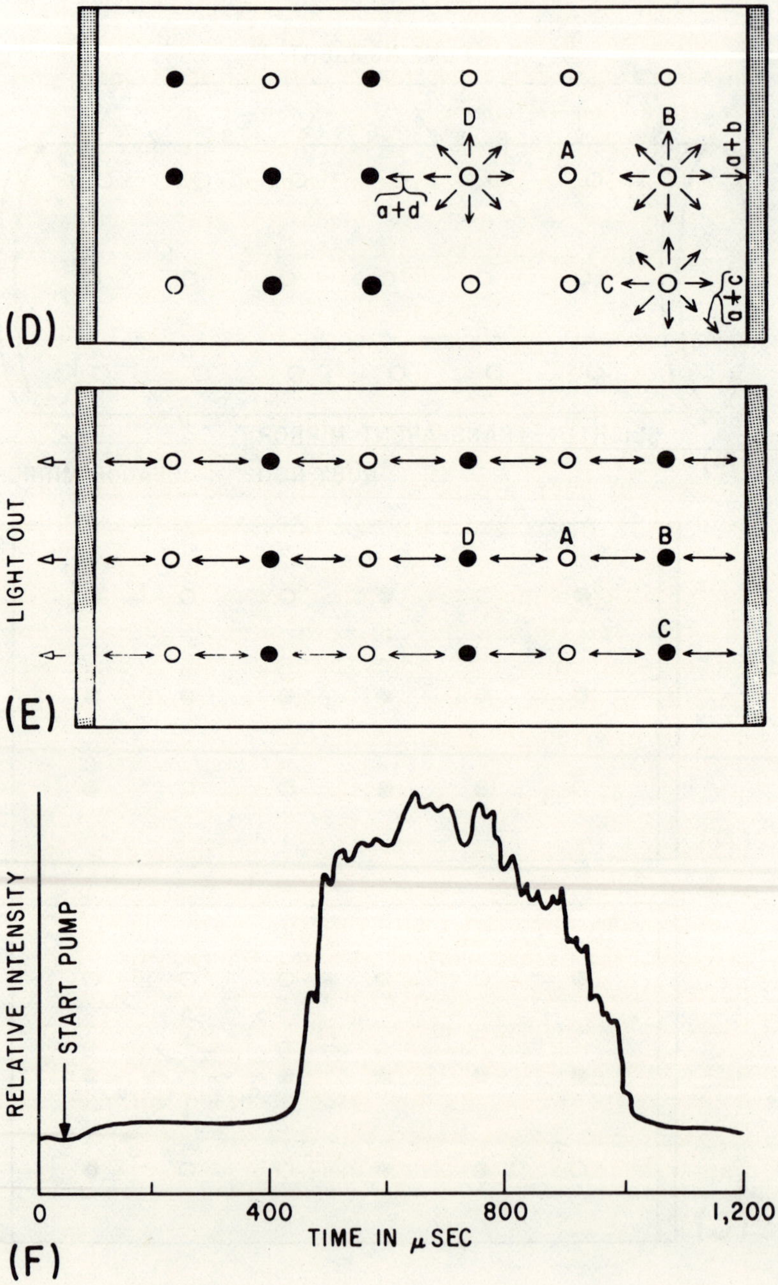
D
B
A
a+b
a+d
C
a+c
(D)
LIGHT OUT
D
A
B
C
(E)
RELATIVE INTENSITY
START PUMP
0
400
800
1,200
TIME IN μ SEC
(F)

tons become more numerous, and consequently the light beam grows more intense as the photons already in the stream trigger still more metastable chromium ions into emitting their radiation.

Eventually the photon stream builds up sufficient intensity so that it bursts from the partially silvered end of the ruby as a single pulse of monochromatic (single color or frequency), spatially coherent light.

PARALLEL RAYS

The light beams coming out of the partially silvered end of the ruby rod are almost exactly parallel, and it is this factor that makes it possible for a laser beam to reach the moon. Conventional light sources such as an incandescent lamp are point sources: their light rays are emitted in a spherical pattern. Conventional rays can be made parallel by use of focusing mirrors and lenses, but such optical systems are far from efficient: the light beam diverges, and consequently loses its intensity at great distances. But since the beams coming from a laser are parallel to begin with, they remain essentially parallel even at exceedingly great distances.

Liquid and Plastic Lasers

The ruby laser was the first laser, but today it is only one member of the class of optically pumped lasers. Furthermore, there are many varieties of ruby lasers. The original ruby lasers worked at room temperature. Later devices have been designed to work at *cryogenic* temperatures, or temperatures close to absolute zero (—273 degrees centigrade). Cryogenic temperatures are usually achieved by immersing the laser in liquid nitrogen or liquid helium. Lasers cooled this way can put out a continuous beam of coherent light instead of a series of flashes.

Other optically pumped lasers include many different crystalline materials, most of which are *doped:* made impure by the infusion of small quantities of some other material—either a rare-earth element, such as europium or neodymium, or an actinide element—a class of heavy metals that includes uranium. Some optically pumped lasers have been made of doped glass (glass to which impurities have been added), of liquid or gas in a quartz cavity or of bundles of plastic fibers.

Gaseous Lasers

The gaseous laser represents a second general class of laser. The working medium is a mixture of helium and neon gas at very low pressure (0.1 millimeter of mercury

Helium-neon gas laser (Raytheon)

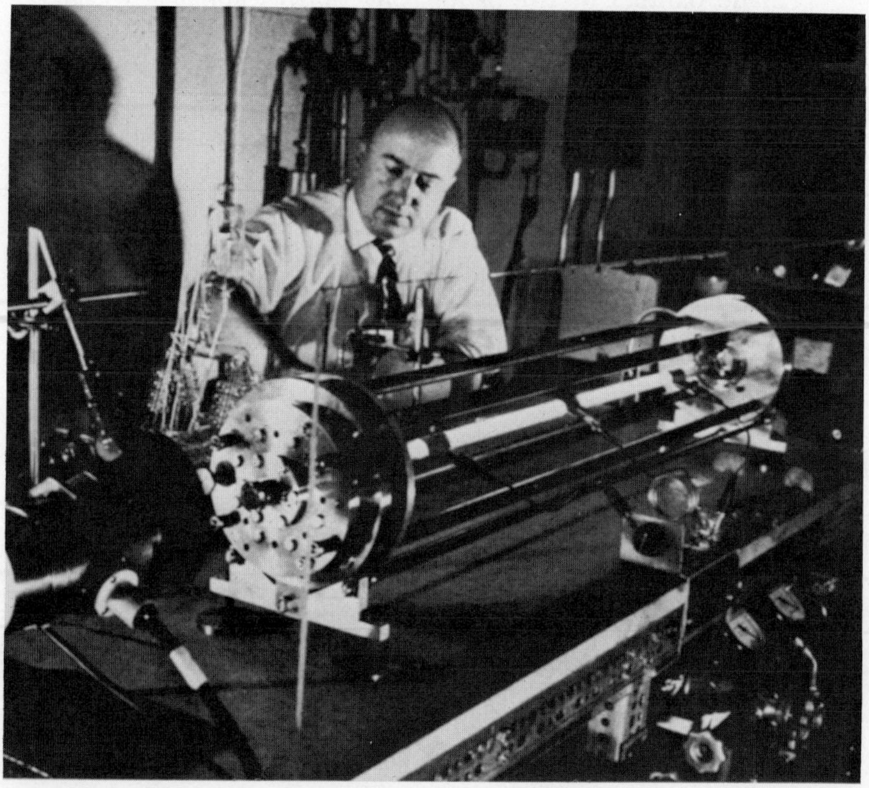

of neon and 1.0 millimeter of mercury of helium). The gas is contained in a cylindrical Pyrex tube about one meter long and 17 millimeters in diameter. At each end of the tube is a quartz plate ground optically flat and with a 13-layer dielectric (or electrically nonconductive) coating on its inner face: this coating produces the same effect as the lightly silvered end of the ruby rod. The spacing of the quartz-plate mirrors can be changed with precision for optimum internal reflection, thanks to an arrangement known as a *Fabry-Perot interferometer.* The laser beam is emitted from both ends of the apparatus.

ELECTRICAL PUMPING

The gas laser is not optically pumped, nor is it pulsed at the rate of three or four times a second as is the ruby laser. Instead it operates in a continuous-wave mode, its excitation supplied by a radio-frequency field—though in some gas lasers, direct current has been used to produce the required discharge. In a typical gas laser the source is a 50-watt transmitter operating on a carrier frequency of 29 megacycles per second. This frequency was selected simply because it lies within a band provided by the Federal Communications Commission for industrial, scientific, and medical use; another frequency would do equally well. The transmitter is coupled to the gas tube by three metal loops.

The radio-frequency generator produces an electrical discharge through the gas that raises the helium gas atoms to an excited state designated as the 2^3S state. This is a metastable state that the helium atoms retain for a finite period of time.

When the helium metastables collide with neon atoms in the ground state, the helium atoms transfer their energy to the neon atoms and drop immediately to the ground

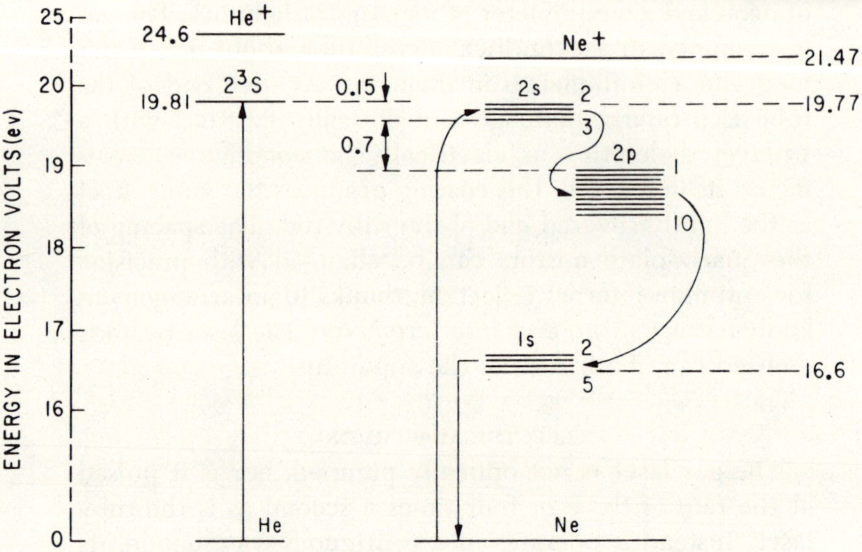

Energy levels in a helium-neon laser *(Electronics)*

state. Simultaneously, the neon atoms are raised to the so-called 2s state because the energy level of the 2s state in neon is nearly equal to the energy level of the 2^3S state in helium.

There are three excited states in neon that are involved in this reaction: the 2s, 2p, and 1s states. We are primarily interested in the transition between the 2s (higher) and 2p (lower) states. The 2s state is a metastable state. Actually, there are four substates in the 2s band and ten substates in the 2p band. Theoretically there are 30 possible transitions, or downward changes in energy level, that could occur, with each giving off radiation at its characteristic wavelength. Actually, only five of these transitions have as yet figured importantly in stimulated emissions; all correspond to wavelengths in the near-infrared region. The strongest of these emissions is one at 11,530 Å.

As in the case of the ruby laser, neon atoms tend to pile

up in the 2s state, and the threshold energy is the amount of input energy that makes the population of neon atoms in the 2s state equal to that in the 2p state. When some random neon atom spontaneously makes the transition from the 2s state to the 2p state, radiation at 11,530 Å stimulates coherent emission.

The photon at 11,530 Å stimulates nearby metastable neon atoms, and they, too, go down the chute and emit their photons at the same wavelength. Photons emitted perpendicular to the Fabry-Perot mirrors bounce back and forth between the mirrors until they acquire sufficient intensity to break out. Photons emitted in other directions are lost through the walls of the tube and do not participate in coherent emission.

When in operation, a gas laser is bathed in an orange-red glow, but this light has nothing to do with its laser action. Most of the coherent output of the gas laser is in the infrared region and is invisible to the eye. The visible glow results from spontaneous transitions of excited neon atoms that do not enter into the stimulated emission of radiation. In fact, the glow is identical to that of any neon sign.

Injection Lasers

The third basic type of laser is the injection laser. An injection laser consists of a semiconductor diode made of gallium arsenide or of gallium arsenide-phosphide.

A diode is an electronic part that has the property of conducting current easily in one direction but almost not at all in the opposite or reverse direction. The injection laser is a forward-biased semiconductor diode. It conducts current in its easy direction.

A semiconductor is a material that does not conduct electricity so well as something like copper does, but does so better than an insulator such as sulphur. The most common semiconductors are the metals silicon and germanium, but some compounds can also be used, and, for the injection laser, gallium arsenide has proved useful. Because gallium is a little better conductor than silicon, and arsenic a little poorer, when mixed together they give roughly the same effect as silicon.

Now, to make a diode out of a block of semiconductor material, it is necessary to dope it. This is done by allowing the two impurities—tellurium and zinc—to diffuse into the block at high temperature. Because the tellurium atom has one more valence (combining) electron than does arsenic, when tellurium atoms replace some of the arsenic atoms in the gallium-arsenic block, there are a few free electrons left over. Since the electron has a negative charge, tellurium-doped gallium arsenide is called N-type, or negative, gallium arsenide.

Because zinc, on the other hand, has one less valence electron than gallium, when some zinc atoms replace a few of the gallium atoms, there are several holes, or electron deficiencies, left over. Therefore, zinc-doped gallium arsenide is called P-type, or positive, gallium arsenide.

The boundary where the regions of N-type and P-type gallium arsenide meet is called the *semiconductor junction*. If you connect the positive terminal of a battery or electronic power supply to the P-type region of a semiconductor diode and connect the negative terminal to the N-type region, the diode will be biased in the forward direction, and current will flow easily across the semiconductor junction. If the power supply is connected with its negative terminal going to the P-region and its positive terminal going to the N-region, the diode will be biased in its reverse direction, and little, if any, current will flow across the semiconductor junction.

HOW DOES IT WORK?

Scientists are not yet sure just what energy transitions occur in the injection laser. But laser action seems to be most pronounced on the P-side of the junction. This might indicate that some energetic electrons making up the current flowing across the junction recombine with holes and give up energy in the recombination process.

The injection laser emits coherent light by passing extremely high current between the terminals of the semiconductor diode, so that light is emitted along the line that defines the semiconductor junction. The light comes out incoherently at first, but as the intensity of the current is increased, the emission becomes coherent. Of course, all

Semiconductor injection laser design as developed by IBM *(Electronics)*

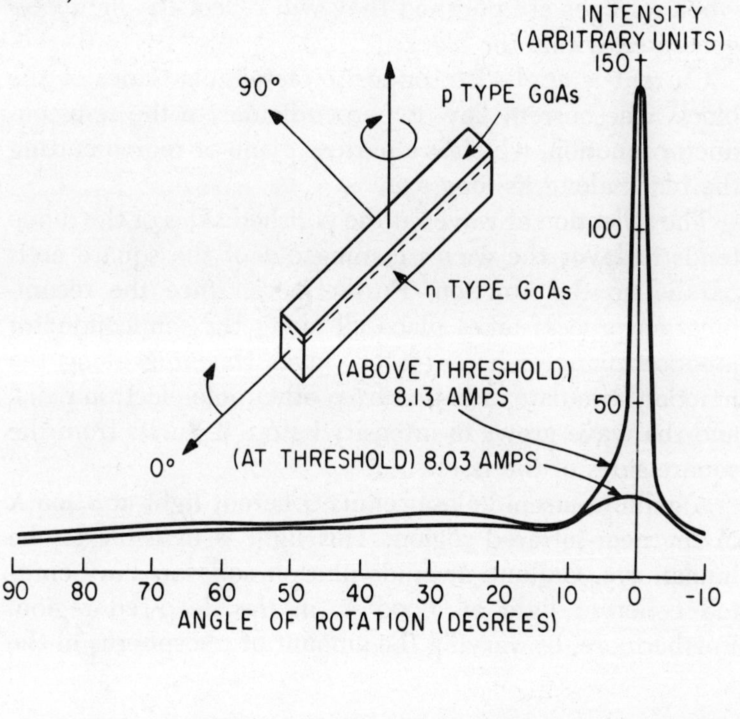

this electrical current passing through the relatively small diode makes the diode heat up rapidly. Since such extreme heating could destroy the semiconductor junction, before the diode is operated it is usually immersed in a cryostat, or double bottle, the inner bottle filled with liquid helium and the outer one with liquid nitrogen. Furthermore, the current is usually pulsed rather than passed continuously.

A typical injection laser is a rectangular parallelopiped (six-sided solid block whose opposite faces are parallel) about ten times as long as it is wide. Dimensions of a typical unit are 1/10 by 1/10 by 1¼ millimeters. The sides are finely polished and tend to reflect light back into the laser so that the emission of coherent light comes out in parallel rays from the square sides of the block. Silvering is not required because the block itself is metallic, and when its sides are polished they will reflect the light rays generated within the block.

Current is applied to opposite rectangular sides of the block. The current flow is perpendicular to the semiconductor junction, which is a narrow plane or region cutting the block along its long axis.

The reflection of waves at the polished sides of the diode tends to favor the waves coming out of the square ends parallel to the junction. Furthermore, since the recombination process takes place all along the semiconductor junction plane, coherent-light waves traveling along the junction stimulate radiation from other hole-electron pairs, and the wave grows in intensity before it bursts from the square sides of the laser.

A gallium-arsenide laser emits coherent light at 8,400 Å in the near-infrared region. This light is invisible to the human eye. Gallium arsenide-phosphide lasers have emitted coherent light at 7,000 Å, in the deep-red region. Furthermore, by varying the amount of phosphorus in the laser, the color can be changed throughout the near-infrared and deep-red regions of the spectrum. Several other intermetallic compounds involving indium and antimony as well as gallium, arsenic, and phosphorus show promise of producing laser action. A silicon-carbide diode was reported to have emitted blue-violet light, but proof of this accomplishment is as yet inconclusive.

The current passed through the particular laser we have described may vary from 10 to 25 amperes or more. At lower currents, the emission is incoherent and involves only a small part of the junction area. As current is increased, the area of incoherent sparkling or sporadic emission of light spreads out along the junction, and coherent emission can be noticed near the center of the junction.

Comparison

Thus there are three main types of lasers: optically pumped lasers, which may be crystalline, glass, liquid, gaseous, or plastic; radio-frequency or direct-current-pumped gas lasers; and semiconductor diode lasers pumped by injection of high current.

GASEOUS LASERS

The gas laser emits coherent light, usually in the infrared region. Gas lasers are used mostly in scientific investigations, such as spectroscopy, and for experiments in space and time, such as verification of some of the consequences of the theory of relativity. The gas laser is useful in these investigations because its output is the most nearly coherent of all lasers and because continuous output is conveniently available from gas lasers even at room temperature.

Because gaseous lasers operate in the continuous wave mode rather than through pulsation, they have proved better than optically pumped lasers for many communications experiments, such as the transmission of speech and music or television pictures.

Furthermore, since gas lasers produce the most nearly coherent output of any laser—the only thing that can cause a helium-neon gas laser to deviate from its 11,530 Å center frequency is mechanical vibration of the apparatus—they have been used for scientific studies, such as checking the experimental evidence of Einstein's theory of relativity and for constructing a precise gyroscope.

OPTICALLY PUMPED LASERS

Optically pumped lasers are used when high energy is required, such as for burning metal, performing delicate eye operations, precision welding or machining. The most used optically pumped laser is still the ruby laser. It is one of the few lasers that can give visible output. Nearly all gas lasers, and most types of optically pumped lasers, work in the infrared region. Most optically pumped lasers emit pulses at a relatively low repetition rate. Continuous output can be achieved only by putting the laser in a cryostat, or double bottle of liquid helium and nitrogen. Although the physical form of a ruby laser is simpler than that of a gas laser, its excitation system is somewhat more complex. The gas laser needs only a simple radio transmitter, while the ruby laser requires an electronic flashgun and either a special xenon flashtube or a carefully designed system of reflectors.

INJECTION LASERS

The injection laser is physically simpler than either the ruby or gas laser. For excitation, it actually needs only a rudimentary direct-current power supply, but it is usually operated in a cryostat. Injection lasers can produce a whole range of coherent output frequencies within the red and infrared regions of the spectrum. They deliver continuous or nearly continuous output, and they, too, have been found useful in communications experiments in which speech, music, or even television pictures have been transmitted. Gallium-arsenide diodes operated at lower current and at room temperature are already being used in portable communications systems. Although the infrared output of these devices is not coherent, they have permitted communications over a range of thirty miles.

Universal Coherence

Sciences have long dreamed of generating coherent emission at all frequencies of the electromagnetic spectrum. Quantum devices have made important contributions toward this end, but a great deal remains to be done. It has been suggested that variations of the word "maser" be coined for all the new devices, including the ones yet to come. There might be rasers (*r*adio-frequency), masers (*m*icrowave), *i*rasers (*i*nfra*r*ed), *l*asers (*l*ight), *uv*asers (*u*ltra*v*iolet), xasers (*X* ray), and gasers (*g*amma-ray). One prominent scientist jocularly suggested the name "daser," standing for "darkness amplification by stimulated emission of radiation."

All this points up the advantage of talking about quantum devices (and specifying whether they are oscillators, amplifiers, or harmonic generators) and designating the wavelength of interest rather than playing with acronyms. It does, nevertheless, seem to be a fact of life that the term "maser" will continue to be used both for amplifiers and for oscillators not only in the microwave region (roughly 1,000 megacycles per second) but perhaps for devices operating at even lower frequencies, when and if such devices are developed.

Likewise, it seems that the term "laser" will continue to

be used to refer both to amplifiers and to oscillators that operate in the near-infrared, visible, and near-ultraviolet portions of the spectrum. Neither extension of laser action into the far-infrared (near microwaves) nor into the far-ultraviolet (near X rays) will result in a change in terminology.

But possibly, when we can successfully generate coherent X rays and gamma rays, another term will be used, for already, as mentioned above, the word "gaser" is being bandied about.

MASERS

Masers are usually true amplifiers instead of the generators that lasers are. This means that they receive a weak signal and pass it on at a higher power level. Masers operate between 300 megacycles per second (100 centimeters or 1 meter wavelength) and 100,000 megacycles per second (3 millimeters).

We might remark parenthetically that there is other millimeter-wave research going on that does not involve masers. One special microwave tube, the Tornadotron, has been reported to have an output of 500,000 megacycles per second, or a wavelength of 0.6 millimeter.

A typical maser consists of a crystal containing chromium that is pumped by the output of a microwave tube operating at a frequency much higher than the one to be received. The microwave signal pumps the chromium ions to an elevated energy level that is metastable.

Incoming signals at a certain lower microwave frequency stimulate the chromium ions to fall from their elevated energy level to an intermediate level before the ground state. In so doing, they emit radiation at the frequency of the incoming signal and thus amplify it.

To avoid the introduction of noise or unwanted signals,

maser amplifiers are placed between the pole pieces of a powerful magnet, and are operated in a double bottle with liquid helium on the inside and liquid nitrogen on the outside.

About a dozen radio astronomical observatories throughout the world use maser amplifiers to pick up radio-frequency emissions from distant planets, stars, and nebulae. Several stations use maser amplifiers for tracking satellites and space probes. So do some of the stations that receive radio and television signals from orbiting communications satellites such as Telstar and Relay. It is possible that maser amplifiers are used in special military radar and communications applications, but if so, the Department of Defense isn't saying!

INFRARED LASERS (IRASERS)

Various kinds of lasers cover the near-infrared spectrum from nearly 13,000 Å right up to visible light. This leaves a gap in the spectrum from 3 millimeters wavelength to 0.013 millimeter. This gap includes the millimeter and submillimeter-wave regions of the radio spectrum and the far-infrared band that encompasses radiation from warm and lukewarm objects.

NEW COLORS IN LASERS

Progress has not been so good in the visible region. Only a few lasers produce visible light, and most of that, as we have noted, is deep red. There is, of course, the ruby laser. Red light has been produced by several other methods as well: by a laser consisting of a crystal of calcium fluoride with the rare-earth samarium dissolved in it; from europium chelate (rhymes with "tea late") embedded in a plastic tube (a chelate is a complex organic or hydrocarbon molecule containing a metal atom, in this case an atom of the rare-earth europium); with the gallium arsenide-phosphide laser; and with some helium-neon gas lasers.

There is a demand for lasers to produce other colors besides red. The Navy would like to have a blue-green laser because blue-green light is best for penetrating seawater and because a blue-green laser could be used as part of an underwater television system to help navigators of nuclear submarines detect the presence of friendly or hostile submarines or other underwater objects.

So far, the only progress in that direction has been the development of "blue-violet lasers," produced by doubling the output frequency of a deep-red laser. (Doubling the output frequency is the same thing as dividing the wavelength by two.) Likewise, there are "green lasers," achieved by doubling the output frequency of lasers operating in the near-infrared region.

But when you double the output frequency of a laser, you lose 8/10 or more of its energy, and what's left will hardly perform the job the Navy has in mind. Therefore the search for different colored lasers continues, with scientists now studying not only rare-earth and actinide metals but even various organic compounds. They feel that, given the right conditions, any substance that will fluoresce can be made to lase. This leaves them with thousands of compounds to investigate.

ULTRAVIOLET LASERS (UVASERS)

So far the story of the ultraviolet laser is short and sweet. One optically pumped laser, using a glass rod in which a small quantity of the rare-earth gadolinium has been dissolved, lases at 3,125 Å in the near ultraviolet.

GAMMA-RAY LASERS (GASERS)

Nothing has been announced officially about X-ray lasers, but certain work is going on with gamma-ray lasers under Navy auspices, though the work has not progressed very far as yet. The Russians have also announced work in this field.

The approach is to use a gamma-ray-emitting isotope of ruthenium to raise a radioactive isotope of rhodium to a higher energy state that is metastable. After a half-life of some 40 days, the level of energy emitted by the ruthenium will drop to that of the metastable state of the rhodium isotope, and trigger emission at roughly 0.3 Å.

There are many problems in the way, however. First, one has to find a way to make a crystal containing the appropriate isotopes without changing their essential characteristics. Next comes the problem of containing the gamma rays (they will penetrate just about anything) so as to achieve spatial coherence. If achieved, a gamma-ray laser would be a death ray in every sense of the word. Gamma rays have several times the burning power of X rays, which are, of course, harmful when improperly applied.

The Future of the Laser

As we have seen, the laser has the advantage of providing a monochromatic or single-color light source. Furthermore, its beam is so collimated that all its energy can be focused on a very small spot. It is also highly directive, with little or no tendency for the beam to bend or spread out even over the astronomical distances of outer space. These properties have suggested a great many uses in national defense, industry, medicine, and science.

Lasers may be developed into devastating antipersonnel weapons for use on the battlefield. They may be sent into space on special platforms to fight intercontinental ballistic missiles or to destroy hostile space stations or satellites. The laser may also be used to modify chemical compounds or even to change the genetic characteristics of the protein molecules of living organisms.

Someday special fiber-optic light pipes or other optical wave guides, such as evacuated tubes with an internal mirror system, may carry laser signals much as coaxial cables now carry telephone conversations and network television programs between cities. A fiber-optic light pipe is a very fine glass, plastic, or arsenic-trisulfide rod polished on the outside; its walls reflect light back inside so that it can bend around corners and still carry a light beam.

One way to put a TV signal on a laser beam is first to impress the complete TV picture and sound (the video signal) on a microwave carrier. The microwave carrier is then used to excite a special crystal situated in a microwave cavity or special metal box. When the laser beam traverses the crystal, entering and leaving the cavity through small side windows, the beam is modulated or made to vary in accordance with the modulated microwave signal. At the receiver, the beam of a microwave traveling-wave amplifier phototube is similarly made to vary in accordance with the variations of the laser light striking the traveling wave tube's photocathode. We now have again the microwave carrier with the video signal riding on it. This signal is demodulated, using conventional electronic circuits to give the original TV picture and sound.

A wideband video channel can be divided into many subchannels, actually some 600, each of which can carry a telephone conversation. Electronic circuits called filters slice up the video channel into so-called voice channels. Each voice channel is about 0 to 2,000 cycles per second wide. Each incoming telephone signal is heterodyned, or moved up, in frequency to fit a specific voice channel at the transmitting end, then moved down in frequency and routed out on its proper telephone line at the receiving end.

A laser communications system would greatly expand the capabilities of our nationwide telecommunications network. Tiny lasers may also function as parts of the memory system of a computer. Such a computer would literally work with the speed of light.

Who knows? You may even one day have a laser ignition system in your automobile!

MILITARY USES

One of the first uses that occurs to most people is to build a big, superpower laser and use it to shoot down ballistic missile nose cones. This would, they reason, make our nation secure from the terrors of thermonuclear war.

But it isn't as easy as all that. Even the most powerful lasers can at present penetrate only ⅛-inch of high-carbon (easily burnable) steel. And the holes they make are mere pinpricks. Furthermore, burning requires that the laser be only a distance of a few feet from the steel. At longer ranges, the water vapor and dust in the atmosphere severely reduce the effective power of the light ray.

Nevertheless, the Air Force is hard at work trying to develop big lasers and figuring out how to deploy them effectively outside the earth's atmosphere: atop mountain peaks, aboard orbiting satellites, or even on antimissile missiles.

Meanwhile, the military and space agencies have other, more prosaic, but none the less vital uses for the laser. When the Apollo lunar capsule carries the first Americans to the vicinity of the moon, the two-man crew aboard the Lunar Excursion Module that will make the actual landing on the moon will probably use a laser altimeter to feel their way onto the lunar surface. Before that, astronauts in Project Gemini will use laser radar to practice rendezvous and docking of satellites in space. Already a large laser at Wallops Island, Virginia, has tracked an orbiting satellite 1,000 miles up. Incidentally, at that range the laser beam was only 200 feet in diameter.

The Army has ordered several laser range finders for use on the battlefield. They will be able to measure the distance to targets far more accurately than their optical or radar counterparts.

During World War II the Army made effective use of sniperscopes and snooperscopes, infrared devices that located targets even at night. But for such devices to be effective, the target had to be a good deal warmer than the background. Now, with an infrared laser, it would be possible to scan the target and get a picture regardless of its temperature.

During World War II the Navy used infrared "Nancy" equipment (usually Nerst tubes or hot filaments enclosed by a black metal hood and placed behind a deep ruby lens) for short-range communications between ships. But the laser affords a much more efficient and less easily detectable source of infrared.

The Armed Forces have a project under way to see just how fast a computer can operate. Some people think that the result will be a new high-speed giant brain for our ballistic missile early-warning system. But a better guess is that such a computer will be used to crack secret enemy codes and ciphers. Anyway, one part of this project is a laser computer, sponsored by the Air Force, in which light pulses would do the counting instead of electrical signals. Such a computer would be faster by several orders of magnitude than any computer now available, since light travels faster than electrical current, which is slowed down by the action of reactive elements, such as capacitors and inductors in the circuit.

INDUSTRIAL APPLICATIONS

Industry is already using lasers to perform delicate machining and welding operations in the manufacture of microelectronic circuits.

A microelectronic circuit is fabricated on a thin wafer of silicon. Sometimes forty circuits are made at one time on a wafer only an inch in diameter. Each circuit can do the work of, say, a five-tube radio or perhaps a computer stage.

The circuits are made by allowing certain selected impurities to diffuse into the silicon wafer in prescribed patterns. These patterns are formed by first allowing a film of silicon dioxide (glass) to grow over the silicon wafer—usually by applying steam to the surface—then selectively etching away portions of the film.

Selective removal of the oxide is accomplished by first coating the oxide with so-called photoresist—a film that becomes tough and acid-resistant when exposed to light—then masking the wafer with a diffusion mask and exposing the unprotected photoresist to light. The wafer is next etched with strong acid, and its silicon-dioxide coat is eaten away except where it is protected by light-hardened photoresist.

Preparation of the diffusion mask is a critical operation, and laser machining of metallic foil is expected to allow making sharper and more precise pattern outlines. Possibly lasers may be used to remove the oxide itself, thus saving several steps in the process of manufacturing microcircuits.

Laser light sources could be valuable in high-speed photography where chromatic aberration or the unequal bending of light of different wavelengths through the camera lens can cause a blurred image.

Since different components of the atmosphere absorb different wavelengths of light to a greater or lesser extent, a bank of lasers used at an airport as a transmissometer could disclose not only the visibility at the end of the runway—as the optical devices already in use do—but also the makeup of the atmosphere at any particular time. Such a laser device could also be useful in air-pollution studies. (Transmissometers are used even though the end of the runway may indeed be visible from the control tower; the view from the tower is not what an approaching pilot sees; besides, the instrument, unlike a human observer, remains on duty around the clock.)

In a chemical process, a laser might be created so that its beam is absorbed to a great extent by the desired product. The laser could be focused permanently through the output pipe, and automatic control equipment could be adjusted so that the product absorbs maximum light from the beam. This would assure that the product in the output pipe has precisely the desired chemical composition.

The ability of a laser beam to carry an almost infinite amount of information has set communications engineers to speculating about its possible use for trunkline or intercity communications. Today, these are handled by coaxial cables or microwave links. One microwave link can carry four television programs simultaneously or replace any one of the television channels with up to 600 telephone conversations. But a laser beam could carry many times this amount of information.

Nevertheless, since dust and water vapor in the atmosphere severely reduce the effective power of a laser beam, a serious problem still remains before lasers can be used for practical communications. Of course, short-distance communications would indeed be possible, as would communications to and from communications satellites. In the latter case, the beam travels in the earth's atmosphere for only a relatively short distance, although the total trip might be 1,000 miles or even more.

One answer to abetting laser communications would be to use light pipes or evacuated tubes with mirrors arranged to conduct the beam around corners where necessary.

A laser telephone exchange has been contemplated. Here the light pulses would be conducted by fiber optic strands. These strands carry light around corners just as copper wires carry electrical current. Though a fiber-optic strand severely cuts down the power of the light being transmitted, in a telephone exchange the length of the interconnecting strands can be kept short by design. The big advantage of a laser telephone exchange would be that there would be no crossed wires or unwanted pickup between adjacent optical fibers so that you would not occasionally hear fragments of another conversation on your line.

MEDICAL USES

Lasers have been regarded as a major boon to medicine. Thousands of Americans suffer each year from a detached retina. In this condition the retina, the light or sensitive area at the rear of the eye, comes loose from the inner surface or choroid coating of the eyeball. The fluid, or humor, with which the eye is filled works in behind the retina and aggravates the condition. Initially, the condition causes distorted vision, but if the retina becomes completely loose from the optic nerve, blindness results. A laser beam can be focused through the lens of the eye so that it makes small scars around the periphery of the retina and thus welds it back into place.

A laser can also burn out small tumors in the eye. In fact, a laser beam can be made as narrow in diameter as the diameter of a single human cell. Some surgeons see the laser, therefore, as a device that can burn out tumors with minimum risk of damage to surrounding healthy tissue. Lasers have also been considered for suturing wounds through heat. The laser would cauterize the wound as it sutured it. It could also be used to disinfect small areas quickly. Dentists have experimented recently with laser drills; they are fast, sure, and painless.

It is conceivable that laser beams can be made even narrower in diameter than a single protein molecule. Such a laser beam might be used to alter the genetic properties of living organisms.

A team of medical scientists has reported that irradiation by a laser beam has altered the electrical conductivity of whole human blood. Just what this means or how it occurs has not yet been made apparent.

SCIENTIFIC APPLICATIONS

Perhaps some of the most far-reaching effects of the laser will be in the fields of pure and applied science. The laser may profoundly affect man's understanding of his natural environment.

Our most basic quantities of measurement are length, mass, and time. Two of these, length and time, are related by a constant, the velocity of light in a vacuum, and yet the value of this constant is only imperfectly known.

Our national standard of frequency is calibrated from the same astronomical observations that give us our measure of time, since the frequency of cycles per second that a wave executes is intimately related to time.

When dealing with radiation in the visible region, scientists measure wavelength instead of frequency. But if the standard radio frequencies could be doubled, redoubled, and then redoubled again as many times as necessary to reach the visible-light region, then length and time would be one and the same thing irrespective of our uncertainty as to the exact speed of light in a vacuum.

Another basic scientific problem is the question of whether ether exists or not. You recall that we explained electromagnetic waves by comparing them to waves in a pond. Many scientists have found it equally hard to conceive of waves without postulating some substance or medium in which the waves could move or propagate.

Accordingly, they postulated ether—a colorless, odorless substance filling all space—in which electromagnetic waves could propagate just as waves propagate in a pond. For years now, scientists have been trying to relegate ether to the same never-never land as phlogiston and other weird substances once postulated by alchemists to explain physical phenomena they could not understand.

The first experiment to disprove the existence of ether was the Michaelson-Morley experiment: If the earth is rotating in a stationary sea of ether, the ether will drift by the earth in a direction counter to the earth's rotation. Now, suppose two light beams are transmitted at right angles to each other in such a way that the ether drift will add to the speed of one beam while the other beam will travel perpendicular to the ether drift and therefore be unaffected by it. Then any difference in velocity caused by ether drift could be detected by measuring the difference in frequency of the two beams. To make the measurement more precise, the apparatus emitting the light beam is next turned around so that the ether drift will oppose the speed of the beam instead of adding to it; the frequency difference (if any) can again be measured. If the sum of the two frequency differences were significant, an ether drift could be said to exist. This experiment has been carried out with the use of gas lasers, but no significant frequency difference has been noticed that could substantiate the existence of an ether.

In the realm of applied science, the laser shows greatest promise in spectroscopy. We have referred at many times to absorption of infrared, light, and ultraviolet frequencies by certain substances. The exact frequencies absorbed depend upon the chemical composition of the substance and the structure of its molecules. The totality of frequencies absorbed or the absorption spectrum of a substance is as individual as your fingerprints. Therefore spectroscopy is a basic tool for physicists and chemists studying the properties of matter. But better discrimination in spectroscopy is needed, and to get it, scientists must know the exact frequencies with which a substance is irradiated. As the number of laser materials increases, and consequently the number of available coherent light frequencies increases, spectroscopists can look forward to more efficient tools that will enable them to gain greater and greater insight into the basic makeup of matter.

Conclusion

In this chapter we have explained the continuum of the electromagnetic spectrum in terms of both frequency and wavelength. We have come to grips with some of the basic concepts of quantum mechanics and have seen how they explain the action of the three basic types of lasers: optically pumped, gaseous electrically pumped, and injection. We have discussed the phenomenon of fluorescence and have seen how laser action is related to fluorescence but differs from it because of (*a*) its frequency coherence or monochromaticity and (*b*) its spatial coherence, or the fact that all wavelets keep in step.

(Incidentally, this last gem of knowledge now makes you smarter than a certain covey of investors with more spare cash than technical knowledge. They lost several kilobucks supporting a glib physicist with a lab full of bottles of fluorescent material that he passed off as lasers completely covering the visible spectrum! Of course, they weren't lasers at all.)

Finally, we have looked at the whole electromagnetic spectrum in terms of how coherent radiation is or might be produced by quantum devices, and have placed a special emphasis on a possible gamma-ray laser. And we have seen the impact of lasers on national defense, industry, medicine, and science.

Now we shall look backward and see how the laser actually came into being.

One basic law rules the operation of all devices that use electric currents. A fine introduction to the study of electricity.

8 A Simple Electric Circuit: Ohm's Law

Albert V. Baez

A chapter from the textbook *The New College Physics, a Spiral Approach.*

WE BEGIN this chapter by considering the operational steps we might take, in an elementary laboratory, in order to learn more about electric current. We shall then try to build up a theory that accounts for our observations.

40.1. A Simple Series Circuit: Measurement of Potential Difference

Figure 40.1 shows what our apparatus looks like: *A*, a six-volt storage battery; *B*, a lamp in a socket; *C*, a knife switch; *D*, a voltmeter; *E*, an ammeter; *F*, some connecting wires. From now on we shall, as much as possible, use the shorthand of conventional diagrams, as in Figure 40.2, which shows battery *A*, lamp *B* (the zigzag line is actually the symbol for an element with resistance), and switch *C* connected in series. When the switch is closed, the lamp lights up. We say that there is an electric current or that there is a flow of electric charge, but we don't, of course, see anything flowing. The fact that the bulb lights up when the switch is closed is the only outward sign that anything flows.

It is not uncommon to begin such an experiment with little or no knowledge of what is inside the magic boxes *A*, *B*, *D*, and *E* (Fig. 40.1). All we

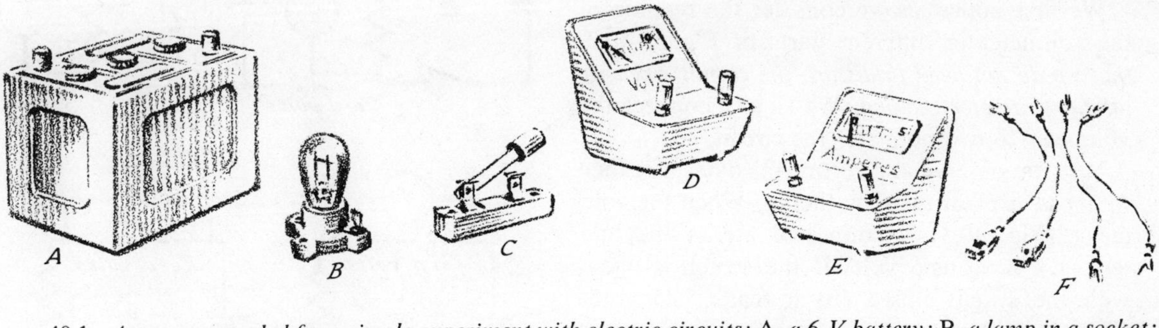

FIG. 40.1. *Apparatus needed for a simple experiment with electric circuits:* A, *a 6-V battery;* B, *a lamp in a socket;* C, *a knife switch;* D, *a voltmeter;* E, *an ammeter;* F, *typical connecting wires, two of the clips on which are called alligator clips.*

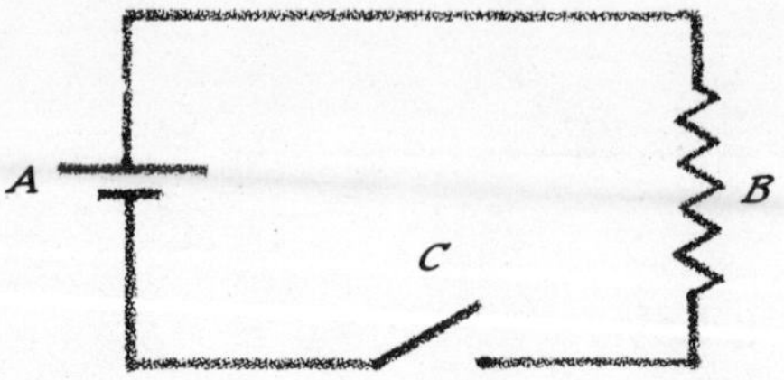

FIG. 40.2. *Schematic diagram of a series circuit including a battery,* A, *connected to a lamp,* B (*shown here as a resistor*), *through a switch,* C.

know is that *D* measures potential difference and that *E* measures current. In this chapter we shall look inside *B*, *D*, and *E*. The battery, *A*, however, will have to remain just an electron pump; I shall leave its inner details out of the discussion because they involve the complicated molecular mechanism by which chemical energy is converted into electrical energy.

We want to understand why the voltmeter readings of Figure 40.3 are what they are at different places. We are going to limit ourselves in this chapter to an understanding of the simple circuit of Figure 40.2. We shall move more slowly than is customary in a chapter on electric circuits, and only when we peek inside the voltmeter and the ammeter shall we see slightly more complicated circuits in series and in parallel. Our immediate objective is limited; but, if you understand all the details of this discussion, you will have a firm grasp of fundamentals.

We first notice, as we consider the reading of the voltmeter in different parts of Figure 40.3, *that we do not need to disturb the circuit when we take a voltmeter reading.* We simply connect the voltmeter to two points of the circuit.

Next we observe and record the data, and then we try to explain them by theory. When the voltmeter (Fig. 40.3) is connected across the battery (A), it reads 6 volts if the switch is open; with the switch closed (B) it reads 5.45 volts. Connected across the lamp, it reads 0 if the switch is open (C) and 5.45 volts if the switch is closed (D). Connected across one of the connecting wires, it reads 0 whether the switch is open (E) or closed (F).

If the voltmeter is telling the truth, the potential difference across the terminals of the battery is 6 volts when there is no current in the circuit (A). The potential difference across the battery drops when there is current (B). There is no potential difference across the terminals of the lamp (C) until the switch is closed (D), and there is never a measurable potential difference across one of the connecting wires. Our theory of what is going on must account for all these readings (and a lot more).

Let's begin our description of what we think is going on. We have already encountered a momentary flow of charge in electrostatic experiments, but something different is obviously happening here, for this current can flow for a long time. Something replenishes the charge; something maintains a potential difference that produces a steady flow of charge. This something, in our experiment, is the battery. The terminals of the battery are *charged* in the very sense in which we used the word in electrostatics. If our battery has only two terminals, an electric field surrounds

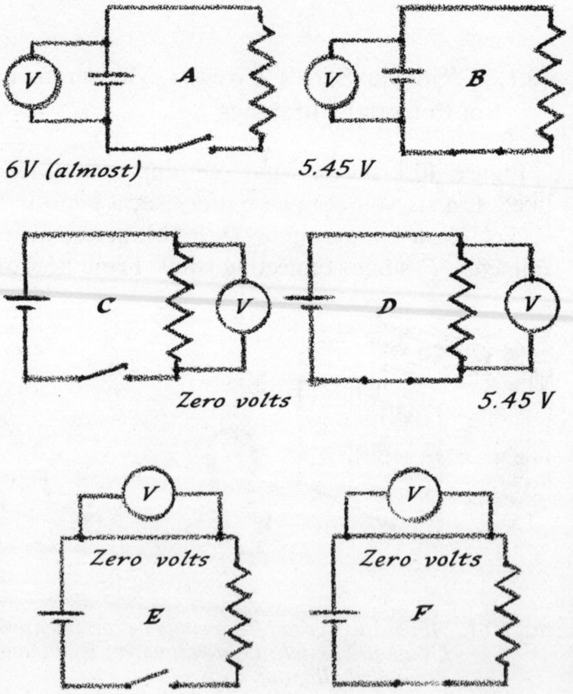

FIG. 40.3. *Readings on a voltmeter as it is connected to different parts of a series circuit that is sometimes open and sometimes closed.*

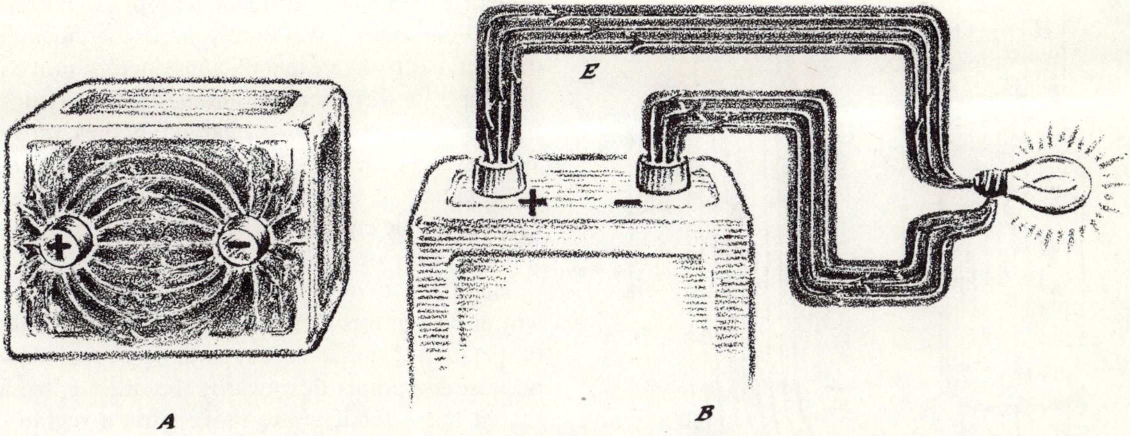

FIG. 40.4. (*A*) *The electric field lines in air surrounding the terminals of a battery.* (*B*) *The electric field lines within a wire connecting the two terminals of a battery through a lamp.*

them as if they constituted an electric dipole. Figure 40.4.A shows the electric field between the two battery terminals. It looks very much like the electric field between two charged metal balls on insulating stands; but there is a difference in what happens to these fields if the terminals are connected with a wire. A wire connecting one charged ball to the other would carry current only for an instant, for the potential difference between them would soon be zero, and the field would vanish. If the terminals of the battery are connected, a large current can exist in the wire for a much longer time, and the field between the terminals will still be like that of Figure 40.4.A after the wire is removed. In Figure 40.4.B we see the electric field lines (**E**) that come into existence *within* the wire that runs from one terminal through the lamp to the other terminal. I said earlier that there can be no electric field within a conductor, but that is true only in the electrostatic case. Charges move in the wire of Figure 40.4.B *because* there is an electric field within it.

Since the lamp gets hot, it is obvious that energy is involved. It looks very much as if something were playing the role that friction plays in mechanics. Something *is* playing that role; it is called *resistance* (defined in § 40.3), and we shall soon consider it in some detail.

Let us now recall the definition of electric field, **E**, as $\mathbf{F}/q$, the force per unit charge (§ 4.4). An electron finding itself in electric field **E** experiences the force $\mathbf{F} = -e\mathbf{E}$. It should experience the acceleration $\mathbf{a} = \mathbf{F}/m$, and it does, but it cannot pick up much speed, for it collides with other electrons. The average behavior of many electrons, starting and stopping, is, nevertheless, a general drift in the direction of $-e\mathbf{E}$. Statistically, the free electrons drift at an average speed determined by the magnitude of the force $-e\mathbf{E}$.

The idea of motion at a constant speed under the action of balanced forces can be perfectly illustrated by the falling of small spheres (such as marbles) through a tall glass beaker containing glycerin (Fig. 40.5.A); balls of the right weight and dimensions achieve a terminal velocity. The force of gravity, $m\mathbf{g}$, pulls them downward, but a viscous frictional force, **f**, pushes them upward. When $m\mathbf{g} = \mathbf{f}$, the acceleration is zero (see § 5.2).

A positive charge, q, in electric field **E** feels the force $q\mathbf{E}$ (Fig. 40.5.B). If it also feels an equal retarding force, **f**, it can move at a constant speed. What happens in a wire is somewhat like this. For two reasons, however, you must not take any such picture literally. First, no electron travels for long without hitting another, and the concept of drift velocity is therefore purely statistical. (It takes a lot of kinetic energy to carry an electron into contact with another, even when the other is anchored to an atom. What I have called hitting just means being decelerated by a force

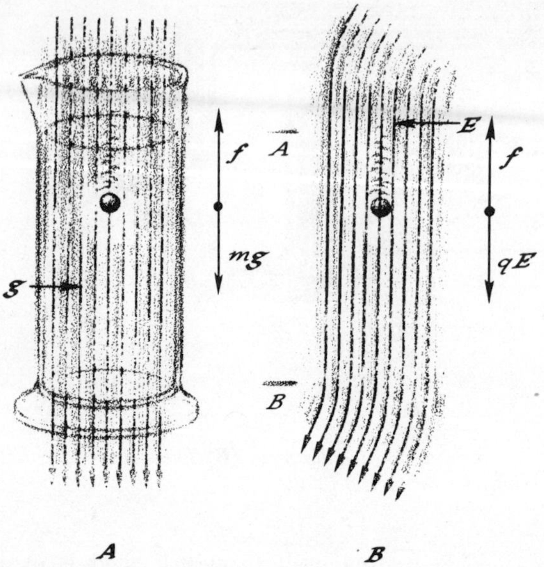

FIG. 40.5. (*A*) *The gravitational field lines running through a tall glass beaker containing glycerin; little spheres fall through it at a constant speed.* (*B*) *The electric field lines in a wire; electric charges move with a constant average speed within the wire.*

field. Here it would pay you to re-read § 3.7, dealing with the concept of contact.) Second, electrons have a negative charge and move opposite to **E**, but this does not damage the model of Figure 40.5.

Traditionally, the direction of current in a wire has been taken as from the positive to the negative pole (in the part of the circuit outside the battery). In this book, since it is now known that in a wire the electrons do the moving, I have broken with tradition by assigning to i the direction of electron flow. But I shall use the symbol $I\,(= -i)$ for the conventional direction (from positive to negative) whenever it can simplify the wording of statements. All the left-hand rules I gave in the study of magnetism relate to i. If we associate the *right hand* with I, similar rules apply. In other words, I is the direction in which positive charges would move in a wire. Since positive charges tend to move from a region of high electric potential to one of low potential, it is convenient to use the traditional symbol for current, I, in these cases. (We simply need to remember that the electrons in metallic conductors move in the opposite direction; in liquids, however, positive as well as negative charged bodies move.) Whenever we use the symbol q without any further specification, it will represent a positive charge. The electronic charge will, of course, be written as $-e$.

There are two ways of expressing the reason why a ball moves downward through the beaker of glycerin. One is to say that it moves down because mg points downward; the other is to say that it has a tendency to move from a region of high potential to one of low potential. The same language applies to *positive* charges in an electric field: they move from A to B in Figure 40.5.B because q**E** points that way, or (since an applied force would do work in moving a positive charge from B to A) they move from a region of high potential to one of low potential.

Potential difference, V, is measured in volts, which we identified earlier (§ 37.2) with joules per coulomb. The work that will move charge Δq through distance x from B to A is (by the formula "work equals force times distance") $\Delta U = (\Delta q)Ex$. The work per unit charge is $\Delta U/\Delta q = Ex$. The left-hand side has the units joules per coulomb, or volts. The right-hand side has newtons per coulomb times meters for units. This equivalence is worth remembering. We may write

$$V = Ex \qquad [40.1$$

or

$$E = V/x \qquad [40.2$$

Now we are getting somewhere. The quantities on the right-hand side of the second equation are measurable, V with a voltmeter (we'd better find out how it works), x with a meter stick.

If we connected a voltmeter across points A and B of Figure 40.5.B, would it show a reading? I said earlier (Fig. 40.3.F) that there is no detectable reading across a wire carrying current. You will have to take my word for it that a certain very sensitive kind of voltmeter would indicate a small potential difference between points A and B if there were a current in the wire.

EXAMPLE 40.1. A sensitive voltmeter indicates a potential difference of 10^{-6} V between points A and B of Fig. 40.5.B. The distance between the points is $x = 2$ m. We wish to know (1) what force, in newtons, an electron feels within the wire; (2) what acceleration it experiences; (3) what the increment in its speed is if it travels for 10^{-7} sec.

1. The force on a charge, q, is $F = Eq$. Since, by equation 40.2, $E = V/x$, we know that $F = Vq/x$. We are given that

$$V = 10^{-6}\ \text{V}$$
$$q = -e = -1.60\times10^{-19}\ \text{coul}$$
$$x = 2\ \text{m}$$

Therefore, if we drop the minus sign,

$$F = \frac{10^{-6} \times 1.60\times10^{-19}}{2}\ \text{nt}$$
$$= 8\times10^{-26}\ \text{nt}$$

2. The acceleration is $a = F/m$. We know that

$$F = 8\times10^{-26}\ \text{nt}$$
$$m = 9.11\times10^{-31}\ \text{kg}$$

Therefore

$$a = \frac{8\times10^{-26}\ \text{nt}}{9.11\times10^{-31}\ \text{kg}}$$
$$= 8.78\times10^{4}\ \text{m/sec}^2$$

3. We know that

$$\Delta v/\Delta t = a$$

Therefore

$$\Delta v = a(\Delta t)$$
$$= 8.78\times10^{4}\ \text{m/sec}^2 \times 10^{-7}\ \text{sec}$$
$$= 8.78\times10^{-3}\ \text{m/sec}$$

There are experimental reasons for believing that this is of the right order of magnitude for the average speed of electrons in a wire.

40.2. Electromotive Force

We can extend the analogy of balls falling through glycerin to a simple electric circuit.

In Figure 40.6.A we see balls rolling and falling under the action of the earth's gravitational field, **g**. If the balls are to keep moving at a constant rate, work has to be done against gravitational force as each ball is lifted from D to A. The energy is supplied by the man, who acquires it by the complicated chemical process that transforms food energy into mechanical energy. Notice that there is a small difference in gravitational potential, $g(\Delta h_1)$, between points A and B, a large differ-

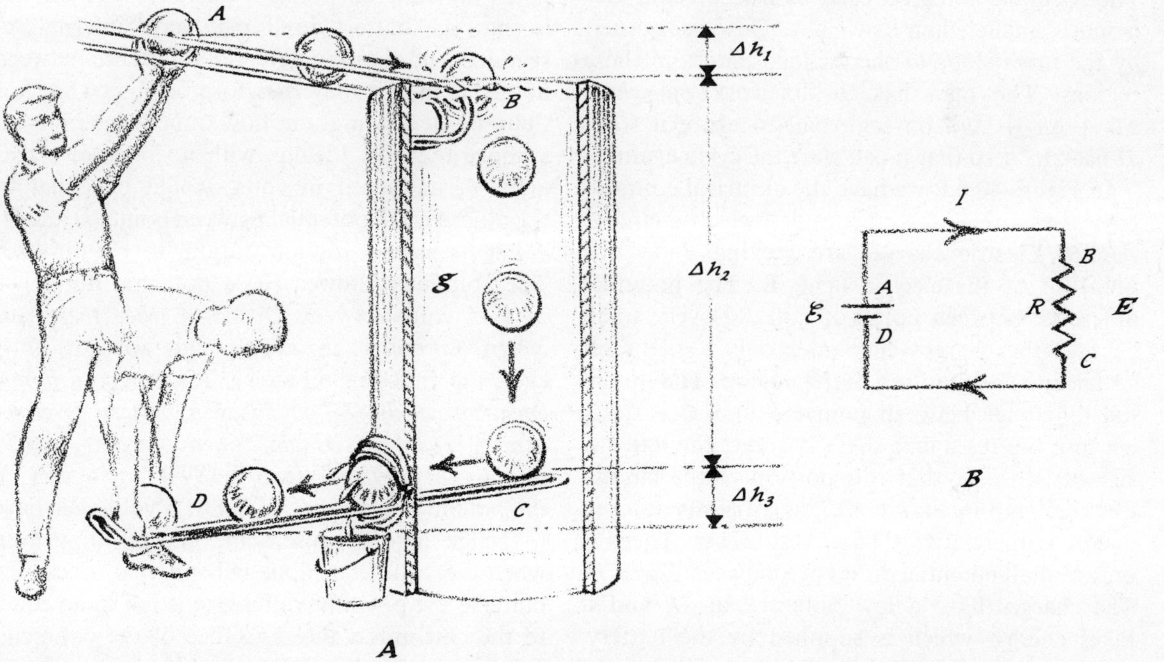

FIG. 40.6. *Analogy between the effect of the earth's gravitational field and that of an electric field.*

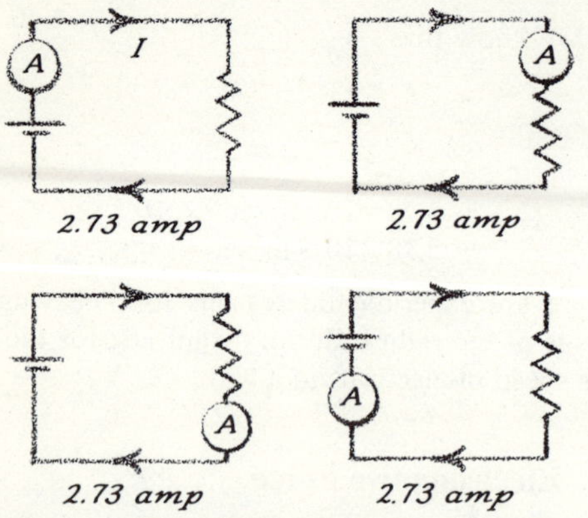

FIG. 40.7. *How an ammeter will read when connected in different parts of a circuit.*

ence, $g(\Delta h_2)$, between points B and C, and a small difference again, $g(\Delta h_3)$, between points C and D. In this arrangement a "potential-difference meter" (analogous to a voltmeter) could consist of an ordinary meter stick.

The frictional force on each ball as it falls in the glycerin from B to C is equal to its weight. This makes the resultant force zero, which is what is required for descent at a constant speed. The frictional force on each ball in AB and CD is much smaller than its weight. This is suggested by the small slope of the inclined planes in these regions. The man has to do work mgh ($h = \Delta h_1 + \Delta h_2 + \Delta h_3$) on each ball to move it from D back to A so that it can start the cycle again.

In Figure 40.6.B we have the electrical counterpart of Figure 40.6.A, a complete electric circuit, $ABCD$. Electric charges are moving under the influence of the electric field, **E**. The potential difference between points A and B is very small because the charges encounter only a slight resistance to their motion in this region. The potential difference between points B and C is great because the resistance there is great; the letter R signifies, in fact, that this portion of the circuit, like the lamp in Figure 40.2, is a **resistor** (a conductor with relatively large resistance). There is only a small potential difference between C and D. The charges have a low potential at D, and it takes energy, which is supplied by the battery, to lift them to a high potential at A. The battery transforms chemical into electrical energy by a complicated process, which I shall not analyze any more than I analyzed the internal workings of the man of Figure 40.6.A.

The ability of the battery to raise positive charges from a low potential at D to a high potential at A is measured by the number of joules per coulomb, $\Delta W/\Delta q$, it needs in order to do this. (It is actually electrons, with negative charges, that are moving—and the other way round; but this poses only semantic problems. We could talk the whole thing out by using different words, but we are here adhering to the classical idea that current consists of positive charges whose potential is *raised* in going from D to A.) The ratio $\Delta W/\Delta q$ is called the **electromotive force** (abbreviated as emf) of the battery and is symbolized as $\mathcal{E}$. It is the work per unit charge done by the battery in moving positive charges against the electric field within the battery. It is not, of course, a force in the Newtonian sense; it is measured in joules per coulomb, or volts, not in newtons; but the word "force" has become firmly established in the vocabulary of electricity. Since $\Delta W/\Delta q$ is measured in volts, you might ask why we do not simply say that $\mathcal{E}$ is the difference in potential between points D and A. The answer is that the battery itself may have internal resistance, and that the potential difference between points D and A may therefore be somewhat less than $\mathcal{E}$, depending on how much internal resistance there is. Ideally, with no internal resistance, $\mathcal{E}$, measured in volts, would be equal to the difference in potential between points D and A.

Let us return, for illustration, to Figure 40.3. The voltmeter showed (B) a potential difference of 5.45 volts between D and A *when there was electric current in the circuit*. This was not, however, the emf of the battery. *The potential difference across the terminals of a battery is never exactly equal to its emf when there is current through the battery*. When the switch is open (A), the potential difference is *almost* 6 volts. We have to hedge here because some charges flow even when the voltmeter alone is connected across the battery; the potential difference is not quite equal to the emf unless the resistance of the voltmeter is infinite—that is, unless the voltmeter draws no current. *A good voltmeter, obviously, has a very high resistance*.

I have been using the term "resistance" in a qualitative way. In order to define it precisely, I have to measure current. Notice that the argument so far has not depended upon current. I have talked only of potential difference ("voltage" in the vernacular of the electrician). But perhaps our rolling-ball analogy (Fig. 40.6) has shown why the reading of the voltmeter in Figure 40.3.F was zero. (It corresponded to a vanishingly small Δh_1.) The *potential rise* ($\mathcal{E} = \Delta W/\Delta q$) within the battery—that is, the emf—must equal the sum of the *potential drops* (ΔV) in the complete circuit or loop. We let V_{AB} mean "the potential difference between points A and B." Since V_{AB} and V_{CD} (Fig. 40.6.B) are both practically zero, the voltmeter readings of Figure 40.3.B,D are practically identical. We now imagine (Fig. 40.6) connecting one terminal of the voltmeter to point A. We then touch points B, C, and D with a wire connected to the other terminal of the voltmeter. We read that $V_{AB} = 0$, that $V_{AC} = 5.45$ volts, and that $V_{AD} = 5.45$ volts. The reason for this is that

$$V_{AD} = V_{AB} + V_{BC} + V_{CD}$$
$$= 0 + 5.45 \text{ V} + 0 = 5.45 \text{ V}$$

Before we can proceed, we need to define resistance in terms of potential difference and current.

40.3. Ohm's Law

We shall now use the ammeter in the circuit of Figure 40.3. *To use an ammeter, you must break into the circuit at some point and allow the current to pass through the ammeter.*† Figure 40.7 shows that the ammeter reads 2.73 amperes in each of four different positions. This simply means that charges are conserved. The number of charges flowing per second past any point in the circuit must be the same as the number flowing per second past any other point; otherwise charges would be either accumulating or leaking away. If, for example (Fig. 40.8), we have a junction, O, where the currents are I_1, I_2, I_3, and I_4, it must be true that $\Sigma I = 0$—that is, that $I_1 + I_2 + I_3 + I_4 = 0$—if we consider "coming into O" as positive and "leaving O" as negative.

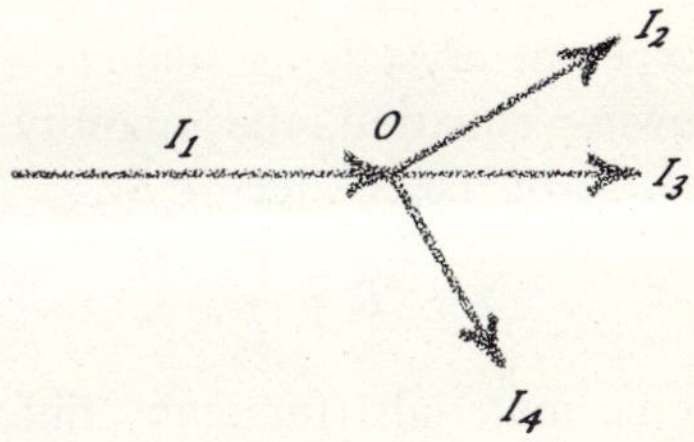

FIG. 40.8. *The sum of inward currents at a junction is equal to the sum of outward currents.*

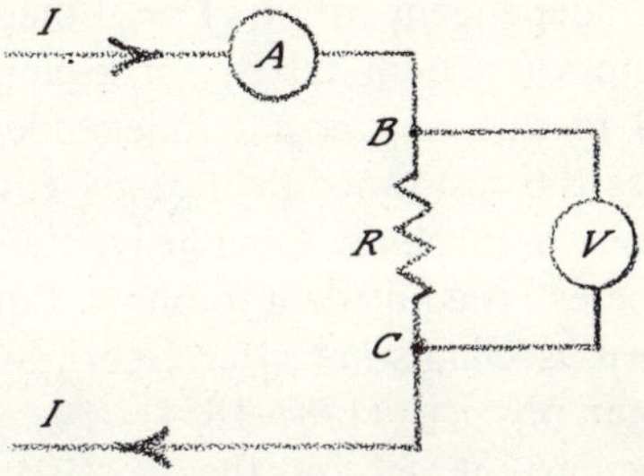

FIG. 40.9. *One way to connect an ammeter and a voltmeter to measure the resistance of a resistor.*

So far Figure 40.7 simply confirms the fact that the current in a single loop is the same everywhere, including the battery. Outside the battery, positive charges tend to flow from regions of high to regions of low potential; inside the battery, the energy supplied by the battery makes it possible for positive charges to flow against the electric field that is naturally there (compare DA in the rolling-ball analogy, Figure 40.6.A).

We now need an experimental fact about metallic conductors. If such a conductor (labeled BC) is connected as in Figure 40.9, the ammeter will show the current in it, and the voltmeter will show the voltage across it. If different currents, I, are made to flow through it, different voltages, V, will appear across it. A plot of V agaiust I is a straight line going through the origin (Fig. 40.10); that is, the ratio of V to I is constant. (This is not true of all kinds of conductors; it is

† Two interesting exceptions to this statement are: (1) a special alternating-current ammeter that just clamps its coil round the current-carrying wire; (2) a special direct-current meter, used by automobile electricians, that works essentially like Oersted's experiment; it is simply clipped onto the battery-charging line.

not true, for example, of vacuum tubes or of certain types of crystals.) I shall now *define*, by the following equation, the quantity called the **resistance,** R, of the conductor BC:

$$R = \frac{V}{I} \qquad [40.3$$

For some materials (for many different kinds of metallic wires, for example) and under certain conditions (at constant temperature, for example) the resistance defined in this way is a constant, independent of I. For other kinds of conductors (vacuum tubes, for example) the R defined in this way is not independent of I. In all cases the resistance defined by equation 40.3 is measured in **ohms.** Obviously, "volts divided by amperes" is equivalent to ohms. Equation 40.3 is known as **Ohm's law** after Georg Simon Ohm, a German physicist (1787–1854).

If the current is I and the cross-sectional area of the wire is A, the **current density, j,** has the magnitude

$$j = \frac{I}{A} \qquad [40.4$$

and is measured in amperes per square meter. For the class of conductors I have been talking about (called **ohmic conductors**) it is an experimental fact that the electric field intensity, **E**, established inside the wire (Fig. 40.11) is proportional to the current density in the wire. In other words, experiments show that

$$\mathbf{E} \propto \mathbf{j} \qquad [40.5$$

(I have written **j** as a vector because **E** is a vector.) There must be a constant of proportionality, ρ, such that

$$\mathbf{E} = \rho\mathbf{j} \qquad [40.6$$

Remembering that E is measured in volts per meter (equation 40.2), let us find the potential difference, V, across a length, l, of wire as follows. Dropping the vector notation, we have

$$El = \rho j l \qquad [40.7$$

Using equation 40.4, we get

$$El = \rho \frac{I}{A} l \qquad [40.8$$

But, according to equation 40.2, $El = V$. Therefore

$$V = \rho \frac{I}{A} l \qquad [40.9$$

FIG. 40.12. *The voltage drop between* P_1 *and* P_2 *is so small that the bird feels no shock.*

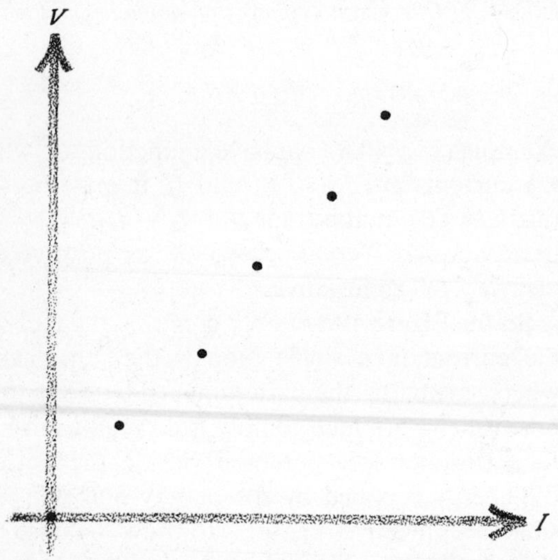

FIG. 40.10. *A plot of voltage against current in an ohmic conductor.*

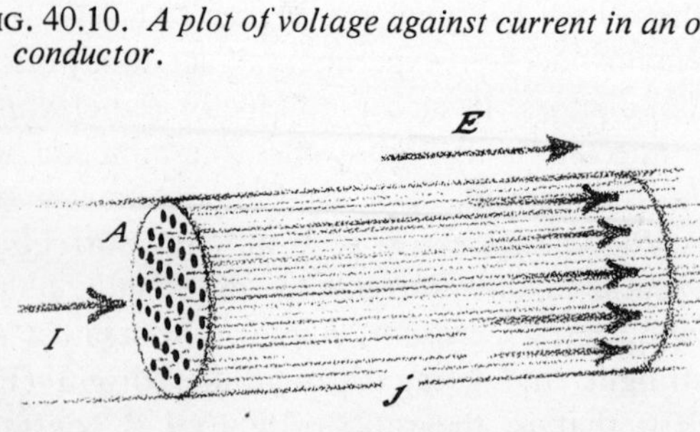

FIG. 40.11. *Illustrating the definition of current density.*

FIG. 40.13. *The voltage drop between* Q_1 *and* Q_2 *might be great enough to kill the bird.*

or

$$\frac{V}{I} = \frac{\rho l}{A} \qquad [40.10$$

But this is the ratio that defines resistance, R (equation 40.3). Hence

$$R = \frac{\rho l}{A} \qquad [40.11$$

That is, the resistance of a wire is directly proportional to its length and inversely proportional to its cross-sectional area. [I could have introduced ρ by means of equation 40.11, but I wanted to emphasize, once again (equation 40.6), the existence of an electric field within a wire carrying a current.] The constant of proportionality, ρ, is called the **resistivity** of the material. Resistivity is the inverse of conductivity. Table 37.2 lists the resistivities of some common substances.

EXAMPLE 40.2. We wish to find the resistance of a piece of copper wire 1 km long and 1 mm in diameter.

We know that

$$\rho = 0.172\times10^{-7} \text{ ohm-meter}$$
$$l = 10^3 \text{ m}$$
$$d = 10^{-3} \text{ m}$$

Therefore

$$A = \frac{\pi d^2}{4} = 7.85 \times 10^{-7} \text{ m}^2$$

and (equation 40.11)

$$R = \frac{0.172\times10^{-7} \text{ ohm-meter} \times 10^3 \text{ m}}{7.85\times10^{-7} \text{ m}^2}$$

$$= 21.9 \text{ ohms}$$

(The filament of an ordinary 100-W light bulb has a resistance of about 100 ohms.)

We can now consider the *voltage drop* in wires carrying current. You have seen birds perched on such wires without being killed and apparently without feeling any shock. Now, one of the harmful things in electric shock, to birds or to people, is the current through the body. This current obeys, approximately, Ohm's law, which implies that we get big currents through the body if we touch points with large potential differences.

There *is* a voltage drop, $V = IR$ (see equation 40.3), in a wire, but the potential difference (Fig. 40.12) between points P_1 and P_2, where the bird's feet rest on the wire, is exceedingly small. In Example 40.2 we saw that the resistance of 1,000 meters of a certain copper wire was 21.9 ohms. The resistance of 10 centimeters would be only 21.9×10^{-4} ohm. Even if the wire carried a current of 100 amperes (very unlikely), the potential drop from P_1 to P_2 would be only 0.219 volt. Such a small potential difference could not send enough current through the bird to do much harm.

A great potential drop might occur (Fig. 40.13) across some distant load—a motor, M, perhaps.

Hence the potential difference between points Q_1 and Q_2 on wires carrying the same current might be very great indeed. If the bird could put one foot at Q_1 and the other at Q_2, it might be killed.

We can now consider our original circuit symbolically. In Figure 40.14 the battery, B, with its internal resistance, r, is enclosed in a dashed line; the lamp, L, has resistance R. The current, I, is the same in both B and L. The potential drop in L is IR (equation 40.3); the potential drop in B is Ir. The charges leave point P at the same potential at which they arrive there. The work per unit charge done by the battery, $\mathcal{E} = \Delta W/\Delta q$, must therefore exactly equal the drop in potential, $IR + Ir$. Hence

$$\mathcal{E} - Ir = IR \qquad [40.12$$

Now Figure 40.3 indicates (D) that $IR = 5.45$ volts and (B) that $\mathcal{E} - Ir = 5.45$ volts. From Figure 40.7 we see that $I = 2.73$ amperes. Therefore

$$R = \frac{5.45 \text{ V}}{2.73 \text{ amp}} = 2 \text{ ohms} \qquad [40.13$$

From Figure 40.3.A we know that $\mathcal{E}$ is almost 6 volts. Therefore, using the equation

$$\mathcal{E} - Ir = 5.45 \text{ V} \qquad [40.14$$

we get

$$\begin{aligned} Ir &= (6 - 5.45) \text{ V} \\ &= 0.55 \text{ V} \end{aligned} \qquad [40.15$$

But $I = 2.73$ amperes. Therefore

$$\begin{aligned} r &= \frac{0.55 \text{ V}}{2.73 \text{ amp}} \\ &= 0.2 \text{ ohm} \end{aligned}$$

We have now accounted for the voltage readings of Figure 40.3, and we have learned something about electric circuits in the process.

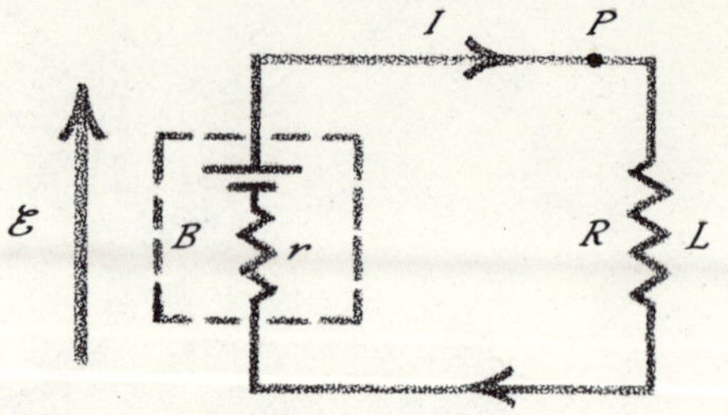

FIG. 40.14. *Our original series circuit treated symbolically. The internal resistance of the battery is shown as* r. *If* $\mathcal{E}$ *is the emf of the battery,* $\mathcal{E} - \text{Ir} = \text{IR}$.

40.4. How the Ammeter and the Voltmeter Work

I have already told how a galvanometer works; it is a coil, mounted between the poles of a magnet, whose dipole moment experiences a torque when it carries current (§ 38.4). If (Fig. 40.15.A,C) a low-resistance conductor, S (called a shunt), is connected across the coil, C, in parallel with it, most of the current flows through S, and we have an **ammeter.** The combination, which has a low resistance, can be designed to measure even a large current, for very little of the current flows through the coil.

The same galvanometer can be converted into a **voltmeter** (Fig. 40.15.B,D). If the coil, C, is connected in series with a resistor, M, of high resistance (called a multiplier), even a large potential difference, V, across the terminals A and B will produce only a small current through the coil, C, since $I = V/R$ and R here includes the resistance of both M and C. The whole device has, as a good voltmeter must have, a high resistance. The details may be clarified by reference to Problems 40.15, 40.16, and 40.17.

It is also left for Problem 40.14 to prove that, when two resistors are connected in series, the resistance of the combination is simply the sum of the two resistances, but that, when they are connected in parallel, the reciprocal of the combination is the sum of the reciprocals of the individual resistors. For resistors in series (as in Fig. 40.15.B,D)

$$R = R_1 + R_2 \qquad [40.16$$

For resistors in parallel (as in Fig. 40.15.A,C)

$$\frac{1}{R} = \frac{1}{R_1} + \frac{1}{R_2} \qquad [40.17$$

40.5. Electric Power Dissipated as Heat

The analogy of balls falling through glycerin (Fig. 40.6.A) is useful, for we see immediately that the loss in potential energy must appear as heat. Similarly, the loss in potential energy of charges moving in the resistor, R, of Figure 40.6.B can appear in the form of heat. The work required to lift a ball in Figure 40.6.A is $W = mgh$. The work per unit mass is $W/m = gh$. Similarly, the work required to move the positive charge Δq from B to A is

$$\Delta W = (\Delta q)V_{BA} = \text{coulombs} \times \text{volts}$$

$$= \text{coulombs} \times \frac{\text{joules}}{\text{coulomb}} = \text{joules}$$

The rate of doing work, P (for power), is

$$P = \frac{\Delta W}{\Delta t} = \frac{\Delta q}{\Delta t} V_{BA} \qquad [40.18$$

But $\Delta q/\Delta t$ is the current, I, in amperes. Hence $P = IV_{BA}$. This must be in joules per second, or watts. If all this power goes into heating the resistor, we may write

$$P_J = IV \qquad [40.19$$

The subscript J is for "joule," to remind us that heat is being generated. Thus "amperes times volts" is equivalent to "watts." Since 4.184 joules = 1 calorie, we may use the expression $IV/4.184$ to compute the calories per second generated in a resistor.

From Ohm's law (equation 40.3) we know that $V = IR$; so we may write

$$P_J = I(IR) = I^2R \qquad [40.20$$

Since, if several resistors are connected in series, they all carry the same I, this form of the equation ($P_J = I^2R$) is useful.

On the other hand, since $I = V/R$, we may write

$$P_J = \frac{V}{R} V = \frac{V^2}{R} \qquad [40.21$$

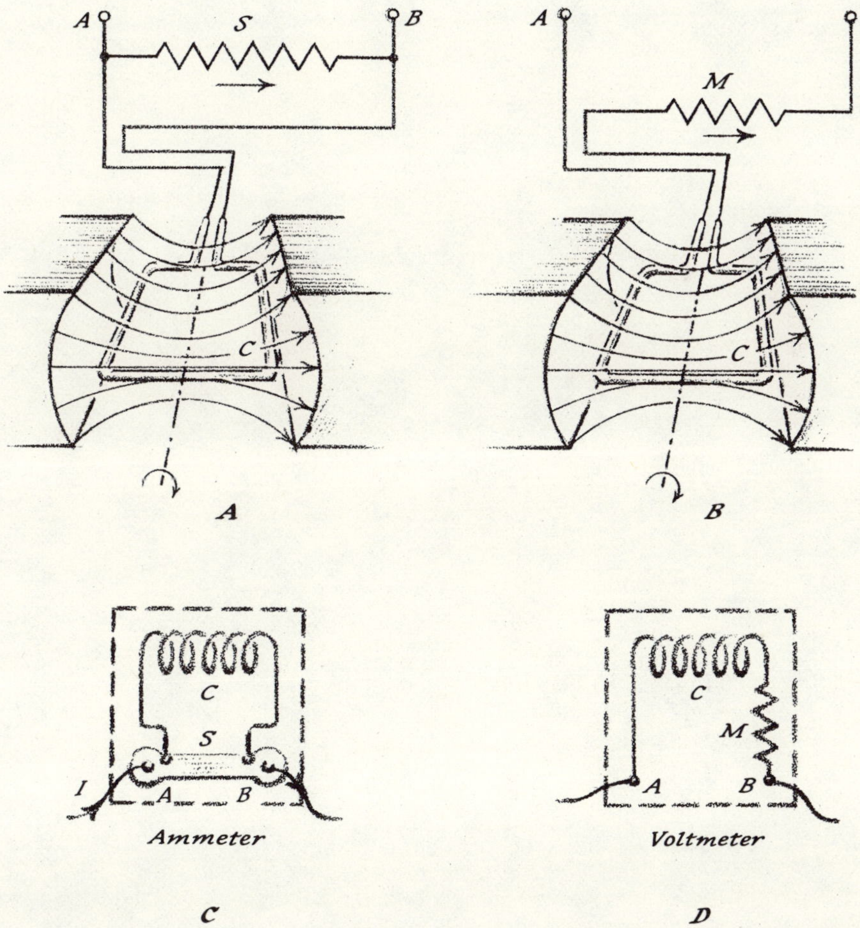

FIG. 40.15. *Symbolic representation of the components (A, C) of an ammeter and (B, D) of a voltmeter.*

Since, if several resistors are connected in parallel, each has the same potential drop as the others, this form ($P_J = V^2/R$) is applicable to such combinations.

40.6. Summary

A battery has the ability to raise positive charges from a low potential to a high potential. Positive charges in an external electric circuit connected to this battery tend to flow from the region of high potential to that of low potential. This flow is called current. Actually, in wires, negative charges (electrons) flow in the opposite direction, but the logic is not affected.

The work per unit charge done by the battery is called its electromotive force, $\mathcal{E}$; it is the ratio $\Delta W/\Delta q$, measured in joules per coulomb, or volts.

The potential rise in the battery must equal the sum of all the potential drops, ΔV, in the whole circuit. The potential drop across an ohmic resistor of resistance R in which there is current I is $V = IR$ (Ohm's law). The resistance of a wire is directly proportional to the product of its length and its resistivity and is inversely proportional to its cross-sectional area.

The flow of charges in a wire is very similar to the flow of a liquid in a pipe. When several wires meet at a point, for example, the sum of the inward currents is equal to the sum of the outward currents.

In both pipes and wires energy can be dissipated in the form of heat. If the potential drop in a wire is V, the work it takes to move charge q across it is qV, and the rate of doing work, or power, is $P = IV$. The power that goes into heating a resistor may be written as I^2R or as V^2/R.

A brief, informal review of the electronic age, past and present.

9 The Electronic Revolution

Arthur C. Clarke

An excerpt from his book, *Voices from the Sky,* originally published in the New York *Times* in 1962.

The electron is the smallest thing in the universe; it would take thirty thousand million, million, million, million of them to make a single ounce. Yet this utterly invisible, all but weightless object has given us powers over nature of which our ancestors never dreamed. The electron is our most ubiquitous slave; without its aid, our civilization would collapse in a moment, and humanity would revert to scattered bands of starving, isolated savages.

We started to use the electron fifty years before we discovered it. The first practical application of electricity (which is nothing more than the ordered movement of electrons) began with the introduction of the telegraph in the 1840's. With really astonishing speed, a copper cobweb of wires and cables spread across the face of the world, and the abolition of distance had begun. For over a century we have taken the instantaneous transfer of news completely for granted; it is very hard to believe that when Lincoln was born, communications were little faster than in the days of Julius Caesar.

Although the beginning of "electronics" is usually dated around the 1920's, this represents a myopic view of technology. With the hindsight of historical perspective, we can now see that the telegraph and the telephone are the first two landmarks of the electronic age. After Alexander Graham Bell had sent his voice from one room to another in 1876, society could never be the same again. For the telephone was the first electronic device to enter the home and to affect directly the lives of ordinary men and women, giving them the almost godlike power of projecting their personalities and thoughts from point to point with the speed of lightning.

Until the closing years of the nineteenth century, men used and handled electricity without knowing what it was, but in the 1890's they began to investigate its fundamental nature, by observing what happened when an electric current was passed through gases at very low pressures. One of the first, and most dramatic, results of this work was the invention of the X-ray tube, which may be regarded as the ancestor of all the millions of vacuum tubes which followed it. A cynic might also argue that it is the only electronic device wholly beneficial to mankind—though when it was invented many terrified spinsters, misunderstanding its powers, denounced poor Röntgen as a violator of privacy.

There is an important lesson to be learned from the X-ray tube. If a scientist of the late Victorian era had been asked "In what way could money best be spent to further the progress of medicine?" he would never by any stretch of the imagination have replied: "By encouraging research on the conduction of electricity through rarefied gases." Yet that is what would have been the right answer, for until the discovery of X rays doctors and surgeons were like blind men, groping in the dark. One can never predict the outcome of fundamental scientific research, or guess what remote and unexpected fields of knowledge it will illuminate.

X rays were discovered in 1895—the electron itself just one year later. It was then realized that an electric current consists of myriads of these submicroscopic particles, each carrying a minute negative charge. When a current flows through a solid conductor such as a piece of copper wire, we may imagine the electrons creeping like grains of sand through the interstices between the (relatively) boulder-sized copper atoms. Any individual electron does not move very far, or very fast, but it jostles its neighbor and so the impulse travels down the line at speeds of thousands of miles a second. Thus when we switch on a light, or send a Morse dash across a transatlantic cable, the response at the other end is virtually instantaneous.

But electrons can also travel *without* wires to guide them, when they shoot across the empty space of a vacuum tube like a hail of machine-gun bullets. Under these conditions, no longer entangled in solid matter, they are very sensitive to the pull and tug of electric fields, and as a result can be used to amplify faint signals. You demonstrate the principle involved every time you hold a hose-pipe in your hand; the slightest movement of your wrist produces a much greater effect at the far end of the jet. Something rather similar happens to the beam of electrons crossing the space in a vacuum tube; they can thus multiply a millionfold the feeble impulses picked up by a radio antenna, or paint a fluorescent picture on the end of a television screen.

Until 1948, electronics was almost synonymous with the vacuum tube. The entire development of radio, talkies, radar, television, long-distance telephony, up to that date depended upon little glass bottles containing intricate structures of wire and mica. By the late 1940's the vacuum tube had shrunk from an object as large as (and sometimes almost as luminous as) an electric light bulb, to a cylinder not much bigger than a man's thumb. Then three scientists at the Bell Telephone Laboratories invented the transistor and we moved from the Paleoelectronic to the Neoelectronic Age.

Though the transistor is so small—its heart is a piece of crystal about the size of a rice grain—it does everything that a radio tube can do. However, it requires only a fraction of the power and space, and is potentially much more reliable. Indeed, it is hard to see how a properly designed transistor can ever wear out; think of little Vanguard I, still beeping away up there in space, and liable to continue indefinitely until some exasperated astronaut scoops it up with a butterfly net.

The transistor is of such overwhelming importance because it (and its still smaller successors) makes practical hundreds of electronic devices which were previously too bulky, too expensive or too unreliable for everyday use. The pocket radio is a notorious example; whether we like it or not, it points the way inevitably to a day when person-to-person communication is universal. Then everyone in the world will have his individual telephone number, perhaps given to him at birth and serving all the other needs of an increasingly complex society (driving license, social security, credit card, permit to have additional children, etc.). You may not know where on Earth your friend Joe Smith may be at any particular moment; but you will be able to dial him instantly—if only you can remember whether his number is 8296765043 or 8296756043.

Obviously, there are both advantages and disadvantages in such a "personalized" communication system; the solitude which we all need at some time in our lives will join the vanished silences of the pre-jet age. Against this, there is no other way in which a really well-informed *and* fast-reacting democratic society can be achieved on the original Greek plan—with direct participation of every citizen in the affairs of the state. The organization of such a society, with feedback in both directions from the humblest citizen to the President of the World, is a fascinating exercise in political planning. As usual, it is an exercise that will not be completed by the time we need the answers.

A really efficient and universal communications system, giving high-quality reception on all bands between all points on the Earth, can be achieved only with the aid of satellites. As they come into general use, providing enormous information-handling capacity on a global basis, today's patterns of business, education, entertainment, international affairs will change out of all recognition. Men will be able to meet face to face (individually, or in groups) without ever leaving their homes, by means of closed circuit television. As a result of this, the enormous amount of commuting and traveling that now takes place from home to office, from ministry to United

Nations, from university to conference hall will steadily decrease. There are administrators, scientists and businessmen today who spend about a third of their working lives either traveling or preparing to travel. Much of this is stimulating, but most of it is unnecessary and exhausting.

The improvement of communications will also render obsolete the city's historic role as a meeting place for minds and a center of social intercourse. This is just as well anyway, since within another generation most of our cities will be strangled to death by their own traffic.

But though electronics will ultimately separate men from their jobs, so that (thanks to remote manipulation devices) not even a brain surgeon need be within five thousand miles of his patient, it must also be recognized that few of today's jobs will survive long into the electronic age. It is now a cliché that we are entering the Second Industrial Revolution, which involves the mechanization not of energy, but of thought. Like all clichés this is so true that we seldom stop to analyze what it means.

It means nothing less than this: There are no routine, noncreative activities of the human mind which cannot be carried out by suitably designed machines. The development of computers to supervise industrial processes, commercial transactions and even military operations has demonstrated this beyond doubt. Yet today's computers are morons compared to those that they themselves are now helping to design.

I would not care to predict how many of today's professions will survive a hundred years from now. What happened to the buggywhip makers, the crossing sweepers, the scriveners, the stonebreakers of yesteryear? (I mention the last because I can just remember them, hammering away at piles of rock in the country lanes of my childhood.) Most of our present occupations will follow these into oblivion, as the transistor inherits the earth.

For as computers become smaller, cheaper and more reliable they will move into every field of human activity. Today they are in the office; tomorrow they will be in the home. Indeed, some very simple-minded computers already do our household chores; the device that programs a washing machine to perform a certain sequence of operations is a specialized mechanical brain. Less specialized ones would be able to carry out almost all the routine operations in a suitably designed house.

Because we have so many more pressing problems on our hands, only the science-fiction writers—those trail-blazers of the future—have given much thought to the social life of the later electronic age. How will our descendants be educated for leisure, when the working week is only a few hours? We have already seen, on a worldwide scale, the cancerous growths resulting from idleness and lack of usable skills. At every street corner in a great city you will find lounging groups of leather-jacketed, general-purpose bioelectric computers of a performance it will take us centuries and trillions of dollars to match. What is their future—and ours?

More than half a century ago H. G. Wells described, in *The Time Machine*, a world of decadent pleasure lovers, bereft of goals and ambitions, sustained by subterranean machines. He set his fantasy eight hundred thousand years in the future, but we may reach a similar state of affairs within a dozen generations. No one who contemplates the rising curve of technology from the Pilgrim fathers to the Apollo Project dare deny that this is not merely possible, but probable.

For most of history, men have been producers; in a very few centuries, they will have to switch to the role of consumers, devoting their energies 100 per cent to absorbing the astronomical output of the automated mines, farms and factories.

Does this *really* matter, since only a tiny fraction of the human race has ever contributed to artistic creation, scientific discovery or philosophical thought, which in the long run are the only significant activities of mankind? Archimedes and Aristotle, one cannot help thinking, would still have left their marks on history even if they had lived in a society based on robots instead of human slaves. In any culture, they would be consumers of goods, but producers of thought.

We should not take too much comfort from this. The electronic computers of today are like the subhuman primates of ten million years ago, who could have given any visiting Martians only the faintest hints of their potentialities, which included the above mentioned Archimedes and Aristotle. Evolution is swifter now; electronic intelligence is only decades, not millions of years, ahead.

And *that*—not transistor radios, automatic homes, global TV—is the ultimate goal of the Electronic Revolution. Whether we like it or not, we are on a road where there is no turning back; and waiting at its end are our successors.

EDISON'S PATENT on the incandescent lamp was accompanied by this drawing. The labeled parts are the carbon filament (*a*), thickened ends of filament (*c*), platinum wires (*d*), clamps (*h*), leading wires (*x*), copper wires (*e*), tube to vacuum pump (*m*).

Thomas Edison's incandescent lamp, generally assumed to be the product of inspired tinkering, was actually only one element in a more far-reaching invention: an entire system of lighting.

10 The Invention of the Electric Light

Matthew Josephson

An article from *Scientific American,* 1959.

"I can hire mathematicians, but mathematicians can't hire me!" By such declarations in the time of his success and world-wide fame Thomas Alva Edison helped to paint his own portrait as an authentic American folk hero: the unlettered tinkerer and trial-and-error inventor who achieved his results by persistence and a native knack for things. He is said, for example, to have tried more than 1,600 kinds of material ("paper and cloth, thread, fishline, fiber, celluloid, boxwood, coconut-shells, spruce, hickory, hay, maple shavings, rosewood, punk, cork, flax, bamboo and the hair out of a red-headed Scotchman's beard") until he hit upon the loop of carbonized cotton thread that glowed in a vacuum for more than half a day on October 21, 1879. Today, in a world that relies for its artificial illumination largely on his incandescent lamp, this invention is not regarded as an especially profound contribution to technology. It rates rather as a lucky contrivance of Edison's cut-and-try methods—of a piece with his stock ticker, mimeograph machine, phonograph and alkaline storage-battery—in the esteem of a public that has come to appreciate the enormous practical significance of higher mathematics and abstruse physical theory.

If Edison's contribution to the light of the world consisted solely in the selection of a filament, this estimate of his person and achievements might be allowed to stand. But the history that is so obscured by legend tells quite another story. Edison's electric light was not merely a lamp but a system of electric lighting. His invention was an idea rather than a thing. It involved not only technology but also sociology and economics. Edison was indisputably the first to recognize that electric lighting would require that electricity be generated and distributed at high voltage in order to subdivide it among a great many high-resistance "burners," each converting current at low amperage (that is, in small volume) with great efficiency into light.

In the 15 months between the time he conceived his invention and the date on which he demonstrated it to the public, Edison and his associates designed and built a new type of electric generator, successfully adapted the then much-scorned parallel or "multiple-arc" circuit that would permit individual lights to be turned on or off separately and, last of all, fashioned a lamp to meet the specifications of his system. The laboratory notebooks of those months of frantic labor show the Wizard of Menlo Park endowed with all the prodigious capacities attributed to him by contemporary legend. They show in addition that this self-taught technologist was possessed of a profound grasp of the nature of electricity and an intuitive command of its logic and power.

It was on September 8, 1878, that Edison was inspired to devote his talents full time to the challenge of electric lighting. On that day he went to Ansonia, Conn., to visit the brass-manufacturing plant of William Wallace, co-inventor with Moses G. Farmer of the first practical electric dynamo in the U. S. Wallace showed Edison eight brilliant carbon-arc lights of 500 candlepower each, powered by a dynamo of eight horsepower. It was with such a system that Wallace and Farmer, as well as Charles Brush of Cleveland, were then beginning to introduce the electric light on a commercial scale, for street-lighting and for illuminating factories and shops. Farmer had made the first demonstration of arc-lighting in this country two years earlier, at the Centennial Exposition in Philadelphia, and John Wanamaker's store in that city was already illuminated with arc lights.

Carbon arcs are still employed in searchlights and in theater floodlights and projectors to produce light of high intensity. The current crossing a small gap between the electrodes creates an arc. Ionization and oxidation of the carbon in the heat of the arc generate a brilliant blue-white light.

In the 1870's Europe was a decade ahead of the U. S. in the technology of arc-lighting. Stores, railway stations, streets and lighthouses in Britain and France were equipped with arc lights. Shedding an almost blinding glare, they burned in open globes that emitted noxious gases, and they could be employed only high overhead on streets or in public buildings. Since they consumed large amounts of current, they had to be wired in series, that is, connected one to another in a single continuous circuit so that all had to be turned on or off together. The multiple-arc circuit, with the lights connected as in the rungs of a ladder between the main leads of the circuit, was not adapted to such systems and was considered prohibitive in cost.

Edison himself had experimented with arc lights, using carbon strips as burners. He had also investigated the

Editor's Note

The author has based this article on material in his biography *Edison*, just published by the McGraw-Hill Book Company.

incandescent light, as had many inventors before him. But the slender rod or pencil of carbon or metal would always burn up, sooner rather than later, upon being heated to incandescence by the current. It would do so though substantially all of the air had been pumped out of the glass envelope in which it was contained. Edison had abandoned the effort to devote himself to a more promising invention: the phonograph.

Now at Wallace's establishment, confronted with the achievements of others in the field, he regained his earlier enthusiasm. As an eyewitness recalled, "Edison was enraptured. . . . He fairly gloated. He ran from the instruments [the dynamos] to the lights, and then again from the lights back to the electric instruments. He sprawled over a table and made all sorts of calculations. He calculated the power of the instruments and the lights, the probable loss of power in transmission, the amount of coal the instrument would use in a day, a week, a month, a year."

To William Wallace he said challengingly: "I believe I can beat you making the electric light. I do not think you are working in the right direction." They shook hands in friendly fashion, and with a diamond-pointed stylus Edison signed his name and the date on a goblet provided by his host at dinner.

From Edison's own complete and explicit notebooks and from the buoyant interviews that he gave to the press at this time we know what made him feel in such fine fettle as he left Wallace's plant. "I saw for the first time everything in practical operation," he said. "I saw the thing had not gone so far but that I had a chance. The intense light had not been subdivided so that it could be brought into private houses. In all electric lights theretofore obtained the light was very great, and the quantity [of lights] very low. I came home and made experiments two nights in succession. I discovered the necessary secret, so simple that a bootblack might understand it. . . . The subdivision of light is all right."

The Subdivision of Light

At this time there flashed into Edison's mind the image of the urban gas-lighting system, with its central gashouse and gas mains running to smaller branch pipes and leading into many dwelling places at last to gas jets that could be turned on or off at will. During the past half-century gas-lighting had reached the stature of a major industry in the U. S. It was restricted, of course, to the cities; three fourths of the U. S. population still lived in rural areas by the dim glow of kerosene lamps or candles. Ruminating in solitude, Edison sought to give a clear statement to his objective. In his notebook, under the title "Electricity versus Gas as a General Illuminant," he wrote:

"Object: E. . . . to effect exact imitation of all done by gas, to replace lighting by gas by lighting by electricity. To improve the illumination to such an extent as to meet all requirements of natural, artificial and commercial conditions. . . . Edison's great effort—not to make a large light or a blinding light, but a small light having the mildness of gas."

To a reporter for one of the leading New York dailies who had shadowed him to Ansonia, Edison described a vision of a central station for electric lighting that he would create for all of New York City. A network of electric wire would deliver current for a myriad of small household lights, unlike the dazzling arc lights made by Farmer and Brush. In some way electric current would be metered and sold. Edison said he hoped to have his electric-light invention ready in six weeks! At Menlo Park, N.J., where his already famous workshop was located, he would wire all the residences for light and hold a "grand exhibition."

Thus from the beginning Edison riveted his attention not so much upon the search for an improved type of incandescent filament as upon the analysis of the social and economic conditions for which his invention was intended. As he turned with immense energy to expanding the facilities at Menlo Park and securing the essential financing, he continued his studies of the gas-lighting industry. In parallel he projected the economics of the electric-lighting system he envisioned.

Gas had its inconvenience and dangers. "So unpleasant . . . that in the new Madison Square theater every gas jet is ventilated by small tubes to carry away the products of combustion." But whatever is to replace gas must have "a general system of distribution—the only possible means of economical illumination." Gathering all the back files of the gas industry's journals and scores of volumes bearing on gas illumination, he studied the operations and habits of the industry, its seasonal curves and the layout of its distribution systems. In his mind he mapped out a network of electric-light lines for an entire city, making the shrewd judgment: "Poorest district for light, best for power—thus evening up whole city." He meant that in slum districts there would be higher demand for small industrial motors. Against tables for the cost of converting coal to gas he calculated the cost of converting coal and steam into electric energy. An expert gas engineer, whose services Edison engaged at this time, observed that few men knew more about the world's gas business than did Edison.

Edison had a *homo oeconomicus* within him, a well-developed social and commercial sense, though he was careless of money and was not an accountant of the type exemplified by his contemporary John D. Rockefeller. Before the experimental work on his invention was under way, he had formed a clear notion, stated in economic terms, of what its object must be. This concept guided his search and determined the pattern of his technical decisions, so that the result would be no scientific toy but a product useful to people everywhere. By his initial calculation of the capital investment in machinery and copper for a whole system of light distribution he was led to define the kind of light he sought and the kind of generating and distributing system he needed.

Backers of the Electric Light

In the crucial matter of financing his inventive work Edison had the generous and imaginative aid of Grosvenor Lowrey, a patent and corporation lawyer well established in the financial community of Wall Street. Lowrey had fallen completely under Edison's spell and regarded him much as a collector of paintings regards a great artist whose works he believes are destined for immortality. Using his extensive connections and the favorable press-notices that he encouraged Edison to secure during late September and early October, 1878, Lowrey assembled a sponsoring syndicate of some of the most important financiers of the time. The underwriters of the Edison Electric Light Company, which was incorporated in mid-October, included William H. Vanderbilt and J. P. Morgan's partner Egisto Fabbri. This was an unprecedented development in U. S. business. Inventors had been backed in the development of inventions already achieved; Edison's financiers were backing him in research that was to lead to a hoped-for invention. In many respects the venture marks the beginning in this country of close relations between finance and technology.

"Their money," Edison said, "was in-

EDISON AND HIS PHONOGRAPH were photographed in 1878 by Mathew Brady. He had worked with electric lights but had turned to the more promising phonograph. In the year that this photograph was made, however, he resumed his work on lighting.

MENLO PARK was depicted in *Frank Leslie's Illustrated Newspaper* for January 10, 1880. The barnlike "tabernacle" of Edison's laboratory is visible at the far right. In its windows passengers on the nearby railroad could see his experimental lights burning.

vested in confidence of my ability to bring it back again." The 31-year-old Edison was by now a well-known figure in Wall Street. His quadruplex telegraph system, by which four separate messages could be transmitted over a single wire, had furnished the pivotal issue in the vast economic war waged between Western Union and the rival telegraph empire of the robber baron Jay Gould. Edison's carbon microphone had transformed the telephone from an instrument of limited usefulness to an efficient system of long-range communication that was now radiating across the country. The shares of gas-lighting enterprises had tumbled on the New York and London exchanges upon Edison's announcement, in the press campaign instigated by Lowrey, that he was now about to displace gas with electricity in the lighting of homes and factories.

The alliance between Edison and his sponsors was nonetheless an uneasy one. The first rift appeared before the end of October, when the rival inventor William Sawyer and his partner Albon Man announced that they had "beaten" Edison and applied for a patent on a carbon-pencil light in a nitrogen-filled glass tube. There was a flutter of panic in the directorate of the Edison Electric Light Company. The suggestion was made that Edison should join forces with Sawyer and Man. Lowrey passed the suggestion on to S. L. Griffin, a former junior executive at Western Union whom Lowrey had hired to help Edison with his business affairs.

Griffin sent back a hasty "confidential" reply: "I spoke to Mr. Edison regarding the Sawyer-Man electric light. . . . I was astonished at the manner in which Mr. Edison received the information. He was visibly agitated and said it was the old story, that is, lack of confidence. . . . No combination, no consolidation for him. I do not feel at liberty to repeat all he said, but I do feel impelled to suggest respectfully that as little be said to him as possible with regard to the matter."

In view of Edison's talent for candid and salty language Griffin's reticence is understandable. After that there was no further talk of consolidation with Sawyer or any other inventor.

The Menlo Park Laboratory

In his belief that he would "get ahead of the other fellows" Edison was sustained by his unbounded confidence in his laboratory, its superior equipment and its staff. The Menlo Park laboratory was still the only full-time industrial research organization in the country, in itself perhaps Edison's most important invention. During this period the physical plant was greatly expanded; a separate office and library, a house for two 80-horsepower steam engines, and a glass blower's shed were added to the original barnlike "tabernacle." Even more important, Edison had collected a nucleus of talented engineers and skilled craftsmen, who were of inestimable help to him in working out his ideas.

The self-taught Edison thought primarily in concrete, visual terms. When he was at work on the quadruplex telegraph, he had even built a model made up of pipes and valves corresponding to the wires and relays of his system, and with running water replacing the electric current, so that he could actually see how it worked. But now he would have to depend far more on theory and mathematics.

One of the happiest effects of Grosvenor Lowrey's personal influence was the hiring of Francis R. Upton, a young electrical engineer who had worked for a year in the Berlin laboratory of the great physicist Hermann von Helmholtz. Edison jocularly nicknamed Upton "Culture," and, according to an oft-told story, put the "green" mathematician in his place with one of his scientific practical jokes. He brought out a pear-shaped glass lamp-bulb and gave it to Upton, asking him to calculate its content in cubic centimeters. Upton drew the shape of the bulb exactly on paper, and derived from this an equation for the bulb's volume. He was about to compute the answer when Edison returned and impatiently asked for the results. Upton said he would need more time. "Why," said Edison, "I would simply take that bulb, fill it with a liquid, and measure its volume directly!"

When Upton joined the staff late in October, Edison had already committed himself to the incandescent light. This, rather than the arc light, was the way to imitate the mildness of gas. But the filament glowing in a vacuum had been sought in vain by numerous inventors for half a century. In choosing the incandescent light rather than the arc light he was "putting aside the technical advance that had brought the arc light to the commercial stage." No one, including himself, had succeeded in making an incandescent lamp that would work for more than a few minutes.

Edison's first efforts in 1878 were not notably more successful. Knowing that carbon has the highest melting point of all the elements, he first tried strips of carbonized paper as "burners" and managed to keep them incandescent for "about eight minutes" before they burned up in the partial vacuum of his glass containers. Turning to the infusible metals, he tried spirals of platinum wire; they gave a brilliant light but melted in the heat. Edison accordingly devised a feedback thermostat device that switched off the current when the heat approached the melting point. The lamp now blinked instead of going out entirely. Nonetheless, with his eye on the problem of financing, Edison filed a patent application on October 5 and invited the press in for a demonstration.

As this discouraging work proceeded in the weeks that followed, Edison turned, with Upton's help, to calculating the current that would be consumed by a lighting system equipped with a certain number of such lamps. They assumed that the lights would be connected in parallel, so their imaginary householder could turn one light in the circuit on or off at will, as in a gas-lighting system. Thinking in round numbers, they assumed that these lamps, when perfected, might have a resistance of one ohm and so would consume 10 amperes of current at 10 volts. Allowing in addition for the energy losses in the distribution system, they found that it would require a fabulous amount of copper to light just a few city blocks. Such a system of low-resistance lights was clearly a commercial impossibility.

This was the gist of the objections which had greeted Edison's first announcements that he would use an incandescent bulb in a parallel circuit. Typical of the scorn heaped upon him was the opinion expressed by a committee set up by the British Parliament to investigate the crash of gas-lighting securities. With the advice of British scientists, the members of the committee declared that though these plans seemed "good enough for our transatlantic friends," they were "unworthy of the attention of practical or scientific men." From Ohm's law, which governs the relationship between voltage, amperage and resistance in a circuit, the report argued that if an electric light of 1,000 candlepower were divided into 10 smaller lights and connected in parallel, each of the smaller lights would radiate not one tenth but "one hundredth only of the original light." In this judgment such figures as Lord Kelvin and John Tyndall concurred. Before the Royal Institution in London the distinguished electrician Sir William Preece declared: "Subdivision of the electric light is an absolute *ignis fatuus*."

Ohm's law does indeed show that the amount of current (amperes) flowing in a circuit is equal to the electromotive force (volts) divided by the resistance (ohms) in the circuit. Edison's contemporaries reasoned that an increase in the number of lights in a circuit would increase the resistance and therefore reduce the flow of current to each. It was thought that the only way to provide these lights with sufficient current was to reduce the resistance in the distribution system. In a parallel circuit this meant increasing the thickness of the copper conductors to an impractical degree. Such were the limits on the operation of arc lights, with their low resistance and huge appetite for current. Upton's calculations showed that this conclusion also applied to Edison's first low-resistance incandescent lamps.

Edison now confounded his collaborator by proposing that he make the same sort of estimates for an entirely different kind of circuit. This time he would assume lights of very high resistance, supplied with current at high voltage and low amperage. In November and December Upton made calculations on the basis of the same number of lights, but lights with the high resistance of 100 ohms each. These lights were to operate on the low current of only one ampere. Their high resistance was to be offset, in accord with Ohm's law, by the high voltage of 100 volts in the circuit. The result was astonishing: A high-resistance system would require only one hundredth of the weight of copper conductor needed for a low-resistance system. And copper was the most costly element involved—the decisive economic factor.

The High-Resistance System

Here was the crux of Edison's insight at Ansonia. He had recognized there that the subdivision of light called for lamps of high resistance which would consume but little current; to balance the electrical equation it would be neces-

INTERIOR OF EDISON'S LABORATORY at Menlo Park was also depicted in the January 10, 1880, issue of *Frank Leslie's Illustrated Newspaper*. At the time of the work on the electric light the laboratory had expanded into several other buildings.

sary to supply the current at high voltage. This was the "necessary secret" that was "so simple." Today every high-school physics student learns that the power lost in transmitting electric energy varies with the square of the current. Thus a tenfold reduction in current meant a decrease of a hundredfold in the energy wasted (or a hundredfold decrease in the weight of the transmission line). It was a conception easily reached by an elementary application of Ohm's law, but it had not occurred to any of Edison's contemporaries. Even Upton did not immediately grasp the full import of Edison's idea. As he said later: "I cannot imagine why I did not see the elementary facts in 1878 and 1879 more clearly than I did. I came to Mr. Edison a trained man, with a year's experience in Helmholtz's laboratory, . . . a working knowledge of calculus and a mathematical turn of mind. Yet my eyes were blind in comparison with those of today; and . . . I want to say that I had company!"

With Upton's figures before him Edison was convinced that a new and strategic invention lay surely within his grasp. It was clear what kind of distributing system he wanted. And he knew what form of incandescent burner would serve his purpose. To offer the necessary resistance to the passage of current it must have a small cross section and so would have a small radiating surface.

By January, 1879, Edison was testing his first high-resistance lamp. It had a spiral of very fine platinum wire set in a globe that contained as high a vacuum as could be achieved with an ordinary air pump. The results were encouraging; these lamps lasted "an hour or two." He then attacked the dual problem of getting a higher vacuum and improving his incandescing element. After another trial with carbon, he returned to metals: platinum, iridium, boron, chromium, molybdenum, osmium—virtually every infusible metal. He thought of tungsten, but could not work it with existing tools. Discouraged by the problem, Edison tried nitrogen in his globe and then resumed his efforts to obtain a higher vacuum. Hearing of the new and efficient Sprengel vacuum pump, which used mercury to trap and expel air, he sent Upton to borrow one from the nearby College of New Jersey (now Princeton University). When Upton returned with the pump late that night, Edison kept him and the other men on the staff up the rest of the night trying it out.

GENERATOR which Edison developed for the needs of electric lighting appears at right in this engraving from SCIENTIFIC AMERICAN for October 18, 1879 (at that time this magazine appeared weekly). The generator was called the "long-waisted Mary Ann."

At this stage Edison made a useful finding: "I have discovered," he noted, "that many metals which have gas within their pores have a lower melting point than when free of such gas." With the aid of the Sprengel pump he devised a method of expelling these occluded gases, by heating the element while the air was being exhausted from the bulb. The platinum wire within the bulb thereupon became extremely hard and could endure far higher temperatures. Edison later said that at this stage he "had made the first real steps toward the modern incandescent lamp."

Meanwhile the spirits of his financial sponsors had begun to droop. Their brilliant inventor, far from having achieved anything tangible, was hinting plainly that he needed more money. The first Brush arc lights were ablaze over lower Broadway, and more were being installed elsewhere with impressive effect. Edison's backers began to have serious doubts as to whether he had pursued the right course. To shore up their morale Lowrey arranged to have Edison give them a private demonstration.

In April, as one of Edison's associates recalled it, "They came to Menlo Park on a late afternoon train from New York. It was already dark when they were conducted into the machine shop where we had several platinum lamps installed in series." The "boss" showed his visitors pieces of platinum coil he was using in the lamps, pointed out the arrangement of the lights and described the type of generator he hoped to build. Then, the room having grown quite dark, he told "Honest John" Kruesi to "turn on the juice slowly."

"Today, I can still see those lamps rising to a cherry-red . . . and hear Mr. Edison saying 'A little more juice' and the lamps began to glow. 'A little more,' . . . and then one emits a light like a star, after which there is an eruption and a puff, and the machine shop is in total darkness. . . . The operation was repeated two or three times, with about the same results."

The platinum coils still consumed a lot of power for the light they gave, and they were costly and short-lived. The temporary Wallace-Farmer dynamos heated up badly, and were not powerful enough to enable Edison to connect his lamps in parallel. Edison admitted that the system was not yet "practical."

It was a gloomy gathering that broke up on that raw April evening. All of Lowrey's abounding faith would be necessary to rally the spirits and funds of Edison's despondent backers. Some rumors of the disappointing demonstration leaked out; the price of Edison stock fell sharply, while that of gas-lighting securities rose. "After that demonstration," Edison's associate relates, "we had a general house cleaning at the laboratory, and the metallic lamps were stored away."

Edison now rallied his staff to efforts on a much broader area of the front "under siege." He followed three main lines of investigation. One group he detailed to the task of developing the dynamo to supply the constant-voltage current required by his high-resistance system. He set another group to pulling down a still higher vacuum in the glass bulbs. The third team, under his watchful eye, carried out the series of experiments in which 1,600 different materials were tested for their worth as incandescent elements.

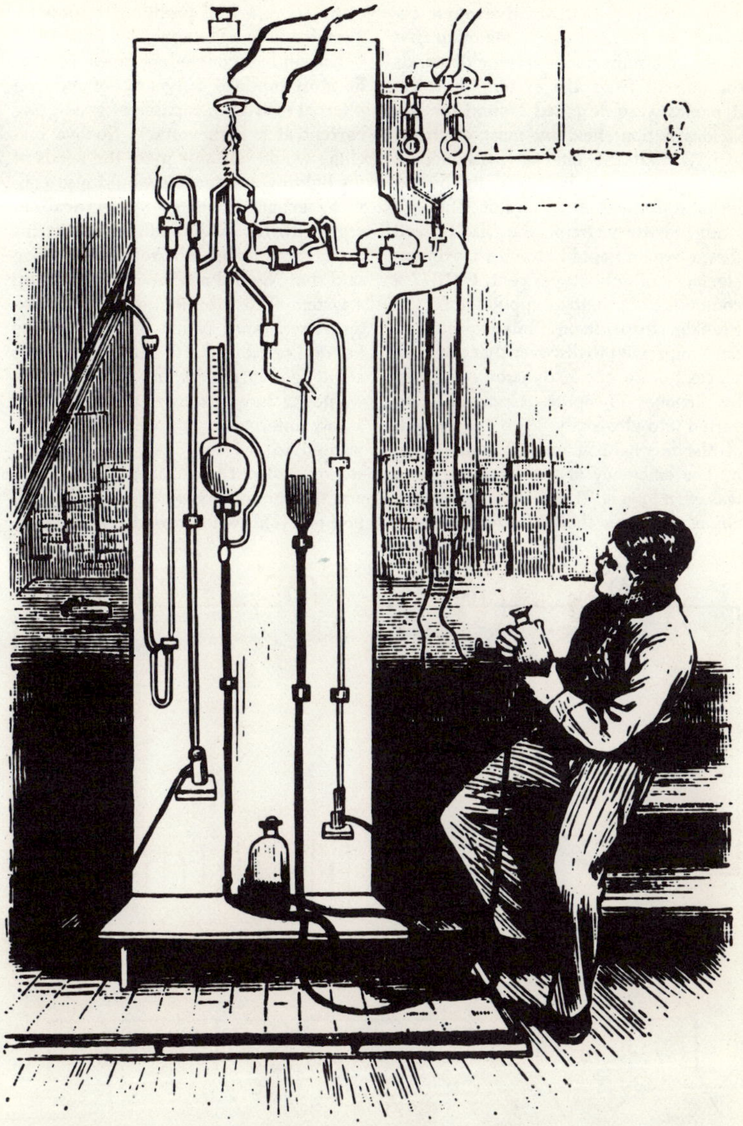

VACUUM PUMP used to remove air from lamp bulbs (*top center*) was of a new type about which Edison had read in a scientific journal. The man is holding a vessel of mercury.

The "Long-Waisted Mary Ann"

To subdivide the electric current for numerous small lights in parallel Edison needed a dynamo which would produce a higher voltage than any dynamo in existence, and which would maintain that voltage constant under varying demands for current from the system. Existing dynamos were designed around the fallacious notion, held by most electrical experts, that the internal resistance of the dynamo must be equal to the external resistance of the circuit. Through study of battery circuits they had proved that a dynamo could attain a maximum efficiency of only 50 per cent. In 1877 a committee of scientists appointed by the Franklin Institute in Philadelphia had been impressed to discover that the most successful European dynamo, designed by Zénobe Théophile Gramme, converted into electricity 38 to 41 per cent of the mechanical energy supplied to it. The efficiency of the Brush dynamo was even lower: 31 per cent. These machines and their theoretically successful contemporaries all produced current at a relatively low voltage.

Edison had concluded, however, that he must produce a dynamo of reduced internal resistance capable of generating current at a high voltage. Such a machine would not only meet the needs of his lighting system but would also convert mechanical energy to electrical energy with far greater efficiency. As his associate Francis Jehl recalled, Edison said that "he did not intend to build up a system of distribution in which the external resistance would be equal to the internal resistance. He said he was just about going to do the *opposite;* he wanted a large external resistance and a low internal resistance. He said he wanted to sell the energy outside the station and not waste it in the dynamo and the conductors, where it brought no profits." Jehl, who carried out the tests

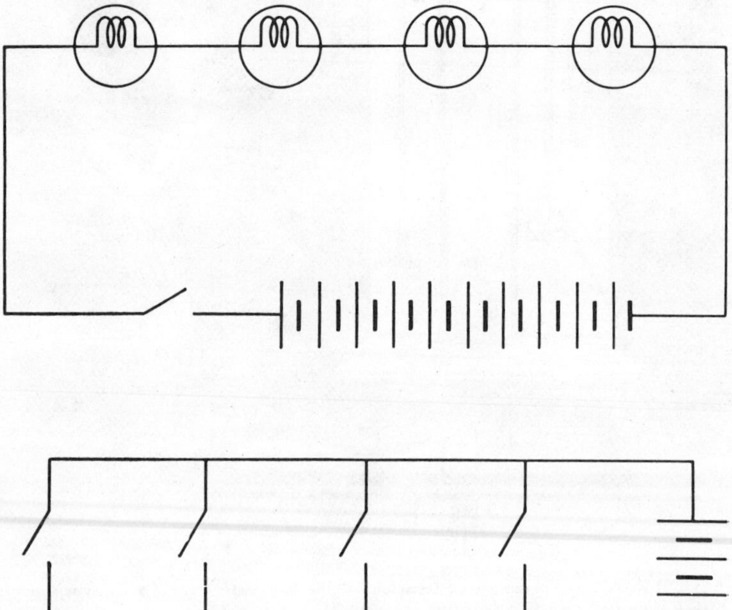

SERIES CIRCUIT (*top*) requires that a number of electric lights (*circles*) be turned on or off at the same time by a single switch (*break in circuit*). Parallel circuit (*bottom*), which was adopted by Edison, makes it possible to turn lights on or off one at a time.

EDISON'S LIGHT.

The Great Inventor's Triumph in Electric Illumination.

A SCRAP OF PAPER.

It Makes a Light, Without Gas or Flame, Cheaper Than Oil.

TRANSFORMED IN THE FURNACE.

Complete Details of the Perfected Carbon Lamp.

FIFTEEN MONTHS OF TOIL.

Story of His Tireless Experiments with Lamps, Burners and Generators.

SUCCESS IN A COTTON THREAD.

The Wizard's Byplay, with Bodily Pain and Gold "Tailings."

HISTORY OF ELECTRIC LIGHTING.

The near approach of the first public exhibition of Edison's long looked for electric light, announced to take place on New Year's Eve at Menlo Park, on which occasion that place will be illuminated with the new light, has revived public interest in the great inventor's work, and throughout the civilized world scientists and people generally are anxiously awaiting the result. From the beginning of his experiments in electric lighting to the present time Mr. Edison has kept his laboratory guardedly closed, and no authoritative account (except that published in the HERALD some months ago relating to his first patent) of any of the important steps of his progress has been made public—a course of procedure the inventor found absolutely necessary for his own protection. The HERALD is now, however, enabled to present to its readers a full and accurate account of his work from its inception to its completion.

A LIGHTED PAPER.

Edison's electric light, incredible as it may appear, is produced from a little piece of paper—a tiny strip of paper that a breath would blow away. Through

FIRST NEWSPAPER ACCOUNT of Edison's brilliant success appeared in *The New York Herald* for December 21, 1879.

of resistance, also remarked that the art of constructing dynamos was then as mysterious as air navigation. All electrical testing was in the embryonic stage. "There were no instruments for measuring volts and amperes directly: it was like a carpenter without his foot rule."

Upton himself had his difficulties in this hitherto unexplored field: "I remember distinctly when Mr. Edison gave me the problem of placing a motor in circuit, in multiple arc, with a fixed resistance; and I . . . could find no prior solution. There was nothing I could find bearing on the [effect of the] counter-electro-motive force of the armature . . . and the resistance of the armature on the work given out by the armature. It was a wonderful experience to have problems given me by him based on enormous experience in practical work and applying to new lines of progress."

The problem of a constant-voltage dynamo was attacked with the usual Edisonian élan. Seeking to visualize every possible structural innovation for his dynamo armature, he had his men lay out numerous wooden dummies on the floor and wind wire around them, spurring them on in their task by laying wagers as to who would finish first.

After Edison had decided upon the form of winding and type of electromagnets to be used, Upton made drawings and tables from which the real armatures were wound and attached to the commutator. Edison eventually worked out an armature made of thin sheets of iron interleaved with insulating sheets of mica; this armature developed fewer eddy currents and so produced less heat than the solid armature cores then used. When the new cores were test-run, it was Upton who made the mathematical calculations from these tests and drew up the final blueprints.

The self-effacing Upton can be given principal credit for interpreting Edison's ideas and translating them into mathematical form. A careful student of contemporary electrical knowledge, he seems to have been conversant with, and to have guided himself by, the design of a German dynamo, made by the Siemens works, that employed an auxiliary source of current to excite its field magnets.

The new Menlo Park dynamo comprised many admirable features for that period. With its great masses of iron and large, heavy wires, it stood in bold contrast to its contemporary competitors. Owing to the two upright columns of its field electromagnets, it was nicknamed "Edison's long-waisted Mary Ann."

When the dynamo was run at the correct speed, the voltage between its armature brushes was approximately 110, and remained fairly constant, falling but slightly when increasing amounts of current were taken out of the machine. Edison and Upton also contrived a simple but ingenious dynamometer by which the torque of a drive belt was used to measure the work output of the steam engine that powered the dynamo. When Kruesi completed the first operating machine, Upton carefully checked the results. To his astonishment—and quite as Edison had "guessed"—the new dynamo, tested at full load, showed 90-per-cent efficiency in converting steam power into electrical energy.

Edison was as jubilant as a small boy. As was usual with him, the world was soon told all about his "Faradic machine." It was described and depicted in SCIENTIFIC AMERICAN for October 18, 1879, in an article written by Upton.

Once more there was scoffing at Edison's "absurd claims." The hectoring of Edison by some of the leading U. S. electrical experts, among them Henry Morton of the Stevens Institute of Technology, now seems traceable to their ignorance. Reading Morton's predictions of failure, Edison grimly promised that once he had it all running "sure-fire," he would erect at Menlo Park a little statue to his critic which would be eternally illuminated by an Edison lamp.

As a matter of fact, this allegedly ignorant "mechanic" was to be found reading scientific journals and institutional proceedings at all hours of the day and night. It was thus that he had learned about the Sprengel vacuum pump. This device enabled him to achieve an increasingly greater vacuum and to test a broad variety of metals, rare earths and carbon compounds under hitherto unexplored conditions.

The globe itself was also much improved, by the inventor's own design, after he had brought to Menlo Park an artistic German glass blower named Ludwig Boehm. Edison one day drew a sketch of a one-piece, all-glass globe whose joint was completely sealed, and late in April, 1879, Boehm, working skillfully with hand and mouth, fashioned it in the small glass blower's shed in back of the laboratory.

"There never has been a vacuum produced in this country that approached anywhere near the vacuum which is necessary for me," Edison wrote in his notebook. After months of effort he could say exultantly: "We succeeded in making a pump by which we obtained a vacuum of one-millionth part of an atmosphere."

In the late summer of 1879 he realized with growing excitement that a key position had been won. He had a dynamo supplying constant high voltage, and a tight glass globe containing a high vacuum. In his mind's eye he saw what might be done with an extremely fine, highly resistant incandescing substance under these conditions. His state of tension is reflected in the laboratory notebooks by such exclamations as: "S - - - ! Glass busted by Boehm!" All that remained for him was to discover a filament that would endure.

The Carbon Filament

In late August or early September—about a year after he first took up his search—he turned back to experimenting with carbon, this time for good. The rods of carbon he had tried earlier had been impossible to handle, as he now understood, because carbon in its porous state has a marked propensity for absorbing gases. But once he had a truly high vacuum and a method for expelling occluded gases he saw that he might achieve better results with carbon than with platinum.

In a shed in back of the laboratory there was a line of kerosene lamps always burning, and a laborer engaged in scraping the lampblack from the glass chimneys to make carbon cake. But lampblack carbon by itself was not durable enough to be made into fine lamp filaments. Edison and Upton had arrived at the conclusion that, given a 100-volt multiple-arc circuit, the resistance of the lamps should be raised to about 200 ohms; this meant that the filament could be no thicker than a 64th of an inch.

Through the summer months Edison and his staff worked at the tantalizing task of making fine reeds of lampblack carbon mixed with tar. His assistants kept kneading away at this putty-like substance for hours. It seemed impossible to make threads out of it; as an assistant complained one day, the stuff crumbled.

"How long did you knead it?" Edison asked.

"More than an hour."

"Well just keep on for a few hours more and it will come out all right."

Before long they were able to make filaments as thin as seven thousandths of an inch. Edison then systematically investigated the relations between the electrical resistance, shape and heat radiation of the filaments. On October 7, 1879, he entered in his notebook a report on 24 hours of work: "A spiral made of burnt lampblack was even better than the Wallace (soft carbon) mixture." This was indeed promising: the threads lasted an hour or two before they burned out. But it was not yet good enough.

As he felt himself approaching the goal Edison drove his co-workers harder than ever. They held watches over current tests around the clock, one man getting a few hours' sleep while another remained awake. One of the laboratory assistants invented what was called a "corpse-reviver," a sort of noise machine that would be set going with horrible effect to waken anyone who overslept. Upton said that Edison "could never understand the limitations of the strength of other men because his own mental and physical endurance seemed to be without limit."

The laboratory notebooks for October, 1879, show Edison's mood of anticipation pervading the whole staff. He pushed on with hundreds of trials of fine filaments, so attenuated that no one could conceive how they could stand up under heat. Finally he tried various methods of treating cotton threads, hoping that their fibrous texture might give strength to the filament even after they had been carbonized. Before heating them in the furnace he packed them with powdered carbon in an earthenware crucible sealed with fire clay. After many failures in the effort to clamp the delicate filament to platinum lead-in wires, Edison learned to mold them together with lampblack and then fuse the joint between them in the act of carbonization.

Then, as Edison later related, it was necessary to take the filament to the glass blower's shed in order to seal it within a globe: "With the utmost precaution Batchelor took up the precious carbon, and I marched after him, as if guarding a mighty treasure. To our consternation, just as we reached the glass blower's bench, the wretched carbon broke. We turned back to the main laboratory and set to work again. It was late in the afternoon before we produced another carbon, which was broken by a jeweler's screwdriver falling against it. But we turned back again and before nightfall the carbon was completed and inserted in the lamp. The bulb was exhausted of air and sealed, the current turned on, and the sight we had so long desired to see met our eyes."

"Ordinary Thread"

The entries in the laboratory notebooks, although bare and impersonal, nonetheless convey the drama and sense of triumphant resolution pervading the laboratory that night: "October 21—No. 9 ordinary thread Coats Co. cord No. 29, came up to one-half candle and was put on 18 cells battery permanently at 1:30 A.M. . . . No. 9 on from 1:30 A.M. till 3 P.M.—13½ hours and was then raised to 3 gas jets for one hour then cracked glass and busted."

As the light went out the weary men

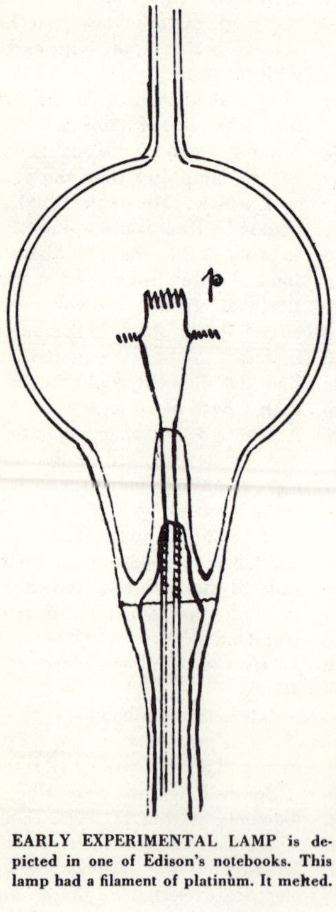

EARLY EXPERIMENTAL LAMP is depicted in one of Edison's notebooks. This lamp had a filament of platinum. It melted.

FRANCIS R. UPTON made invaluable calculations for Edison's system. An electrical engineer who had studied with Hermann von Helmholtz, he was named "Culture" by Edison.

waiting there jumped from their chairs and shouted with joy. Edison, one of them recalled, remained quiet and then said: "If it can burn that number of hours I know I can make it burn a hundred." Yet all the workers at Menlo Park—Edison, Upton, Kruesi, Boehm and the rest—were completely astonished at their success. They had become accustomed to laboring without hope. "They never dreamed," as one contemporary account put it, "that their long months . . . of hard work could be ended thus abruptly, and almost by accident. The suddenness of it takes their breath away."

For once Edison tried to be discreet and keep his momentous discoveries a secret until he could improve upon his lamp filament. At length, after experimenting with various cellulose fibers, he found that paper, in the form of tough Bristol cardboard, proved most enduring when carbonized. Edison was exultant when this filament burned for 170 hours, and swore that he would perfect his lamp so that it would withstand 400 to 1,000 hours of incandescence before any news of it was published.

On November 1, 1879, he executed a patent application for a carbon-filament lamp. Its most significant passage was the declaration: "The object of the invention is to produce electric lamps giving light by incandescence, which lamps shall have high resistance, so as to allow the practical subdivision of the electric light. . . . The invention consists in a light-giving body of carbon wire . . . to offer great resistance to the passage of the electric current, and at the same time present but a slight surface from which radiation can take place." The specifications called for a distinctive one-piece all-glass container, lead-in wires of platinum that passed through the glass base and were fused to the carbon filament, and joints that were sealed by fusing the glass.

Here were the essential features of the basic Edison carbon-filament lamp, in the form that was to be known to the world during the next half century. It was not the "first" electric light, nor even the first incandescent electric lamp. It was, however, the first practical and economical electric light for universal domestic use.

Edison had spent more than $42,000 on his experiments—far more than he had been advanced by his backers. Now he asked for more money so that he might complete a pilot light-and-power station at Menlo Park. But the directors were still uncertain about the future of the invention. Was it "only a laboratory toy," as one of them charged? Would it not need a good deal of work before it became marketable? Grosvenor Lowrey stoutly defended his protégé. He got no results until he prematurely, and over Edison's objections, made the secret of the electric lamp public.

Rumors had been spreading for several weeks. New Jersey neighbors told of brilliant lights blazing all night at Menlo Park, and railroad passengers between New York and Philadelphia also saw the bright lights with astonishment from their train windows. In Wall Street there was a flurry of speculation in Edison stock; the price rose briefly to $3,500 a share.

Then came a front-page story in *The New York Herald* on Sunday, December 21, 1879. There followed an exclusive article about the inventor's struggles for the past 14 months, told to the world, *con amore*, by Marshall Fox, who had written much of Edison before. The detailed treatment of such an adventure in applied science as a feature story was something of an innovation. Also somewhat unusual in the journalism of the time was its relative accuracy of detail, owing to help provided by Upton, who also supplied drawings for the *Herald's* Sunday supplement. The writer did his best to explain how this light was produced from a "tiny strip of paper that a breath would blow away"; why the paper filament did not burn up but became as hard as granite; and how the light-without-flame could be ignited—without a match—when an electric current passed through it, giving a "bright, beautiful light, like the mellow sunset of an Italian autumn."

In the week following Christmas hundreds of visitors made their way to the New Jersey hamlet. Edison hurried with his preparations for an announced New Year's Eve display as best he could, but was forced to use his whole staff of 60 persons to handle the crowds. He could do no more than put on an improvised exhibition, with only one dynamo and a few dozen lights.

The closing nights of the year 1879 turned into a spontaneous festival that reached its climax on New Year's Eve, when a mob of 3,000 sight-seers flooded the place. The visitors never seemed to tire of turning those lights on and off.

The inventor promised the sight-seers that this was but a token of what was in store. He was awaiting the completion of a new generator, he said, and intended to illuminate the surroundings of Menlo Park, for a square mile, with 800 lights. After that he would light up the darkness of the neighboring towns, and even the cities of Newark and New York.

Hi-fi is a field in which erroneous ideas abound. Both human and electronic factors are involved in the accurate reproduction of sound.

11 High Fidelity

Edgar Villchur

Two chapters from his book *Reproduction of Sound* published in 1962.

IT MIGHT APPEAR that following a discussion of the nature of sound, the logical subject to consider would be the criteria for reproducing this sound with "high fidelity" to the original. One other element, however, should be covered first—the way in which we hear.

Perception of Sound

We have already seen, in examining units of measurement for pitch and power—the octave and the decibel—that our perception of sound does not necessarily correspond directly to the objective reality. The illusion is consistent, however, so that a given sound always has the same effect on a normal ear.

An important element in the perception of sound was discovered by Fletcher and Munson in 1933. These investigators demonstrated that our impression of loudness did not depend solely on the amplitude of the sound wave, but on other things as well. Specifically, they showed that sound in the lower treble range of the frequency spectrum—the 3500-cps region—appeared to be much louder than sound of the same amplitude at any other part of the spectrum. Thus, if the frequency scale was swept by a tone which continuously rose in frequency but kept exactly the same amplitude, the *loudness,* or apparent amplitude, would increase to a maximum at about 3500 cps and then fall off again.

This fact does not have much practical interest for the person listening to reproduced music, except as it describes the relative nuisance value of different types of noise. No matter how lop-sided our interpretation of acoustic reality, we make the same interpretation in the concert hall as in our living room, and the craftsmen who designed musical instruments (who worked to satisfy their ears, not sound-level meters) perceived sound in the same way.

Fletcher and Munson made a second discovery, however, that does bear directly on the reproduction of sound. They found that the effect described above took place in varying degree, depending on the over-all level of the sound. For very high amplitude sound the drop in loudness with frequency below 3500 cps hardly occurred at all, while for very soft sound the effect was maximum. Above 3500 cps the effect remained constant, within 2 or 3 db, no matter what the over-all sound level.

The well-known "equal loudness contours," also referred to as the Fletcher-Munson curves, are reproduced in *Fig.* 2–1. Each curve plots the sound amplitude required to produce the same perceived loudness at different frequencies of the scale. It can be seen that normal hearing losses in the bass end become progressively greater as the over-all sound level is decreased.

This means that if an orchestra plays a musical passage at the sound level represented by 90 db, and if this music is reproduced at the 60 db level, we will hear the bass with less *relative* loudness than we would have heard it at the concert itself. If you follow the 90- and 60-db curves, shown superimposed in *Fig.* 2–2, you will see that there is approximately a 14 db perceived loss at 50 cps–it takes 14 db more of actual amplitude, in the lower curve, to produce the same relative loudness at 50 cps as it does in the upper curve.

In order to re-create the original balance of perceived frequencies at low volume levels, it has become customary to introduce bass boost which is related to the setting of the volume control, either automatically or otherwise.

A volume control tied to automatic bass boost is called a *loudness control.* (Some loudness controls also boost the treble spectrum appreciably at low volume settings. There is no justification for this in the Fletcher-Munson curves.)

High Fidelity to What?

The assumption will be made here that the purpose of high fidelity equipment is to reproduce as closely as possible the experience of the concert hall, not to transcend or improve it.

I remember an exhibition at New York's Museum of Modern Art, during the late thirties, of "high fidelity" reproductions of water color paintings. Lifesize reproductions were hung side by side with the originals, and it was often difficult or impossible to tell them apart. There was no question in anyone's mind about how to judge the quality of these prints. The only criterion was accuracy. The public that visited the exhibit was used to looking at paintings, and was able to make an immediate comparison

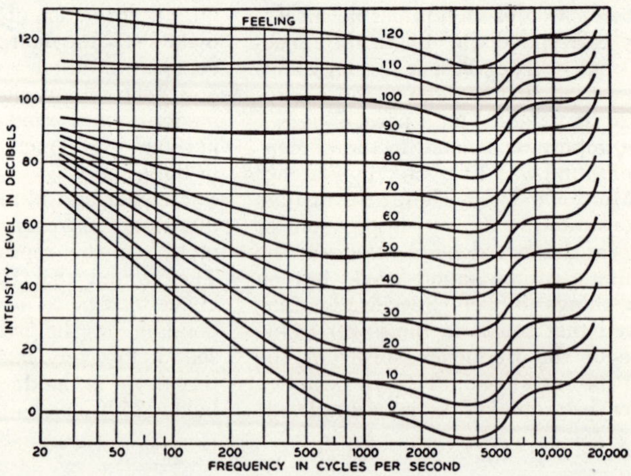

Fig. 2-1. The Fletcher-Munson equal loudness contours. For each curve, the height at any point represents the sound amplitude required to produce the same subjective loudness as at 1000 cps. (After Fletcher and Munson)

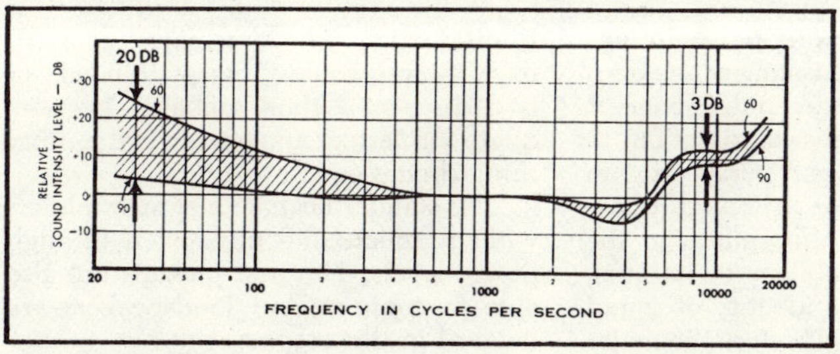

Fig. 2-2. The 60 and 90 db Fletcher-Munson curves superimposed. The shaded area represents the difference in normal hearing loss from one sound level to the other.

between the copy and the original. No one thought of the prints as entities in themselves, with qualities independent of the qualities of the originals.

This point of view does not always hold in the field of high fidelity musical reproduction. Only a minority of today's high fidelity public are concert-goers. Many have never attended a live concert; they know the sound of the orchestra or of individual musical instruments only as it is reported by amplifiers and loudspeakers. They may know what they like in reproduced sound, but they have no way of evaluating the realism of reproduction.

This partly explains why so much variation is tolerated in audio equipment. The same record may sound very different when played through different brands of equipment, each brand equally acceptable in the market place. The evaluation of high fidelity components is popularly thought of as an entirely subjective matter, like comparing the tone of one violin to that of another rather than like holding a facsimile up to its original.

For similar reasons high fidelity demonstrations such as the annual Hi-Fi shows can get away with a lot of sound that is startling but essentially non-musical. Some of the "reproduced" sound that greets the show visitor is necessarily unfamiliar because it has no live counterpart. A harmonica blown up in volume to the dimensions of a theatre organ is a new and different instrument. A crooner whispering into a microphone an inch away invents a new sound; his unamplified voice is never heard in public. A combination of Bongo drum, chimes and electric guitar creates a *tutti* which one may like or dislike, but for which there is no equivalent in one's memory to serve as a live standard.

Such sound can only be accepted as a self-sufficient entity, like an old calendar chromo. Any resemblance to live music or to painting is purely coincidental, and the science and/or art of *re*production is not really involved.

High fidelity has undoubtedly increased rather than decreased the ranks of music lovers, and there are probably more people than ever who are unimpressed with gimmick sound. Many designers and manufacturers in the field work only for naturalness of reproduction. The designer of integrity avoids like the plague those exaggerations that sometimes attract the novice—over-emphasized bass for "depth," over-emphasized mid-range for "presence," over-emphasized treble for "brilliance." These distortions are more properly called, respectively, boominess, nasality or "honkiness," and harshness.

Many demonstrations are not, fortunately, of the gimmick type, and use musical material played at musical levels. There have also been concerts staged with live musicians, in which direct comparisons of reproduced sound to the sound of the live instruments could be made, in the same way that direct comparisons of prints to original paintings were made at the Museum of Modern Art. The live vs. recorded public concert is one method of giving direction to equipment designers and perspective to high fidelity consumers. Although transferring concert hall atmosphere to the home has special problems of its own, success in creating an identity of sound in the concert hall itself solves the major part of the problem. Even more vital to maintaining balance and perspective in the high fidelity world is live concert attendance.

We are now prepared to discuss the technical standards of quality that may be applied to a sound reproducing system. There will be no dividing lines proposed, at which low fidelity becomes medium, high, or super.

Frequency Response

The frequency response of a sound reproducing system, or of one of its components, describes its relative handling of parts of the input signal which differ in frequency. "Handling" may refer to electrical amplification, as in an amplifier, to conversion of mechanical to electrical energy, as in a pickup, or to conversion from electrical to acoustical energy, as in a loudspeaker.

There are two aspects of frequency response: the *range* of frequencies handled, and the *uniformity* with which the unit or system responds to different frequencies. Knowledge of the first of these is useless without knowledge of the second. Let us therefore pass over the question of range for the moment, and determine what uniformity will be required for the range we finally decide on.

Uniformity of Response

Although the trained ear can usually perceive a change of sound level of a db or less in test signals, the average observer is probably less sensitive to a change of sound level in a particular frequency range of a musical passage. Reproduction which remains constant over its frequency range within one or two db would thus probably be adequate for perfect apparent fidelity, other things being equal.

This standard can be met in amplifiers without much difficulty, even at high power levels. The best pickups are also able to conform, but loudspeakers are laggard in this respect.

The results of non-uniform reproduction are several. Undue volume in a particular section of the sound spectrum can produce stridency or boominess as opposed to natural musical sound. More particularly, the existence of sharp peaks in the response curve, usually representing a resonant condition, mean that *hangover* or *ringing* may be present—the speaker cone or section of cone will continue to vibrate after the signal has stopped. This is perceived as a "rain-barrel" effect, a muddying up of the sound and impairment of the distinctness of the different instrumental voices. Such an effect is also indicated when the listener is unable to distinguish clearly the pitch of low-frequency tones.

Another important effect of peaked frequency response is the exaggeration of unwanted noise components such as turntable rumble or record surface scratch. This effect was not given its due recognition in the earlier days of high fidelity, when the existence of rumble and surface noise was proudly displayed as evidence of extended frequency range.

The amount of surface noise in a good quality modern LP record and the amount of rumble from a good record player are such that there will not be much significant noise produced in a system with uniform frequency response, even though the frequency range be extended to the limits of the present state of the art. In a comparison test conducted recently between two tweeters, the one which was able to reproduce almost an octave more of treble (into the inaudible region) showed a dramatic decrease of surface noise, due to its extreme evenness of response. There was no selective reproduction of discrete frequency regions, and the switch to the superior speaker produced a fuller, more natural treble simultaneously with the reduction in surface noise.

A similar situation exists with regard to turntable rumble. A peaked system whose response falls off rapidly below 60 cps may exhibit more turntable rumble than a smooth system whose full response extends an octave lower.

Tell-tale evidence of the existence of peaked reproduction in the bass may be gathered from listening to the reproduction of speech. The male speaking voice ordinarily contains no sound components whose frequency is below 100 cps, and the reproducing system should give no hint (by a boomy, resonant quality in the voice) that it is also capable of speaking in the tones of the double bass.

Range of Response

It is generally agreed among acoustics authorities that the range of 40 to 15,000 cps is sufficient for perfect or near-perfect apparent fidelity in the reproduction of orchestral music. The phrase "near-perfect" is meant to imply that when such a range has been achieved the designer should direct his attention to inaccuracies of reproduction more gross than are associated with the frequency limitations indicated.

For the pipe organ enthusiast, however, there is significant intelligence (significant, that is, from the point of view of the emotional impact of the music) down to 32 cps or lower. 32.7 cps is three octaves below middle C relative to *A*-440, and is the lowest note of the average pipe organ, although many larger organs reach down an octave lower. These low organ tones are distinguished by the fact that they contain a strong fundamental component. The lowest tones of the piano, on the other hand, contain no fundamental energy that significantly affects the quality of the sound. Even though the lowest key on the piano strikes 27.5 cps, response down to this frequency is not required for the reproduction of piano music.

Probably no characteristic of audio components is so freely booted about by advertising copywriters as frequency range. Any numerical range of frequencies listed is totally meaningless unless accompanied by a description of the decibel tolerance above or below reference that is being used, and, for a loudspeaker, by a description of off-axis response as well. A 3-in. speaker made for portable radios will "respond" when stimulated by a 30-cps signal—perhaps by having its cone tear loose and fly out into the air—and almost any speaker, even a woofer, will make some kind of sound when stimulated by a high-powered 15,000-cps signal. A frequency response rating must mean something more than that a signal of given frequency makes a speaker move audibly, or that it makes an amplifier show an electrical output of some sort at its terminals. It must mean that within a stated frequency range, and, for power devices, within a stated range of power, the fundamental output of a given device is uniform to a stated degree.

Treble Dispersion

The on-axis response of a loudspeaker may be very deceiving, because the higher frequencies tend to be directed in a beam which continually narrows as the frequency is raised. Good sound dispersion must therefore be a qualifying factor for any treble response curve.

A speaker which has relatively uniform treble output both on-axis and off-axis (over a reasonably large solid angle—perhaps 45 degrees in any direction from the axis) will reproduce music with a "spaciousness" that does not exist when there is more concentrated beaming of the treble. Furthermore, severely attenuated off-axis response in the treble means that the total sound power radiated at treble frequencies is considerably less than that implied by the on-axis response curve. It is this total radiated power, rather than the on-axis pressure, that determines whether a speaker will sound dull, natural, or over-bright in a normally reverberant room.

Transient Response

Transient response refers to the accuracy of reproduction of the wave envelope, and is concerned with the reproduction of attack and decay characteristics of the sound. We have seen that uniform frequency response predicts the absence of ringing; if the steady-state frequency response curve does not have peaks, the reproduced sound will die away just as in the original.

Consider, for example, the tone represented in (A) of *Fig.* 2–3. Perfect reproduction would produce an identical wave form, differing perhaps only in amplitude, while poor transient response would be indicated by the hangover that is apparent in (B). The continuation of the reproduced signal after the original

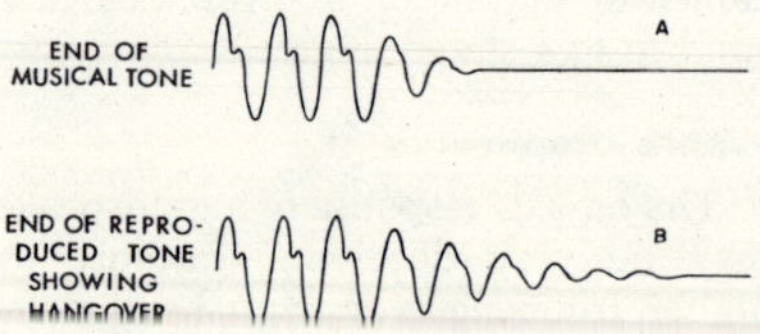

Fig. 2-3. Poor transient response.

has ended may be compared to a color smear on a reproduced painting.

Attack time involves the reproduction of frequencies higher than the fundamental. Although a percussive tone may have a low fundamental pitch, the frequency components associated with its steep attack characteristic may be very high. Natural reproduction of a drum beat through a two-way speaker system may thus be accomplished by the "woofer" handling the fundamental tone and its proper decay, while the "tweeter" contributes the sound components that make up the sharp attack.

Harmonic and Intermodulation Distortion

Reproducing devices have a characteristic way of performing with less than perfect accuracy. In addition to the frequencies at which they are asked to vibrate mechanically (or alternate electrically) they introduce new modes of oscillation of their own—and these new frequencies are harmonics, integral multiples of the original frequency. This inaccuracy is called *harmonic distortion.* It is measured as the ratio of the amplitude of the spurious harmonics to the true signal, in per cent.

We have seen that harmonics of fundamental frequencies are produced in any case by musical instruments. Yet small amounts of harmonic distortion produce very unpleasant effects. The sound becomes harsh, unmusical; the bass is wooden and the treble painful.

The primary reason for this is that with harmonic distortion comes an attendant evil—intermodulation distortion. Intermodulation distortion can be described as the introduction of new sound components, at sum and difference frequencies, when tones of two or more frequencies are passed through a non-linear system—that is, a system which creates harmonic distortion. These sum and difference frequencies are harmonically unrelated to the original musical tones. They are musically discordant, and they serve to create raucous, unmusical sound in a degree proportional to their relative strength. The formation of intermodulation products is illustrated in *Fig.* 2–4.

The primary importance of low distortion has always been recognized by audio authorities. It has also become in-

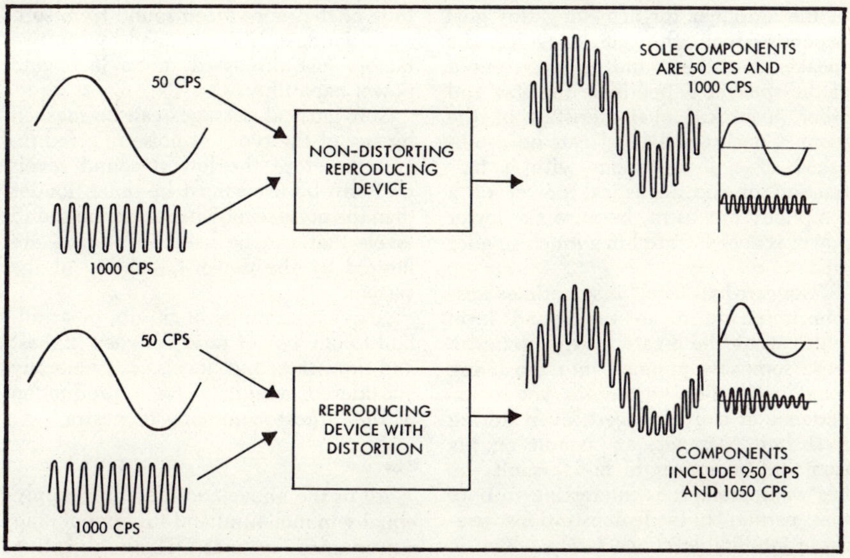

Fig. 2-4. Intermodulation distortion as a result of harmonic distortion of the low-frequency wave form. Note that the wave envelope of the high-frequency tone is "modulated."

creasingly recognized by the high fidelity public in recent years, after the first flush of excitement over reproducing regions of the frequency spectrum previously untouched. Amplifier manufacturers now feature distortion data over frequency response data; unfortunately it is very rare for loudspeaker specifications to make any quantitative reference to distortion at all. The reason lies in the fact that while both harmonic distortion and intermodulation distortion (the latter is usually greater by a factor of 3 or 4) can be kept to extremely low values in high quality amplifiers—a small fraction of one per cent at rated power—the corresponding values for loudspeakers are much higher. In the octave below 60 cps it is a rare speaker indeed which can hold harmonic distortion, at any appreciable sound level, below the 5 per cent mark over the entire octave, and many speakers produce percentages of distortion in this frequency region ten times as great. But the listening results are not as bad as might appear at first glance: speaker response is normally severely attenuated in this lower range, which helps, and there is comparatively little musical material of such low frequency to be distorted.

When the reproducing system has a minimum of low frequency distortion, very low bass tones of high power, such as might be produced by organ pedal pipes, not only remain pure in timbre themselves but do not create intermodulation with the rest of the music; they do not destroy the purity of the treble by introducing false tones.

Power Capability

The power capability of a high-quality reproducing system should be such as to be able to establish an intensity level of sound in the living room equal to the level at a good seat in the original concert hall. The electrical power required of the amplifier for achieving this goal depends upon the efficiency of the speaker, and the sound power required of the speaker depends on the size and other acoustical characteristics of the room. Concert-hall level can be established in a living room with a tiny fraction of the acoustical power of a symphony orchestra, because the lower power is concentrated in a much smaller area.

"Concert-hall level" is sometimes misinterpreted to mean the sound level which would be created if the orchestra were somehow jammed into the living room itself. The writer has yet to experience at a live concert, even during *fortissimo* passages, an assault on his ears that compares to hi-fi assaults he has weathered. It is interesting to note that certain hi-fi demonstrations preclude intelligible conversation which is not shouted, while whispered conversations in a concert hall are liable to prove extremely distracting and annoying to one's neighbors. It is the sound intensity level at the ear, not the power of the orchestra, that we are trying to reproduce.

Noise Level

Any sound component not present in the original program material, other than distortion products, is referred to as noise, even though it may be periodic and not conform to our strictly scientific definition. Hum, rumble, surface scratch, tube hiss or other circuit noise and similar disturbances tend to destroy the auditory illusion, and must be kept to a minimum.

A standard for satisfactorily low noise has been established by the FCC for FM broadcast stations. It is that the power ratio of the maximum signal to the noise must always be at least 60 db; this represents a ratio of one million to one.

Dynamic Range

The dynamic range, or range of amplitude of the reproduced sound from softest to loudest, is determined by the two factors just discussed, noise level and power capability.

Soft musical passages can be masked by any of the types of noise referred to, and therefore the lowest sound levels that can be used must be much louder than the noise level. The maximum sound levels that can be used, of course, are limited by the power capability of the system.

A dynamic range of 60 db, or a million to one power ratio between highest and lowest sound levels, is generally considered adequate for reproduction of the largest symphony orchestra.

Stereo

All of the above considerations apply equally to monaural and to stereophonic reproducing systems. These objective elements of equipment fidelity—low distortion, adequate frequency response, dynamic range, etc.—are able, in stereo, to contribute more to the subjective illusion of musical reality than in a monaural system.

A stereo record-reproduce system has in effect two parallel and complete monaural systems. The work of each component along the way is done twice. The sound is picked up by two separate microphones; the output of each microphone is recorded on a separate track of the tape; the record groove, although not doubled, is cut in such a way as to independently contain the record of each signal channel; the pickup contains two separate generating elements which independently sense and transmit each signal channel; the two signal outputs of the pickup are sent through independent amplifiers and fed to two independent loudspeakers. There are variations on this ideal scheme, but the above describes the basic concept of stereo.

The purpose of this dual-channel reproduction is, in the simplest terms, to help recreate the acoustical atmosphere of the concert hall. In the old-fashioned stereopticon each visual channel gave a slightly different perspective view of the subject. Similarly, in stereo recording, each microphone gets a slightly different auditory perspective. It is important to note that this auditory perspective is of the orchestra or soloists *in the hall* in which they are performing, not merely of the musical performers in the abstract. This is important because a good part of the sound that reaches our ears at a concert does not come directly from the orchestra, but is reflected from the walls and ceiling of the concert hall.

The channels of a stereo system are identified as "right and "left." This does not mean that one microphone picks up the sound of the right section of the orchestra only, and that the other microphone picks up the sound from the left section of the orchestra. It does mean that one microphone has a right-oriented perspective of the total sound in the recording hall, and that the other microphone has a left-oriented perspective of the total sound. When these two recorded channels (which, like the two photos on a stereopticon card, are very similar to each other) are reproduced through two separate loudspeakers they create, although not perfectly, the illusion of the acoustical environment and sense of space of the concert hall. There is an increased awareness of the physical position of different instruments, but this is very much less important than the general increase in realism and the consequent increase of clarity, particularly from the point of view of the distinctness of the different musical voices.

There is an approach to stereo recording, commonly referred to as "ping-pong" stereo, which provides an exaggerated separation between the right and left channels. If only the left side of the orchestra were playing during a particular passage, there would be practically no sound from the right recording channel. The left-right orientation of the different instruments is the primary goal in this case, rather than reproduction of the original acoustical environment. The degree to which one's attention is directed to the physical position of the instruments in "ping-pong" stereo is often much greater than that at the live concert itself.

The greatest benefit of good stereo recording and reproduction is that it frees us, to a greater extent than was possible previously, from the acoustical environment of the listening room, and transports us to some extent to the acoustical environment of the hall in which the recording was made. The normal living room does not provide the proper acoustical atmosphere for a musical concert, particularly of a large orchestra. Musical instrument designers worked in terms of the tonal qualities that would be produced in the type of concert hall with which they were familiar.

THE SOUND REPRODUCING SYSTEM

THE PHONOGRAPH is a classic example of an invention that cannot be credited wholly to one man. In 1877 Edison directed his assistant, John Kruesi, to construct the first complete record-reproduce system, but sound recorders were sold on a commercial basis as early as 1860, and Thomas Young's "A Course of Lectures on Natural Philosophy" described and illustrated a crude but practical sound recorder in 1807.

Young's recorder consisted of a sharp metal stylus held by spring tension against a revolving cylinder, the cylinder coated with wax and turned by a governor-controlled gravity motor. When a vibrating body such as a tuning fork was held against the stylus, a wavy line was cut into the wax. This line represented the wave form of the vibrations, and it could be studied and analyzed at leisure. The recorder was a mechanical draftsman, that could sense very small motions and record pressure changes that took place within a period of a very small fraction of a second.

By 1856 Léon Scott de Martinville had constructed the "phonautograph" (self-writer of sound) illustrated in *Fig.* 3–1. The sound wave form was scratched by a hog-bristle stylus on the surface of a cylinder coated with lampblack, but the big advance over Young's machine was the fact that the phonautograph could record directly from the air. The force of the acoustical vibrations

Fig. 3-1. The phonautograph of Léon Scott de Martinville — a commercial sound recorder of the eighteen sixties. (Courtesy Smithsonian Institution)

was concentrated by a horn onto a diaphragm, and the stylus was attached to the diaphragm, so that the recording needle did not have to actually touch the vibrating source of sound. This device, which corresponds in function to the modern oscilloscope, was a catalogue item of the Paris firm of Koenig, and was sold as a measuring instrument to acoustical laboratories.

The phonautograph which is at the Smithsonian Institution at Washington would undoubtedly reproduce music if a proper record were placed on its revolving cylinder. The theoretical possibility of playback was understood then, too, but the lampblack records were useless for playback, as their grooves were not rigid enough to direct the vibrations of a playback needle. About half a year before Edison got his brainstorm Charles Cros conceived a method for bringing the groove sinuosities back to life as sound. The lampblack recording was to be photo-engraved on a metal cylinder, and running a needle through the hard groove would then cause the needle to vibrate from side to side, in the same time pattern as the hog bristle stylus that first inscribed the line.

For reasons which may be related to nineteenth century differences in tradition between the scholar and the industrial engineer, Cros didn't even construct a working model, but merely filed a complete, sealed description of his system with the *Académie des Sciences*. On the other hand, less than a month after Edison first conceived of a reproducing phonograph the country was reading about a working unit in newspaper headlines. There was a great stir of excitement over this amazing tonal imitator, (see *Fig.* 3–2) with public demonstrations, lectures before august scientific bodies, and a visit to the White House.

Fig. 3-2. Edison with his tin-foil phonograph. (Photograph by Brady — courtesy Smithsonian Institution)

The excitement soon died down, as the Edison machine was an impractical toy, with neither permanent records nor usable fidelity. The recorded groove was indented into a semi-hard material, tin foil; it was only able to retain its shape partially, and that for very few playings. Subsequent technical improvements, however, made the phonograph a popular device by the turn of the century. It is curious that our modern recording system, in which the record is a mechanical copy of the original master, is more closely related to Cros' system than to Edison's. Emil Berliner, the father of the moulded or cast record, began his research work by successfully carrying out Cros' proposals.

The Mechanical or "Acoustic" Phonograph

It would be useful to consider the design of the non-electric phonograph, as illustrated in (A) of *Fig.* 3–3. A better insight can thereby be gained into the function of the various components of a modern electronic system.

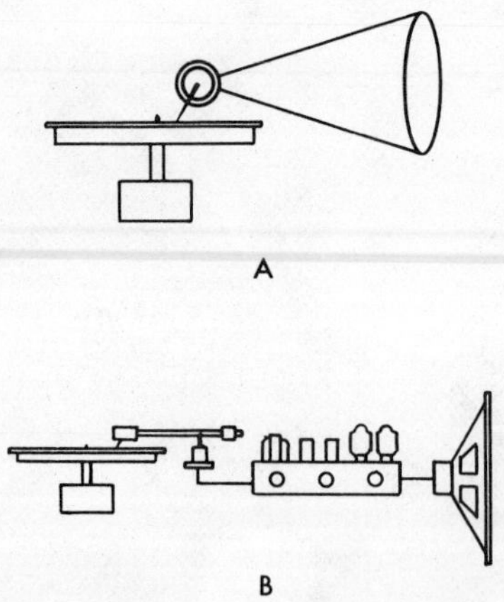

Fig. 3-3. (A) The mechanical phonograph. (B) The electric phonograph.

The wave forms frozen into the record groove control the vibrations of the playback stylus when the groove is dragged past the stylus by a revolving turntable. These stylus vibrations, although they contain a fairly large amount of mechanical energy, engage practically no air, like the revolutions of a bladeless electric fan. The needle is therefore attached to a diaphragm, which vibrates in sympathy with the stylus and has a much larger surface area in contact with the air of the room.

But even the reproducing diaphragm doesn't get a sufficient bite of the air for practical purposes. Therefore the diaphragm is placed at the narrow throat of an acoustical horn, and the actual usable sound emerges into the room from the much larger mouth of the horn. The system works somewhat as though the diaphragm area were really that of the horn's mouth.

It can be seen that all of the energy radiated by the horn is taken from the mechanical vibrations of the needle, and the forces between needle and record groove are necessarily great. This has obvious implications for record wear, but perhaps more important, the demands for power placed on the "sound box" or "speaker" (old-fashioned terms for the needle-diaphragm-head assembly) place a severe limitation on musical fidelity. High distortion and peaked and severely limited frequency response are to be expected.

The Phonograph Amplifier

The solution to this problem lies in changing the function of the phonograph pickup, from the primary generator of sound power to a device which controls an outside source of power. If the power from the outside source is made to oscillate in imitation of the needle vibrations, two benefits can result:

1. The final output sound derived

from the record groove can be much louder.

2. The power demands on the pickup itself are no longer heavy. The pickup can be designed for quality rather than loudness; the problems of achieving uniform, extended frequency response and low distortion are considerably lessened. So, incidentally, is the required weight on the pickup and the grinding away of the record groove.

The control of an outside source of power to conform to given oscillations is called *amplification*. The first phonograph amplifier was pneumatic: the needle was made to actuate an air valve, which periodically throttled a flow of compressed air. Most of the work of radiating sound power was thus performed by the air compressor, and the stylus was relieved of part of its burden.

All modern sound reproducing systems use amplifiers, but unlike the first pneumatic systems these amplifiers are electronic. The phonograph pickup is no longer a sound generator but an electric generator. It produces small alternating voltages at its terminals, whose wave forms conform to those of the groove and of the recorded sound. The pickup has to generate very little power, because the output voltage can be amplified to almost any desired degree. The amplified electrical power must finally, of course, be converted back into sound by a loudspeaker. The two types of reproducing system, electrical and purely mechanical, are shown in *Fig.* 3–3.

The Modern Sound Reproducing System

The purpose of the historical approach used above has been to furnish the reader with an appreciation of the reason for the modern audio system being designed as it is. With the electronic amplifier supplying the brute force, so to speak, the mechanical components—pickup and loudspeaker—can be built in such a way as to suppress the natural resonant tendencies inherent in mechanical vibratory systems.

Before discussing each of the audio components in detail, it would be useful to make a brief survey of the entire reproducing system. A complete monaural system is illustrated in *Fig.* 3–4.

First of all the disc record must be revolved by a *motor* and *turntable*. The chief operational requirements of this part of the system are that it revolve at the correct speed, that the speed be constant, and that extraneous vibrations do not communicate themselves to the pickup.

The first of these requirements is for the purpose of keeping the reproduced music at the same absolute pitch at which it was recorded: too fast a turntable speed will make the pitch sharp, and too low a speed will make it flat. The second condition listed, constant speed, is required in order to avoid pitch variations, or "wow." The third requirement, lack of extraneous vibrations, keeps low-frequency noise called "rumble" out of the final sound.

The groove variations are sensed by the *needle*, or *stylus*, which in high-quality systems is jewel tipped; it is usually diamond. The needle must have an unmarred, smooth surfaced, hard tip, normally of spherical shape.

The *pickup* is an electric generator (usually either of the piezo-electric, variable reluctance, or moving-coil type) whose function is to translate the mechanical vibrations of the needle into electrical oscillations of the same wave form. It must do this with minimum distortion of the wave form, and must not allow resonances of its own to influence its output voltage significantly. It is also an advantage for the pickup to impose as little work as possible on the needle. The greater the force required for the groove to displace the needle from side to side, the greater the vertical bearing force will have to be to

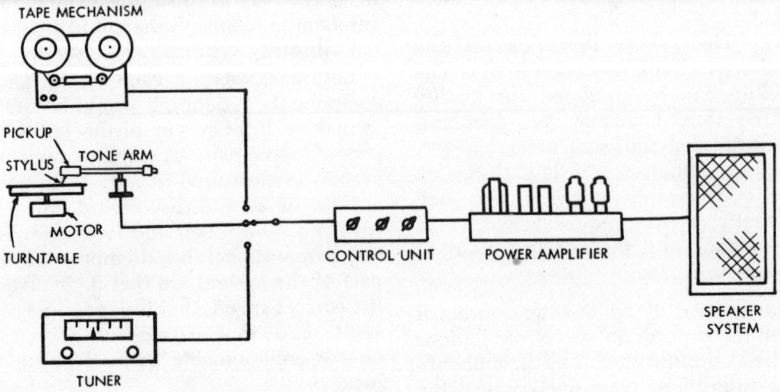

Fig. 3-4. Diagram of a complete monaural sound reproducing system.

maintain proper and constant stylus-groove contact, and the greater the wear of both record and needle.

The *tone arm* holds the pickup in place over the groove, and must provide sufficient freedom of motion so that the pressure of the groove walls alone can make the needle move across the record, following the recorded spiral. It must also be free enough to follow warp and eccentricity of the disc easily. The tone arm must hold the pickup approximately tangent to the groove being played, must provide the proper vertical force for the pickup, and must not allow its own resonant behavior to influence the system.

The electrical output of one type of pickup, the piezo-electric, is usually fed directly to the amplifier. It is of the order of ½ volt or more, and is a fairly accurate replica of the recorded sound. This is so because the characteristic frequency response of the pickup is more or less the inverse image of the frequency characteristics "built in" to the record. (This last subject will be taken up in detail later.)

The reluctance and moving-coil pickups, however, produce a much smaller amount of electrical energy. The output voltage of these pickups (which are classed together as *magnetic* types) may be as low as a few thousandths of a volt. Furthermore the characteristic frequency response of the magnetic pickup does not compensate for the way in which the frequency characteristics of the recorded sound has been doctored. Therefore the pickup output must be passed through a *preamplifier* before it enters the amplifier proper.

The preamplifier is normally combined with the main amplifier control sections (volume and tone controls). Its functions are to increase the output voltage of the pickup, and to compensate accurately for the frequency characteristics of the record so that the sound is not deficient in bass and heavy in the treble. Since different record companies have made records with different characteristics the preamplifier may allow the operator to choose between several types of frequency compensation. The need for such control, which is called variable record equalization, has disappeared with modern records, which are standardized on the RIAA recording characteristic.

The control section of the amplifier allows the operator to regulate the volume, and, in most cases, to either accentuate or attenuate ("boost" or "cut") the bass and treble portions of the reproduced sound independently. The primary function of tone control is to compensate for deficiencies in associated equipment or program material, and to compensate for acoustical conditions of the room in which the music is heard. When the control section and phonograph preamplifier are combined on one chassis, the entire unit is commonly referred to as a preamplifier.

The power amplifier receives the electrical signal as it is finally shaped, and releases another signal, ideally identical in all respects except power. The power amplification may be tens of millions of times, from a fraction of a microwatt (one millionth of a watt) to dozens of watts.

Although the demands on the amplifier are very great, and although it appears to be the most complicated of the system components, it is the least imperfect of these components. The percentages of harmonic and intermodulation distortion, the irregularities of frequency response, and the extraneous noise introduced by an amplifier built according to the best current design practice, and without regard for cost, are such that they are not limiting factors in the fidelity of the reproduced sound.

The final component of the sound system is the loudspeaker system, which consists of the speaker mechanism itself and the speaker enclosure. The loudspeaker converts the alternating electrical output of the amplifier into mechanical vibrations of a cone or diaphragm. But the cone vibrating by itself cannot, for reasons that will be discussed further on, produce adequate bass energy. It must be mounted in an enclosure or baffle of some sort, which gives the vibrating surface the "bite" of air that it needs to radiate low-frequency sound.

The speaker and its enclosure, like the amplifier, should introduce as little distortion and frequency irregularity into the signal as possible. Typical speaker deficiencies are irregular frequency response, poor transient response (hangover), and harmonic and intermodulation distortion.

Two other components are shown in *Fig.* 3–4. The *tuner* is a device which converts AM or FM radio signals to audio signals that can be handled by the audio amplifier; the tape transport mechanism, with its associated pream-

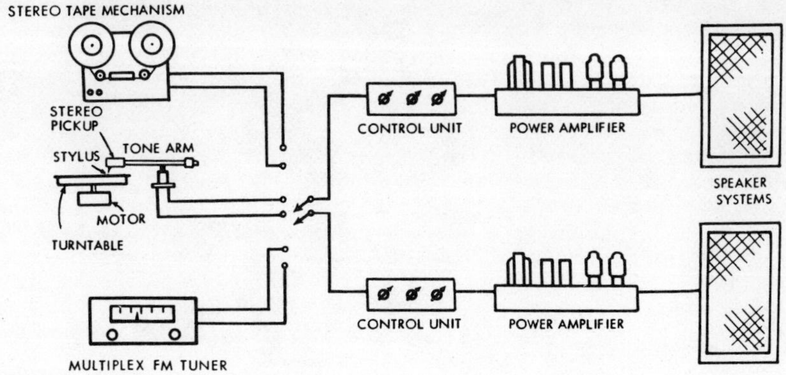

Fig. 3-5. A stereo reproducing system.

plifier, provides a signal of the same nature as that coming from the tuner or phonograph pickup.

Fig. 3-5 shows the basic elements of a stereo reproducing system. The stereo tape mechanism has two heads which independently reproduce each channel that is recorded in parallel on the tape; the stereo pickup provides two separate output signals from the two channels recorded in the groove (the turntable and pickup arm do not have to be duplicated); the stereo tuner receives the "multiplex" FM stereo signal and separates it into two separate channels, which it feeds independently to each of the control units. Each control unit and each power output is shown duplicated. The two control units and power amplifiers may be separate, or they may be combined on one chassis, or all four units may be combined on one chassis, but in any case they must provide independent amplification for each channel.

Since the Niagara power plant was built, commercial electric power has been almost entirely alternating current. Now new consideration is being given to the advantages of direct current for long distance power transmission.

12 The Future of Direct Current Power Transmission

N. L. Allen

A popular article published in 1967.

The history of technology provides many examples of unexpected turns of fortune, and electrical technology is no exception. It frequently happens that a principle or technique, originally the basis of a well-established system, is superseded by a device making a significant advance, only to reappear in a different guise as the 'last word' in the state of the art. An obvious example is the crystal of the early radio receiver. This was superseded by the thermionic valve, but it has now developed into the more sophisticated form of the transistor. Not many years before the era of the crystal receiver, an appreciable proportion of electrical energy was generated, transmitted, and used in the form of direct current. At that time, generation and consumption usually took place in the same locality, distribution was simple, and the quantities of energy transmitted were small by modern standards. However, serious limitations appeared as it became necessary to distribute electrical energy more widely, and direct current as the distributing medium gave way to alternating current.

In many countries, the economic advantages of being able to concentrate power generation in large stations have led to the adoption of a comprehensive network of power lines that interconnect generating plant and the areas where the power is used. As the length of a power line increases, the current passed, for minimum power loss, decreases: the economic operating voltage for transmission of a given power therefore increases. The transmission of larger quantities of energy at high voltages and low currents is greatly facilitated by the ease with which alternating current can be transformed to the voltage most appropriate for the power lines. In the receiving areas of the system, the voltage can equally easily be transformed to lower values suitable for distribution, and a system of far greater flexibility can be set up than is the case with direct current. Further, it is difficult to switch and, particularly, to interrupt direct current. The interruption of an alternating current by circuit breakers is relatively easy because the current passes through zero twice in every cycle.

This combination of circumstances made alternating current the natural choice as power systems increased in size. The main links operated initially at 132 kilovolts, but the need for increased power during the post-war years has led to the adoption of 275 kilovolts and, more recently, 400 kilovolts as the operating voltages of the principal links in Britain. The power is distributed locally at lower voltages. During this period, the remaining direct current distribution systems have been reduced or eliminated.

Transmission over long distances

What, then, is the place of direct current? There is certainly no good reason for turning away completely from alternating current distribution. But there have always been some situations in power distribution practice in which direct current has distinct advantages over alternating current, and it is worth while considering what these situations are.

One basic factor in power system design is the need to find the simplest and most efficient means of transferring power from one point to another. Figure 1(a) shows the basic three-phase alternating current system and figure 1(b) a favoured direct current system, which has positive and negative polarities on the two lines, and is linked by convertors to alternating current for generation at one end and distribution at the other. In both cases, the maximum voltage to earth is E, but for alternating current, it is the root-mean-square value $E/\sqrt{2}$ that determines the power transmitted. This is $3EI_A \cos \varphi/\sqrt{2}$, where I_A is the current in each conductor, lagging behind the voltage in phase by φ degrees. In the direct current system, the power transmitted by each line is EI_D, where I_D is the current. For transmission of equal power by the two systems, therefore, it can be shown that each alternating current line has $4/(3 \cos^2 \varphi)$ times the cross sectional area of the corresponding direct current line, a factor which is always greater than 1·33. Moreover, the alternating current system requires three cables rather than two, so that the amount of copper required is $2/\cos^2 \varphi$ times that in the direct current system, a factor which is always greater than 2.

Direct current, then, reduces the cost of the cables. This may appear trivial compared with the other capital costs in electrical systems, but over great distances, as in the United States and the Soviet Union, the saving in cable, and in the means of supporting the cable, becomes a very significant factor that can outweigh the cost of providing the convertor stations at each end of the system.

Great distances bring further problems in alternating current transmission that do not occur with direct current. These problems arise from the relationship between the wavelength of the oscillation and the dimensions of the system. The quarter-wavelength of a 50 cycles per second wave in air is about 900 miles, and the transmission of energy through a conductor can be regarded as due to an influx of energy along its length from the electromagnetic field that surrounds it. Over short distances, this field is very nearly the same at all points, since electromagnetic energy is conveyed with the velocity of light. But at distances greater than 900 miles, the fact that the velocity of light is finite results in significant differences, at any instant, in the phase of the current at the two ends.

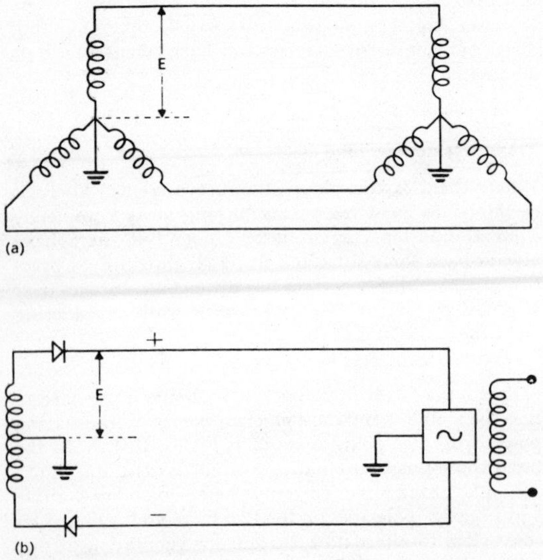

Figure 1 Simplified distribution systems: (a) alternating current, (b) direct current.

This situation leads to difficulties where two parts of a power circuit, joined by a long alternating current link, are out of phase and where a loop is formed through another part of the network of different length. Large circulating currents will be set up unless some form of compensation is applied. A direct current link obviates these difficulties; as a corollary, it may be noted also that if a direct current line is used to link two alternating current systems, they need not be synchronized with each other.

Transmission over short distances

For long-distance transmission, overhead lines, supported by towers, are used. The virtues of direct current are most clearly shown when the current is carried by underground or underwater cables. Here, the central core of the cable, which is at the transmission voltage, is surrounded by an insulant, the exterior of which is at earth potential. This constitutes a coaxial capacitor, and the capacitance per mile of a cable rated at 200 kV is typically about 0·3 microfarads. In an alternating current circuit, this capacitance is charged and discharged, through the inductance and resistance of the cable itself, once every half-cycle. Additional generating capacity is needed to supply this charging current. In the example quoted, at 200 kV, the charging current requires about 5000 kilovolt-amperes per mile of cable; at 400 kV the figure is about 15 000 kilovolt-amperes per mile. For appreciable lengths of cable, the losses become such that the charging currents must be supplied at intermediate points. At 200 kV, these points are about 25 miles apart for 50-cycle alternating current; at 400 kV, only 15 miles. Thus, alternating current transmission becomes impracticable in cables over long distances. Further, the cost of the generating capacity needed to supply the charging current is significant. Taking a rough figure of £50 per kilowatt of installed capacity at the generating station, this extra cost is £250 000 per mile for a 200 kilovolt cable. By contrast, with direct current in the steady state, there is no charging current. It may well be worthwhile, therefore, to accept the cost of converting to direct current to avoid having to provide this charging current. Direct current is also advantageous in that there are no dielectric losses due to reversal of the electric stress in the insulant.

The balance between the two systems

To summarize, direct current has significant advantages for the transmission of bulk power over great distances by overhead lines, and over short or long distances by cable. In addition to the technical advantages already examined, direct current may be valuable in linking two alternating current systems that need not then be synchronized. Alternatively, a very large alternating current system may be divided by direct current links into two or more smaller systems: this is a possible future development as power systems continue to increase in size. It is necessary, however, to examine some disadvantages of direct current, and some relevant non-technical factors, to demonstrate the balance affecting the final choice of system.

The most obvious drawback to the use of direct current is the need for conversion at each end of the link in order to integrate it with established alternating current systems. The technical details are outlined later, but it may be mentioned here that the cost of the conversion equipment is about twice that of the alternating current equipment required for the termination of a power line of corresponding size and output. These costs must be set against the savings inherent in the direct current system. There is therefore, a limit to the length of a line, below which the capital outlay on a direct current system is higher than that of an alternating current system. Estimates of the critical length for a long overhead line naturally vary, depending mainly on the power to be transmitted and the voltage to be employed, but figures of more than 300 miles have frequently been quoted. This approach is unlikely to be favoured, therefore, in the British Isles, but such systems are being developed in the United States and in the Soviet Union. For underground or submarine cables, where dielectric losses and charging currents are so important, the 'critical length' is reduced to about 30 miles, and it is in short submarine links and in urban transmission lines that direct current finds its second important application. Indeed, where large amounts of power have to be introduced into large cities, legal and social considerations may predominate over technical and economic factors. It is frequently extremely difficult to obtain permission to erect overhead lines in urban areas, and the disturbance to local amenities caused by the towers for high-tension cables may not be justifiable. Underground cables become necessary, and it is preferable to use direct current for distances greater than about 30 miles.

In choosing between the systems, the fact that there can be no direct current transformer and that there is no satisfactory circuit breaker ensures that alternating current maintains its general superiority for distribution purposes. The use of direct current is thus confined to the bulk transmission of high power between discrete parts of a system or between two separate systems.

The Reader for Unit 3 contained the first part of Newman's biography of this outstanding mathematician and physicist. This final part covers primarily his work on electromagnetic theory.

13 James Clerk Maxwell, Part II

James R. Newman

A biographical essay published in 1955.

In February, 1858, Maxwell wrote a letter to his aunt, Miss Cay, beginning, "This comes to tell you that I am going to have a wife." "Don't be afraid," he added, "she is not mathematical, but there are other things besides that, and she certainly won't stop mathematics." His engagement to Katherine Mary Dewar, daughter of the principal of Marischal College, was formally announced the same month, and in June they were married.

Their union became very close: they enjoyed doing things together — horseback riding, reading aloud to each other, traveling — and he even found useful tasks for her in his experimental work. The marriage was childless, but this very fact increased the couple's dependency and devotion. Maxwell regarded the marriage tie in an "almost mystical manner." The published letters to his wife overflow with religiosity.*

The Aberdeen appointment terminated in 1860 when the two colleges, King's and Marischal, were fused into a new university and Maxwell's chair in physics at Marischal was eliminated. He was not long at liberty. In the summer of the same year he became professor of natural philosophy at King's College, London, a post he retained until 1865. The teaching schedule at King's was long and arduous; in the evenings there were lectures to be given to "artisans" as part of his regular duties. Living in London offered him the opportunity to see something of Faraday, with whom, up to this time, Maxwell had had only correspondence, to make the acquaintance of other scientific men and to renew old friendships. He was no solitary. "Work is good, and reading is good, but friends are better," he wrote to his friend Litchfield.

Yet despite academic duties and social distractions, the five years in London were the most productive of his life. The paper "On the Theory of Three Primary Colors," the two articles in the *Philosophical Magazine* on "Physical Lines of Force" and the culminating electrical memoir "A Dynamical Theory of the Electromagnetic Field," the Bakerian lecture "On the Viscosity or Internal Friction of Air and other Gases," and the celebrated paper "On the Dynamical Theory of Gases," all belong to this period. He also performed important experimental work during these years. At his house in Kensington,

* He did not write in this vein to others and it is a little puzzling why he found it necessary in corresponding with her to quote Scriptures, to express the fervent hope that the Lord would protect her from evil, and that she would get her eyes off "things seen and temporal and be refreshed with things eternal."

in a large garret, he measured the viscosity of gases and obtained practical confirmation of the theoretical work I have described. (For example, he found that the viscosity of air at 12 millimeters of mercury measured the same as at normal pressure of 760 millimeters, thus proving that viscosity is independent of density.) To maintain the necessary temperature, a fire had to be kept up in the midst of very hot weather and kettles kept boiling to produce steam, which would be allowed to flow into the room. Mrs. Maxwell acted as stoker. Another investigation dealt with the ratio of the electromagnetic to the electrostatic unit of electricity and led to one of Maxwell's greatest discoveries. But I must postpone explaining this work, even though to do so means abandoning the strict chronology of events in Maxwell's life, until I have sketched the development of his ideas on electricity.

To gain an appreciation of Maxwell's stupendous contribution to this branch of science it is necessary first to describe very briefly the position of electrical theory when he embarked on his studies.

In the eighteenth century, Charles Augustin de Coulomb established the fundamental facts of electrostatic attraction and repulsion. He showed that an inverse-square law, resembling that of gravitational forces, applied to electric charges: attraction or repulsion between charged bodies is directly proportional to the product of the charges and inversely proportional to the square of the distance between them.* (The same discoveries, and others going beyond them, were made earlier by the brilliant English recluse Henry Cavendish, but his researches remained unpublished until 1879.) The next major advance was that of Hans Oersted, who in 1819 found that the flow of electric current through a wire parallel to a magnetic needle makes the needle swing to a position at right

* $F = k\frac{qq'}{r^2}$, where F equals the force; k, a constant; q and q', the charges; r, the separating distance.

angles to the current. In other words, a current produces a magnetic field.

A complementary series of advances was made early in the same century by the French physicist and mathematician André Ampère, whom Maxwell called the Newton of electricity. The accolade was not undeserved, but there is a special reason for Maxwell's conferring it: Ampère was the first to explain the relationship of electric currents in terms of mechanical action,* an approach later perfected by Maxwell himself. By experiment Ampère learned that a coil of wire carrying an electric current behaves like a magnet, and he suggested that a magnet owes its property to tiny electrical currents inside the steel molecules. Thus a great conceptual link was forged; for magnetism was shown to be not distinct from electricity, but rather a name we give to some of the effects of moving electric currents.

The crown of these fundamental researches was the immortal work of Michael Faraday. He found that an electric current flowing in one circuit can cause ("induce") a current to flow in another circuit; that there is a magnetic field between two currents; that a current can also be induced to flow in a wire by use of a magnet — in other words, as a symmetric counterpart to the phenomena discovered by Oersted and Ampère, that changes in a magnetic field produce an electric current.

Faraday's explanation of these phenomena is of central importance to understanding Maxwell's work. He imagined *lines of force* running through space as the instrumentality of electric and magnetic actions.

These lines, it should be emphasized, were not conceived as mere mathematical makeshifts, but as entities possessing physical properties. The lines spread out in every direction from an electric charge or magnetic pole; every electric line of force

* He showed how to calculate the mechanical forces between circuits carrying currents, from an assumed law of force between each pair of elements of the circuit.

starts from a positive charge and ends on a negative charge; the more powerful the source, the more lines emanate from it. Along the lines there is tension, a kind of pull, so that each line behaves like an elastic thread trying to shorten itself; lines of force repel each other sideways; the ends of a line of force, representing charges, can move freely over the surface of a conductor but are anchored on an insulator.

This hypothetical system, for which Faraday was convinced he had found experimental evidence, was the starting point of Maxwell's studies. He believed in it; he sought to develop it.

However, it must not be supposed that everyone accepted Faraday's hypothesis. In fact, the majority of electricians — I use the term in its older sense — regarded lines of force as a concept much inferior to that of "action at a distance." They likened electricity to gravitation. They imagined a charge (or mass) situated at one point in space mysteriously influencing a charge (or mass) at another point, with no linkage or connection of any kind, however tenuous, bridging the distance between the charges (or masses). Where Faraday sought to assimilate the behavior of electricity to that of a mechanical system, in which all parts are joined by levers, pulleys, ropes and so on, the others held electricity to be a special case, to which mechanical analogies were inapplicable. Maxwell admirably summarized the cleavage between the two views: "Faraday, in his mind's eye, saw lines of force traversing all space, where the mathematicians saw centres of force attracting at a distance; Faraday saw a medium where they saw nothing but distance; Faraday sought the seat of the phenomena in real actions going on in the medium, they were satisfied that they had found it in a power of action at a distance impressed on the electric fluids."

Maxwell's first electrical paper "On Faraday's Lines of Force" was delivered at Cambridge in 1855, within a few months after he had finished reading Faraday's *Experimental Researches*. What he tried to do was imagine a physical model embodying Faraday's lines, whose behavior, like that of any machine, could be reduced to formulae and numbers. He did not suggest that the model represented the actual state of things; on the other hand, he had no confidence in what mathematical manipulations alone would reveal about the actual state of things. It was important, he said, so to balance the method of investigation that the mind at every step is permitted "to lay hold of a clear physical conception, without being committed to any theory founded on the physical science from which that conception is borrowed."* Such a method will neither lead

* The opening paragraph of the paper is worth giving in full. "The present state of electrical science seems peculiarly unfavorable to speculation. The laws of the distribution of electricity on the surface of conductors have been analytically deduced from experiment; some parts of the mathematical theory of magnetism are established, while in other parts the experimental data are wanting; the theory of the conduction of galvanism and that of the mutual attraction of conductors have been reduced to mathematical formulae, but have not fallen into relation with the other parts of the science. No electrical theory can now be put forth, unless it shows the connection not only between electricity at rest and current electricity, but between the attractions and inductive effects of electricity in both states. Such a theory must accurately satisfy those laws the mathematical form of which is known, and must afford the means of calculating the effects in the limiting cases where the known formulae are inapplicable. In order therefore to appreciate the requirements of the science, the student must make himself familiar with a considerable body of most intricate mathematics, the mere attention of which in the memory materially interferes with further progress. The first process therefore in the effectual study of the science, must be one of simplification and reduction of the results of previous investigation to a form in which the mind can grasp them. The results of this simplification may take the form of a purely mathematical formula or of a physical hypothesis. In the first case we entirely lose sight of the phenomena to be explained; and though we may trace out the consequences of given laws, we can never obtain more extended views of the connections of the subject. If, on the other hand, we adopt a physical hypothesis, we see the phenomena only through a medium, and are liable to that blindness to facts and rashness in assumption which a partial explanation encourages. We must therefore discover some method of investigation which allows the mind at every step to lay hold of a clear physical conception, without being committed to any theory founded on the physical science from which that conception is borrowed, so that it is neither drawn aside from the subject in pursuit of analytical subtleties, nor carried beyond the truth by a favorite hypothesis. In order to obtain physical ideas without adopting a physical theory we must make ourselves familiar with the existence of physical analogies. By a physical analogy I mean that partial similarity between the laws of one science and those of another which makes each of them illustrate the other. Thus all the mathematical sciences are founded on relations between physical laws and laws of numbers, so that the aim of exact science is to reduce the problems of nature to the determination of quantities by operations with numbers."

into a blind alley of abstractions, nor permit the investigator to be "carried beyond the truth by a favorite hypothesis."

Analogies are, of course, the lifeblood of scientific speculation. Maxwell gives a number of examples, among them the illuminating suggestion of William Thomson comparing the formulae of the motion of heat with those of attractions (such as gravitation and electricity) varying inversely as the square of the distance. To be sure, the quantities entering into heat formulae — temperature, flow of heat, conductivity — are distinct from a quantity such as force entering into the formulae of inverse-square attraction. Yet the mathematical laws of both classes of phenomena are identical in form. "We have only to substitute *source of heat* for *center of attraction, flow of heat* for *accelerating effect of attraction* at any point, and *temperature* for *potential*, and the solution of a problem in attractions is transformed into that of a problem of heat."*

Maxwell proposed a hydrodynamical analogy to bring before the mind in "convenient and manageable form those mathematical ideas which are necessary to the study of the phenomenon of electricity."† The analogy was combined with Faraday's lines of force, so that they were converted from lines into "tubes of flow" carrying an incompressible fluid such as water. He was then able to show that the steady flow of particles of this fluid would give rise to tensions and pressures corresponding to electrical effects. The fluid moving through a system of such tubes represented electricity in motion; the form and diameter of the tubes gave information as to strength and direction of fluid (electric) flow. The velocity of the fluid was the equivalent of electrical force; differences of fluid pressure were analogous to differences of electrical pressure or potential. Since the tubes were flexible and elastic, and ar-

* "On Faraday's Lines of Force," *Transactions of the Cambridge Philosophical Society*, vol. X, part I, included in *The Scientific Papers of James Clerk Maxwell, op. cit.*

† *Ibid.*

ranged so as to form surfaces — each tube being in contact with its neighbors — pressure transmitted from tube to tube furnished an analogy to electrical induction.

One of Faraday's key concepts deals with the effect on space of lines of magnetic force. A wire introduced into ordinary space remains inert; but if magnetic lines of force are introduced into the space, a current flows through the wire. Faraday explained this by saying that the introduction of the magnet threw the space into an "electro-tonic state." This concept could not be fitted into the hydrodynamical analogy; Maxwell admitted that while he could handle Faraday's conjecture mathematically, the electro-tonic state at any point of space being defined "as a quantity determinate in magnitude and direction," his representation involved no physical theory — "it is only a kind of artificial notation."*

It was a wonderful paper, and Faraday, to whom Maxwell sent a copy, appreciated how much it advanced the "interests of philosophical truth." "I was at first almost frightened," he wrote Maxwell, "when I saw such mathematical force made to bear upon the subject, and then wondered to see that the subject stood it so well."† Other students, however, thought the subject stood it not at all well. Electricity was mysterious enough without adding tubes and surfaces and incompressible fluids. But Maxwell, who had good training in being considered queer, went on with the task of extending Faraday's ideas.

The second great memoir, *On Physical Lines of Force*, a series of three papers published in the *Philosophical Magazine* in 1861 and 1862, was an attempt to describe a more elaborate mechanism that would not only account for electrostatic effects but also explain magnetic attraction and Faraday's concept of

* For a discussion of Maxwell's use of physical analogy, see Joseph Turner, "Maxwell on the Method of Physical Analogy," *The British Journal for the Philosophy of Science*, vol. VI, no. 23, November, 1955.

† Campbell and Garnett, *op. cit.*, p. 519.

electromagnetic induction. Again, Maxwell used a concrete, mechanical image to exhibit and develop his theory.* For, as he said, "scientific truth should be regarded as equally scientific whether it appears in the robust form and vivid colouring of a physical illustration or in the tenuity and paleness of a symbolic expression."

In the new model a magnetic field is produced by the rotation in space of what Maxwell called "molecular vortices." These may be thought of as slender cylinders (Maxwell himself had a disconcerting way of modifying the image as he went along) that rotate round the lines of magnetic force. The lines, traced by the pattern of iron filings about a magnet, are the axes of rotation of the cylinders; the velocity of rotation depends on the intensity of the magnetic force. Two mechanical effects are associated with the cylinders: *tension* in the direction of the lines of force, and *pressure*, exerted in the "equatorial" direction (i.e., lateral pressure), arising from the centrifugal force produced by the rotating cylinders. Combined, these effects mechanically reproduce magnetic phenomena: magnetism is a force exerted both along the axis and outward from the axis.

It may now be asked how this curious arrangement fitted in with the known facts that an electric current produces a magnetic field, and changing magnetic forces produce an electric current. Step by step Maxwell answers this question.

The first point to clarify is the structure of a uniform magnetic field. Maxwell supposed this to consist of a portion of space filled with cylinders rotating at the same velocity and in the same direction "about axes nearly parallel." But immediately a puzzle confronted him. Since the cylinders are in contact, how can they possibly rotate in the same direction? For

* As Turner (*op. cit.*) points out, Maxwell employed two analogies. One bridged a stationary field and a solid under stress. The other is between electricity and fluid motion, "with its suggestion that Ampère's laws be modified to satisfy the equation of continuity."

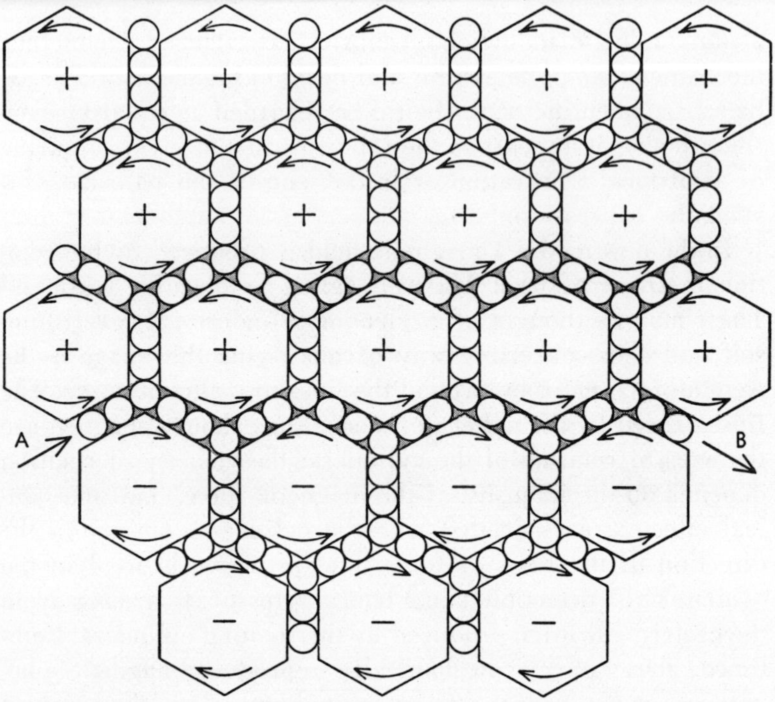

Model of an electromagnetic field used by Maxwell visualized "Molecular vortices" rotating in space. In this illustration the vortices are slender cylinders seen from the end. (Maxwell gave the cylinders a hexagonal cross section to simplify the geometry of the model.) Between the vortices are small "idle wheels." If a row of the idle wheels is moved from A *toward* B, *they cause the adjacent vortices to rotate in the opposite direction.* (Scientific American)

as everyone knows, a revolving wheel or cylinder causes its neighbor to revolve in the opposite direction; thus one would expect the rotation of the cylinders to alternate in direction from one to the next. Maxwell hit upon a pretty idea. He supposed the cylinders to be separated by rows of small spheres, like layers of ball bearings, which acted as gears (in Maxwell's words, "idle wheels"). This arrangement, resembling a device envisaged a century earlier by John Bernoulli, the younger, fulfilled the requirement. The spheres rotate in an opposite sense to that of each of the cylinders with which they are in contact, and so the cylinders all rotate in the same direction.

And now, as just reward for his ingenuity, Maxwell found that the spheres could be made to serve another, even more valuable, purpose. Think of them as particles of electricity. Then by purely mechanical reasoning it can be shown that their motions in the machine of which they are a part serve to explain many electrical phenomena.

Consider these examples. In an unchanging magnetic field the cylinders all rotate at the same constant rate; thus they maintain a constant magnetic field. The little rotating spheres keep their position; there is no flow of particles, hence no electric current, a result that tallies with the properties of a uniform magnetic field. Now suppose a change in the magnetic force. This means a change in the velocity of rotation of the cylinders. As each cylinder is speeded up, it transmits the change in velocity to its neighbors. But since a cylinder now rotates at a slightly different speed from its neighbor, the spheres between them are torn from their positions by a kind of shearing action. In other words, they begin to move from their centers of rotation, in addition to rotating. This motion of translation is an electric current; again, a result that tallies with the properties of a changing magnetic field.

Observe now how the model begins to live a life of its own. It was designed, as J. J. Thomson has pointed out,* to exhibit Faraday's great discovery that magnetic changes produce electric currents. It suggested to Maxwell the no less striking converse phenomenon that changes in electric force might produce magnetism.† Assume the spheres and cylinders are at rest. If

* Sir J. J. Thompson, "James Clerk Maxwell," in *James Clerk Maxwell, A Commemoration Volume, op. cit.*

† Ampère, of course, had already demonstrated that currents *in wires* produced accompanying magnetic fields.

a force is applied to the spheres of electricity, they begin to rotate, causing the cylinders of magnetism with which they are in contact to rotate in the opposite direction. The rotation of the cylinders indicates a magnetic force. Moreover, the model holds up even as to details. Take a single illustration. Magnetism acts at right angles to the direction of flow of current. If you will examine the diagram of Maxwell's model, you will see that the cylinders will rotate in the direction perpendicular to the motion of the spheres, thus bearing out the observation that a magnetic force acts at right angles to the flow of a current.

"I do not bring it forward," Maxwell wrote of his system, "as a mode of connection existing in Nature. . . . It is, however, a mode of connection which is mechanically conceivable and easily investigated, and it serves to bring out the actual mechanical connection between the known electromagnetic phenomena.* Certain aspects of these "mechanical connections" have already been discussed — rotations, pressures, tensions — which account for the reciprocal relations between currents and magnetic forces.† The connections also serve to explain the repulsion between two parallel wires carrying currents in opposite directions, an effect produced by the centrifugal pressures of the revolving cylinders acting on the electrical particles between them. The induction of currents is similarly elucidated: this phenomenon, says Maxwell, is simply "part of the process of communicating the rotary velocity of the vortices [cylinders] from one part of the field to another." In other words, whenever electricity (Maxwell's particles) "yields to an electromotive force," induced currents result. His diagram and the accompanying text make this beautifully clear.

Maxwell was not done with his model. It had helped in the

* "On Physical Lines of Force," *op. cit.*

† The model explained, for example, why a current of electricity generated heat: for as the particles (or spheres) move from one cylinder to another, "they experience resistance, and generate irregular motions, which constitute heat."

interpretation of magnetism, the behavior of electric currents, the phenomenon of induction; it had yet to pass the supreme test: that is, to supply a mechanical explanation of the origin of electromagnetic waves. To orient ourselves in this matter we must examine briefly the question of condensers and insulators.

An electric condenser is a device for storing electricity. In its simplest form it consists of two conducting plates separated by an insulating material, or dielectric as it is called. The plates can be charged, after which the charges attract each other through the dielectric and are thus said to be "bound" in place. Faraday in his experiments had come upon a curious fact. He found that two condensers of the same size, fed by the same electric source and with insulation of equal thickness, differed markedly in their capacity to take or to hold a charge if the insulating material (dielectric) was different. This was difficult to understand if all dielectrics were equally impermeable to an electric current. Moreover, if it were true, as Maxwell already was beginning to suspect, that light itself is an electrical phenomenon, how could light pass through certain dielectrics, empty space among them? With the help of his model, Maxwell advanced a bold hypothesis. Conductors, he said, pass a current when the electrical particles they contain are acted upon by an electric force. Under such an impulsion, the little particles move more or less freely from cylinder to cylinder, and the current flows as long as the force persists. Not so in a dielectric. The particles are present but an easy passage from cylinder to cylinder is impossible. This fact may be taken as the characteristic attribute of a dielectric, having to do with its physical structure. Yet it was known that "localized electric phenomena do occur in dielectrics." Maxwell suggested that these phenomena also are currents — but of a special kind. When an electric force acts on a dielectric, the particles of electricity are "displaced," but not entirely torn loose; that is, they behave like a ship riding at anchor in a storm. The medium in which they are located, the magnetic cylinders, is elastic; under pressure the particles move, a limited distance, until the force pushing them is balanced by the stresses due to the elastic reaction of the medium. Thus a state of equilibrium is attained. But as soon as the impelling force ceases to act, the particles snap back to their original positions. The net effect of these mechanical actions is twofold. First, the initial displacement of the electric particles constitutes a current that passes through the dielectric. A current of this type is called a *displacement current* to distinguish it from currents that flow through conductors and are therefore known as conduction currents.* Wherever there is an electric force, said Maxwell, there is displacement; wherever there is displacement, there is a current.

Second, whenever the pressure displacing the particles is released, and they snap back, they overshoot and oscillate briefly about their fixed positions. The oscillation is transmitted through the magnetic medium (the insulator) as a wave. This wave is the return phase of the displacement current.† (Maxwell suggested this disturbance on analogy to the displacement of an elastic solid under stress.)

Maxwell next arrived at an epoch-making conclusion. The velocity of the displacement wave, or current, depends on the electrical properties of the medium in which it moves. Moreover, this velocity, as he showed, was "within the limits of experimental error, the same as that of light." Hence, he in-

* The contrast between displacement currents and currents through conductors was vividly expressed by Henri Poincaré. A displacement current, he said, is an elastic reaction like the compression of a spring: it can only be effected by pressure against resistance. Equilibrium is reached when resistance balances pressure. When the pressure is removed the spring regains its original form. A conduction current, on the other hand, is like a viscous reaction such as is encountered in moving a body immersed in water. It can be effected only by pressure; the resistance depends on velocity; the motion continues as long as the motive force acts, and equilibrium will never be established. "The body does not return to the starting point, and the energy expended in moving it cannot be restored, having been completely transformed into heat through the viscosity of the water." (*Maxwell's Theory and Wireless Telegraphy*, New York, 1904.)

† If the electric force applied to the insulator is varied continually, it will produce a continually varying displacement wave: in other words, a continuing current.

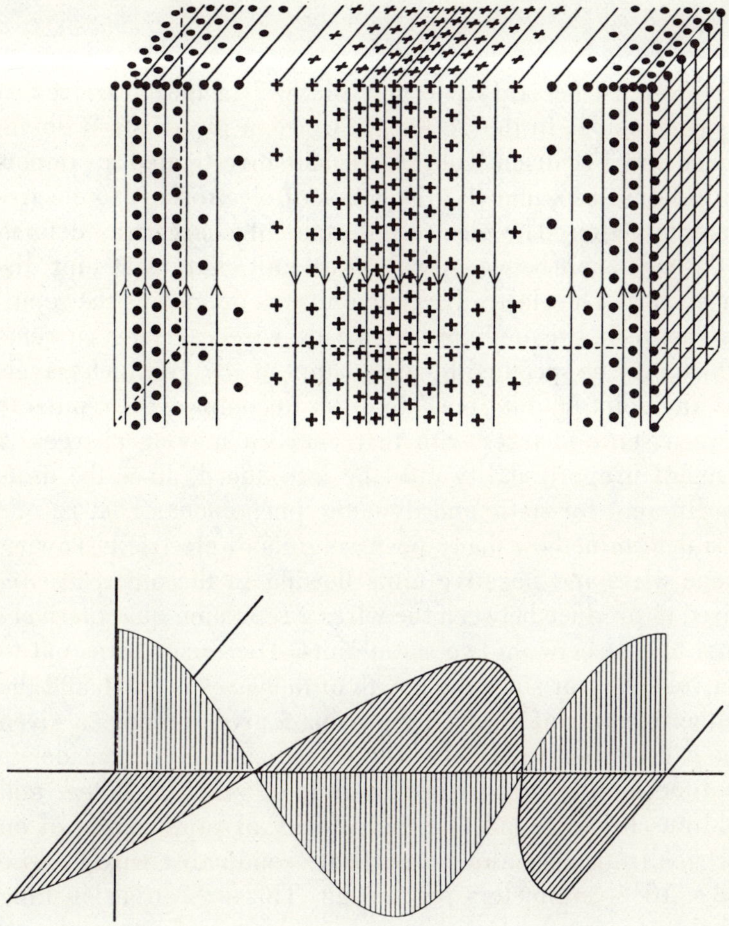

Electromagnetic wave as visualized by Maxwell is a moving disturbance which tends to separate positive (plus sign) and negative (dot) charges. In the drawing at the top, magnetic lines of force (arrows) lie at right angles to the direction in which the disturbance is moving. The drawing at the bottom depicts the two components of the electromagnetic wave. The electrical component is shown in black, the magnetic component in color. **(Scientific American)**

ferred, "the elasticity of the magnetic medium in air is the same as that of the luminiferous medium, if these two coexistent, coextensive and equally elastic media are not rather one medium."

More must be said as to how Maxwell actually arrived at this conclusion. In the 1850s the German physicists Wilhelm Weber and Friedrich Kohlrausch had investigated an important relationship, namely, the ratios of electrostatic to electrodynamic action. The electrostatic unit of charge was defined as the repulsion between two (like) unit charges at unit distance apart. The electrodynamic unit was defined as the repulsion between two definite lengths of wire carrying currents "which may be specified by the amount of charge which travels past any point in unit time." In order to compare the repulsion between static charges with that between moving charges, a factor of proportionality must be introduced, since the units are different for static and dynamic phenomena. That is, one must determine how many positive units of electricity flowing in one wire, and negative units flowing in the other, are required to produce between the wires a repulsion quantitatively equal to that between two static units. The factor turns out to be a velocity; for since the length of the wires is fixed, and the number of units of electricity passing a given point in a given time can be measured, what the investigator must consider is the dimension length divided by time or velocity. Weber and Kohlrausch had found that the velocity of propagation of an electric disturbance along a perfectly conducting wire is close to 3×10^{10} centimeters per second. This was an astonishing coincidence, for the figure was about the same as the velocity of light as it had been determined a few years earlier by the French physicist Hippolyte Fizeau.

Kirchhoff remarked the coincidence, but did not pursue it; Maxwell did. In 1860 he attacked the problem experimentally, using an ingenious torsion balance to compare the repulsion between two static charges and two wires carrying currents. The Weber-Kohlrausch results were roughly confirmed. Also, at about the same time (he said, in fact, that the pencil and paper work was done before seeing Weber's memoir), he calculated the velocity of displacement currents in empty space or in any other dielectric. The resulting values tallied closely.

In other words, currents in a wire, displacement currents in a dielectric, and light in empty space (which of course is a dielectric) all traveled with the same velocity. With this evidence at hand, which he communicated in a letter to Faraday in 1861, Maxwell did not hesitate to assert the identity of the two phenomena — electrical disturbances and light. "We can scarcely avoid the inference," he said, "that light consists in the transverse undulations of the same medium which is the cause of electric and magnetic phenomena."

"On Physical Lines of Force," despite its cogwheels and other gross mechanical adjuncts, may be regarded as the most brilliant of Maxwell's electrical papers. If it did not claim to give a picture of the true state of things, it was at least enormously enlightening as to how electricity and magnetism could interact in a purely mechanical relationship. Maxwell himself summarized the achievements of the theory as follows. It explained magnetic forces as the effect of the centrifugal force of the cylinders; induction as the effect of the forces called into play when there is a change of angular velocity of the cylinders; electromotive force as an effect produced by stress on the connecting mechanism; electric displacement as a result of the elastic yielding of the mechanism; electromagnetic waves as an accompaniment of displacement. The paper is one of the rare examples of scientific literature in which one may glimpse the play of imagination, the actual exercise of inductive power, the cultivation of nascent ideas.

None of the basic concepts unfolded in this memoir was discarded in the more mathematical writings that followed. But Maxwell now had to outgrow his model. In "A Dynamical Theory of the Electromagnetic Field," published in 1864,* Maxwell, in Sir Edmund Whittaker's words, displayed the architecture of his system "stripped of the scaffolding by aid of which it had first been erected."† The particles and cylinders

* *Royal Society Transactions*, vol. CLV.

† *History of the Theories of Aether and Electricity: The Classical Theories*, London, 1951.

are gone; in their place is the field — "the space in the neighborhood of the electric or magnetic bodies" — and the aether, a special kind of "matter in motion by which the observed electromagnetic phenomena are produced." The matter composing the aether has marvelous properties. It is very fine and capable of permeating bodies; it fills space, is elastic and is the vehicle of "the undulations of light and heat." Yet for all its refinements and subtleties, the medium is no less a mechanical rig than the cylinders and spheres of its predecessor. It can move, transmit motions, undergo elastic deformations, store potential (mechanical) energy and release it when the deforming pressures are removed. Though susceptible to the action of electric currents and magnets, it is nonetheless a mechanism that, as Maxwell said, "must be subject to the general laws of Dynamics, and we ought to be able to work out all the consequences of its motion, provided we know the form of the relation between the motions of the parts." In the preceding paper Maxwell already had devised a set of equations that described the possible mechanical basis of electrical and magnetic phenomena, and, in particular, how certain changes in electric and magnetic forces could produce electrical waves. He now elaborated the hypothesis of displacement currents and obtained the expressions that are in substance the famous Maxwellian equations of the electromagnetic field.

In their most finished form the equations appear in the *Treatise on Electricity and Magnetism* (1873), the culmination of Maxwell's researches, which he wrote at Glenlair in the years following his resignation from King's College. This celebrated work deals with every branch of electric and magnetic science and presents the results of twenty years of thought and experiment. Maxwell remained faithful to Faraday, whose point of view is emphasized throughout the *Treatise*. Characterizing his own part as that of an "advocate," Maxwell makes no attempt to describe the hypotheses propounded by Weber, Gauss, Riemann, Carl and Franz Neumann, or Ludwig Lorenz, the foremost cultivators of the theory of action at a distance.

The task Maxwell set himself was, first, to formulate mathematically electromagnetic phenomena as observed experimentally, and, second, to show that these mathematical relationships could be deduced from the fundamental science of dynamics; or to put it another way, that the laws of electricity in motion could be derived from the laws applicable to any system of moving bodies. As always, Maxwell was very cautious in expressing himself about the nature of electricity. It behaves, he said, like an incompressible fluid; "wherever there is electric force there is electric displacement." These, as J. J. Thomson observed, are the only definite statements about electricity to be found in the treatise, which led Hertz to say that Maxwell's theory is Maxwell's equations, and caused Helmholtz to comment that "he would be puzzled to explain what an electric charge was on Maxwell's theory beyond being the recipient of a symbol."

What are the Maxwellian equations? I cannot hope to give an easy answer to this question, but at the cost of deliberate oversimplification I must try summarily to explain them, for they are the heart of the theory.

Maxwell based the equations on four principles: (1) that an electric force acting on a conductor produces a current proportional to the force; (2) that an electric force acting on a dielectric produces displacement proportional to the force; (3) that a current produces a magnetic force (i.e., a moving electric charge is surrounded by a magnetic field) at right angles to the current's lines of flow and proportional to its intensity; (4) that a changing magnetic force (or field) produces a current proportional to the intensity of the force. The third and fourth principles exhibit a striking symmetry. The third is Faraday's law of electromagnetic induction, according to which "the rate of alteration in the number of lines of magnetic induction passing through a circuit is equal to the work done in taking unit electric charge round the circuit." Max-

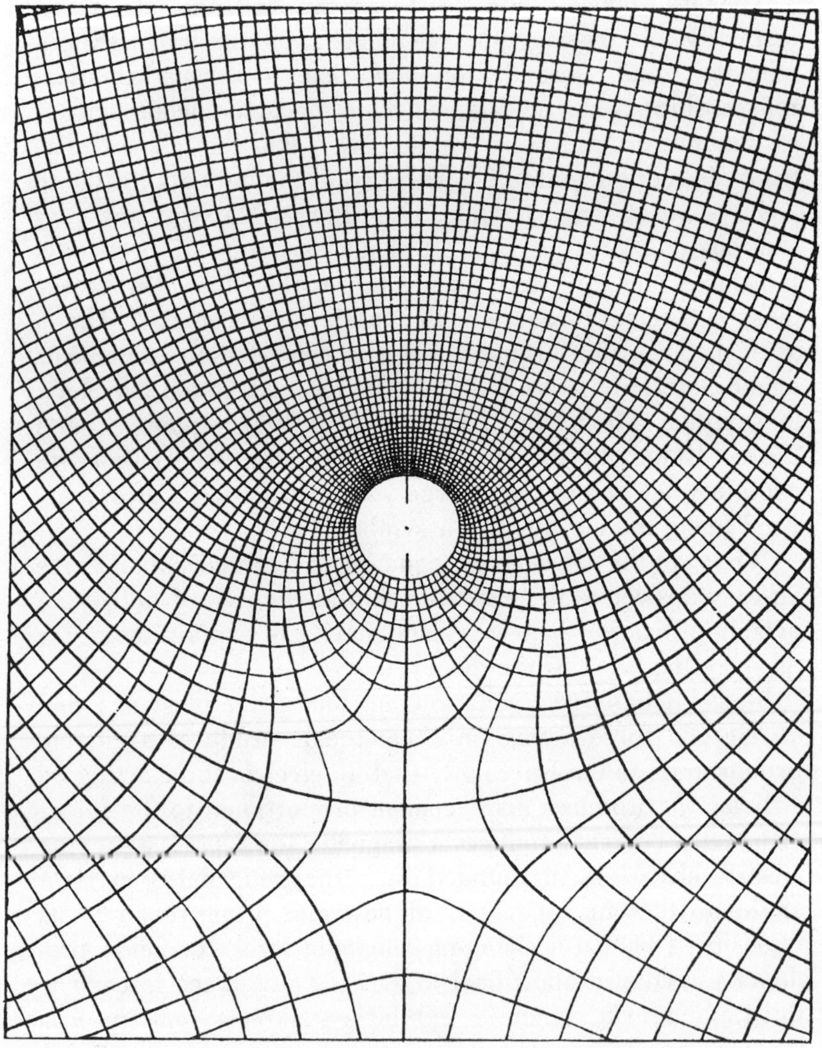

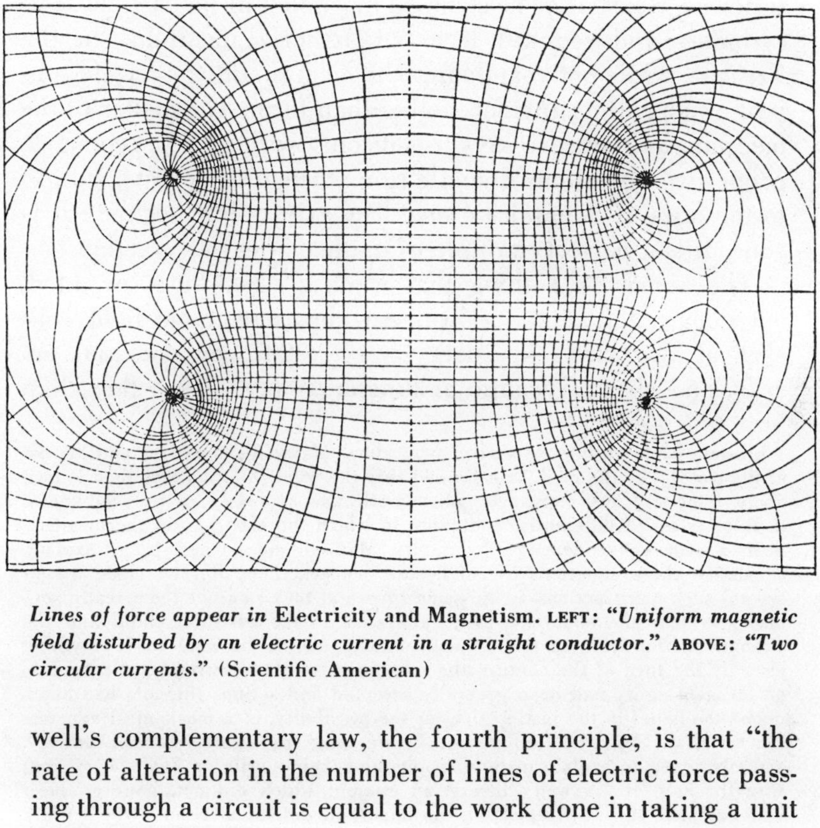

Lines of force appear in Electricity and Magnetism. LEFT: *"Uniform magnetic field disturbed by an electric current in a straight conductor."* ABOVE: *"Two circular currents."* (Scientific American)

well's complementary law, the fourth principle, is that "the rate of alteration in the number of lines of electric force passing through a circuit is equal to the work done in taking a unit magnetic pole round it."

On this foundation two sets of symmetrical equations can be erected. One set expresses the continuous nature of electric and magnetic fields; the second set tells how changes in one field produce changes in the other. In these formulations the mechanical aspects of the theory are retained, perfect continuity is preserved by treating electricity as if it were an incompressible fluid, and wave phenomena are deduced as the consequences of displacement in a dielectric.

How does the concept of the field enter the theory? We have

followed Maxwell as he stripped his model of its particles and cylinders and reduced it to an aetherial medium. In the *Treatise*, while not abandoning the medium altogether, he robs it of almost all its attributes other than form. The matter of the medium, as Poincare says, is left only with purely geometric properties, the atoms dwindle to mathematical points, subject to the laws of dynamics alone. The grin is left but the cat is gone. It is a perfect example of mathematical abstraction.*

The aether is a thing that wiggles when it is prodded, but does nothing on its own. An electromagnetic field consists of two kinds of energy, electrostatic or potential energy, and electrodynamic or kinetic energy. The aether, like

* Einstein made an interesting comment about Maxwell's equations and his use of the concept of the field. "He showed that the whole of what was then known about light and electromagnetic phenomena was expressed in his well-known double system of differential equations, in which the electric and the magnetic fields appear as the dependent variables. Maxwell did, indeed, try to explain, or justify, these equations by intellectual constructions. But he made use of several such constructions at the same time and took none of them really seriously, so that the equations alone appeared as the essential thing and the strength of the fields as the ultimate entities, not to be reduced to anything else. By the turn of the century the conception of the electromagnetic field as an ultimate entity had been generally accepted and serious thinkers had abandoned the belief in the justification, or the possibility, of a mechanical explanation of Clerk Maxwell's equations. Before long they were, on the contrary, actually trying to explain material points and their inertia on field theory lines with the help of Maxwell's theory, an attempt which did not, however, meet with complete success. Neglecting the important *individual* results which Clerk Maxwell's life work produced in important departments of physics, and concentrating on the changes wrought by him in our conception of the nature of physical reality, we may say this: — before Clerk Maxwell people conceived of physical reality — insofar as it is supposed to represent events in nature — as material points, whose changes consist exclusively of motions, which are subject to partial differential equations. After Maxwell they conceived physical reality as represented by continuous fields, not mechanically explicable, which are subject to partial differential equations. This change in the conception of reality is the most profound and fruitful one that has come to physics since Newton; but it has at the same time to be admitted that the program has by no means been completely carried out yet."

I am puzzled as to what Einstein meant in saying that Maxwell's equation eliminated the notion of mechanism in explaining electromagnetic phenomena. Similar views have been expressed by many other physicists and philosophers. Maxwell himself would not have agreed with this position. His writings refute it. The inference was drawn by his successors. But there is a more important a universal condenser, may be conceived as storing energy — in which case, being elastic, it is deformed. Since the aether fills all space and therefore penetrates conductors as well as dielectrics, it no longer makes any difference whether we deal with a conduction current or a displacement current; in either case the aether is set in motion. This motion is communicated mechanically from one part of the medium to the next and is apprehended by us as heat, or light, or mechanical force (as in the repulsion between wires) or other phenomena of magnetism and electricity. The ruling principle of all such phenomena, it should be observed, is that of least action. This is the grand overriding law of the parsimony of nature: every

---

point that requires clarification; namely, do the equations justify the inference? It is true that a field is not the same as a material particle, and that the motion of a particle is not the same as a change in a field. It is true also that the concept "material particle" was long held to be intuitively clear, while the concept "field" has never been so regarded. This makes it easier to say mysterious things about fields, which no one would dream of saying about particles. But a more careful definition of these concepts, as physicists actually use them, raises serious question as to whether a field is any less suited to a "mechanistic" explanation than a system of material particles; indeed, whether a mechanistic explanation fits either or neither case. In modern physics material particles are not what they once were. They are pale abstractions, quite incapable of anything so robust as a collision. But then what is a collision? One thinks of billiard balls knocking together, as a pristine example. This, however, is a plain man's way of thinking. The modern physicist has rid his mind of such seductive images. (As far back as the eighteenth century, the Italian physicist Boscovich proposed the idea that the heart of an atom is not solid substance but a mere center of immaterial force.) As particles fade, the field becomes more substantial. Properties are now ascribed to it that make it seem more real and more potent than a billiard ball or a boulder. Of course the field is hard to describe in homely terms. Yet it is quite capable, as physicists tell us, of doing homely things. It produces and undergoes changes — now as if it were a cloud, now an engine, now an ocean. In short it has mechanical effects. By this I mean effects of a kind produced by what used to be called material particles. Moreover, it has mechanical properties. By this I mean properties of a kind produced by what we call a machine. The field can do things no system of particles or machine yet conceived can do. Since it can also do all they can do, it is a supermachine. Is there any point in saving the name? I think there is, to keep our thinking straight. We ought to keep it to describe both fields and particles or we ought to discard it entirely. If the word "mechanism" has any meaning in the universe of refined observation, it has as much meaning in relation to fields as to particles. At the same time I am quite prepared to believe that it has as little meaning in one case as the other; for that matter, no meaning in either.

action within a system is executed with the least possible expenditure of energy. It was of the first importance to Maxwell that electrical phenomena should satisfy the principle, for otherwise his mechanical explanation of the phenomena would not have been possible.

With these points in mind, we may examine a set of Maxwell's equations in a form that describes the behavior of an electromagnetic field under the most general conditions, i.e., a field moving in empty space. No conductors are present, no free charges, and the medium is a vacuum. The equations then read

$$1)\ div\, E = 0$$

$$2)\ div\, H = 0$$

$$3)\ curl\, E = -\frac{1}{c}\frac{\partial H}{\partial t}$$

$$4)\ curl\, H = \frac{1}{c}\frac{\partial E}{\partial t}$$

The meaning of the symbols is as follows: *E* and *H* represent electric and magnetic field strength; since they vary in time, and from place to place, they are functions of the space coordinates *x*, *y*, *z* (not shown) and of the time coordinate, *t*. *C* is the velocity of light and enters the equations as the rate of propagation; *div* (an abbreviation for divergence) and *curl* (an abbreviation for rotation) represent mathematical operations whose physical meaning is explained below.

Divergence is essentially a measure of rate of change. In words, then, equation 1

$$div\, E = 0$$

says that in a moving field the electric intensity is the same at every point, i.e., the rate of change is zero at every point. More loosely, this equation extends to the field the classical principle that electric lines of force can be neither created nor destroyed. Thus the equation says that the number of electric lines of force, representing the field strength, that enter any tiny volume of space must equal the number leaving it. Making use of still another analogy, if one conceives of electricity in Maxwell's idiom, as an incompressible fluid, equation 1 states that as much fluid flows out of a tiny volume of space in a given time as flows in.*

* For the reader interested in a little more detail, the following explanation may be helpful. Equation 1 states that the divergence of the electric field intensity is zero at any point in space and at any instant of time. The meaning of the equation may be visualized as follows. It is customary to represent E at a given instant of time by a series of lines whose relative density in space is proportional to E. These lines have direction because E is a vector. Consider a point P and a sphere surrounding P. Let us suppose that the intensity of the electric field on the left hemispherical surface of the sphere is uniform over the surface and is directed at each point perpendicular to the surface.

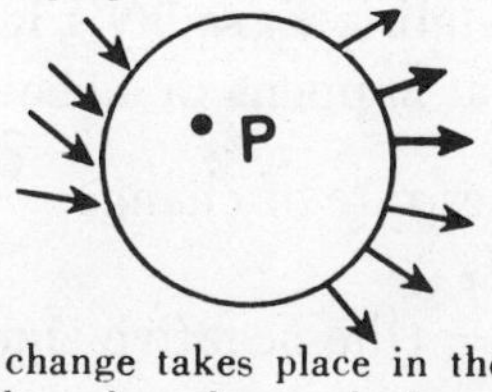

Suppose further that some change takes place in the electric field intensity E in the region occupied by the sphere but such that on the right hemispherical surface the field E is again uniform and perpendicular to the surface but stronger than on the left portion. We would indicate this increase in the intensity of E by having more lines leave the sphere on the right than enter on the left. Using the number of lines as a measure of E, we count the lines entering the spherical surface and multiply this number by the area of the hemisphere, and regard this product as negative. Let us next form the analogous product of the area and the number of lines leaving the surface, and regard this product as positive. The algebraic sum of these two products, that is, the positive plus the negative, is called the net electric flux through the spherical surface. This net flux is the divergence of E over the volume of the sphere. In our illustration the net flux of E has increased as E passes through the sphere. Hence we should say in this case that the divergence of E through the sphere is positive. If we now divide this net flux through the sphere by the volume of the sphere, we obtain the next net flux per unit volume. We now imagine that the sphere becomes smaller and smaller and contracts to the point P. Of course the net flux per unit volume changes and approaches some limiting value. This limiting value, which is a mathematical abstraction, is *div E* at the point P. Thus *div E* is essentially a measure of the spatial rate of change of E at the point P. Since equation 1 says that for electric fields $div\ E = 0$ at each point P, we may say that the net spatial rate of change of E is zero in empty space. More loosely stated, this equation says that electric field lines are neither created nor destroyed at the point P. It is to be noted that the phrase "spatial rate of change" is intended to emphasize that the divergence is concerned with the way in which E changes from point to point in space at the same instant of time. This spatial rate must be distinguished from the rate at which some quantity, for example, E itself in equation 4, may change during some interval of time.

Equation 2

$$div\, H = 0$$

makes the same assertion for magnetic lines as equation 1 makes for electric lines.

Equation 3

$$curl\, E = -\frac{1}{c}\frac{\partial H}{\partial t}$$

is Maxwell's way of stating Faraday's law of induction. The equation describes what happens in a changing magnetic field. The right side expresses rate of change, $\frac{\partial H}{\partial t}$, multiplied by a very small factor, $-\frac{1}{c}$ (the negative sign before the fraction is purely a matter of algebraic convenience); the left side expresses the fact that an electric field is created by a changing magnetic field. But the equation is more than analytic; thanks to the sign *curl*, it actually gives a picture of the event. A simple diagram may help make this clear. Suppose the existence of a magnetic field uniform over a region of space. We draw a circle

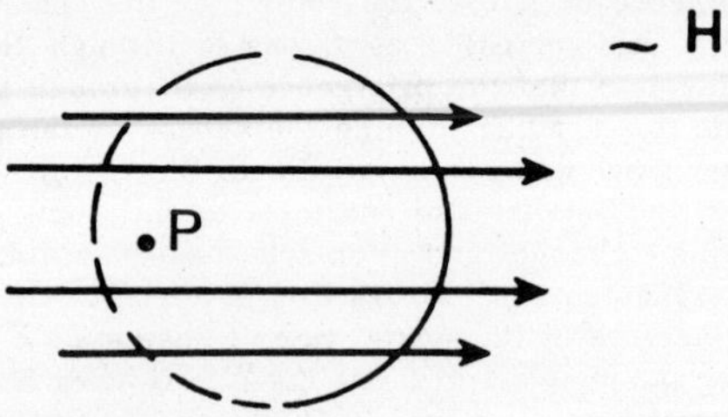

surrounding a bundle of parallel lines, which represent the intensity and direction of the magnetic field. The circle lies in a plane perpendicular to the lines. If the field is changed (by motion or by increase or reduction of strength), it produces an electric field that acts in a circle around the lines of magnetic force (though it may also act in other directions). By summing the work done in moving unit electric charge around the circle, we obtain what is called the net electromotive force around the circle.* If the circle were made of wire, the changing magnetic lines would of course induce the flow of a current; but even without a wire — and therefore no current — a force would be induced. Dividing this force by the area enclosed by the circle gives the net electromotive force (per unit area) which "curls" around the circle. Now imagine the circle growing smaller and smaller and shrinking finally to the point *P*. By this limiting process we obtain a limiting value of the net electromotive force per unit area: this is *curl E* at *P*. Thus equation 3 says that the limiting value of electromotive force per unit area equals the rate of change of *H* at the point *P*, multiplied by the tiny negative fraction, $-\frac{1}{c}$.† Or, again, more loosely stated, a changing magnetic field creates an electric field whose electromotive force per unit area at any given point and instant of time equals the time rate of change of the magnetic field at that point and instant.

Equation 4

$$\mathit{curl}\, H = \frac{1}{c}\,\frac{\partial E}{\partial t}$$

says that, except for the change in algebraic sign (which has to do with the directions of the fields), the roles of *E* and *H* in

* In physical terms, we obtain the net capacity of the electric field to move current along the circle.

† The symbol *c*, which here stands for the ratio of the electrostatic to the electromagnetic units of electricity, is required to translate *E* (an electrostatic phenomenon) and *H* (an electromagnetic phenomenon) into the same system of units. The equation explains how Maxwell was able to connect electrical and magnetic phenomena with the velocity of light, for *c* is in fact that velocity.

equation 3 may be reversed. At any given point and instant the magnetomotive force (the analogue for magnetic fields of electromotive force) per unit of area created by a changing electric field is equal to the time rate of change of the electric field multiplied by the tiny positive fraction $\frac{1}{c}$. Now, the reader who has followed this discussion will perceive that the time rate of change of E, $\frac{\partial E}{\partial t}$, is none other than Maxwell's displacement current. For since the changes are taking place in the dielectric known as empty space, the only currents that can flow are displacement currents.* Prior to Maxwell, it was thought that the magnetic field H could be produced only by currents that flowed in wires passing through the circle. If no wires were present, the law thought to be applicable was $curl\ H = 0$. It was Maxwell's great discovery, deduced mechanically from his model and expressed mathematically in this equation, that a time-varying electric field produces (or must be accompanied by) a net "curled" magnetic force even in an insulator or empty space.†

According to Maxwell's theory, the introduction of a time-varying electric force in a dielectric produces displacement waves with the velocity of light. To put it another way, it is the surge and ebbing of the force that produces the periodic displacement waves; a static charge would merely create an instantaneous displacement, which would be fixed, but not a

* Equation 4 assumes the existence of this current and relates it quantitatively to the magnetomotive force generated by the existent magnetic field. Physically we may regard the magnetic field as creating the displacement current or, conversely, regard the displacement current as creating the accompanying magnetic field and magnetomotive force.

† Maxwell called $\frac{\partial E}{\partial t}$ the displacement current, the term "displacement" meaning that the electric field intensity E was being altered or displaced as time varies, and the term "current" suggesting that $\frac{\partial E}{\partial t}$ had the properties of a current flowing in a wire even though $\frac{\partial E}{\partial t}$ existed in empty space.

wave. Now, an electric current, as we have seen, whether in a dielectric or in a conductor, is accompanied by a magnetic force; and similarly a periodic wave of electric displacement is accompanied by a periodic magnetic force. The wave front itself, as Maxwell showed, comprises electric vibrations at right angles to the direction of propagation and a magnetic force at right angles to the electric displacement. The compound disturbance is therefore called an electromagnetic wave. A light wave (which is a displacement wave) is, as Henri Poincaré later elaborated, "a series of alternating currents, flowing in a dielectric, in the air, or in interplanetary space, changing their direction 1,000,000,000,000,000 times a second. The enormous inductive effect of these rapid alternations produces other currents in the neighboring portions of the dielectric, and thus the light waves are propagated from place to place."

The electromagnetic theory of light was testable experimentally, and indeed stood up remarkably well in laboratory trials. But this was only a limited confirmation of Maxwell's system, for if his reasoning was correct, there must be other electrical waves produced by initial disturbances of differing intensity. These waves would differ from light in wave length and would therefore not be visible, yet it should be possible to detect them with appropriate instruments. How to find them, not to say generate them, was now the crucial problem. Maxwell did not live to see it solved. Not until ten years after his death were his prophecies fulfilled and the skepticism of his most distinguished contemporaries refuted. As late as 1888 Lord Kelvin referred to Maxwell's waves as a "curious and ingenious, but not wholly tenable hypothesis"; but a year later Helmholtz's greatest pupil, Heinrich Hertz, nosed out Oliver Lodge in the race to demonstrate their existence. In a series of brilliant experiments he showed how electric waves could be "excited" (i.e., generated) by oscillation and detected by a circular conductor provided with a small gap; and how they could be polarized, reflected, refracted, made to form shadows and to interfere with each other. The connection, he said, "between light and electricity . . . of which there were hints and suspicions and even predictions in the theory, is now established. . . . Optics is no longer restricted to minute aether waves, a small fraction of a millimetre in length; its domain is extended to waves that are measured in decimetres, metres and kilometres. And in spite of this extension, it appears merely . . . as a small appendage of the great domain of electricity. We see that this latter has become a mighty kingdom."

The *Treatise*, written while Maxwell was "in retirement" at Glenlair, drew only part of his energy. As a "by-work" during the same period he wrote a textbook on heat, which appeared in 1870, and a number of papers of considerable importance on mathematics, color vision and topics of physics. He maintained a heavy scientific and social correspondence, enlarged his house, studied theology, composed stanzas of execrable verse, rode his horse, went on long walks with his dogs, visited his neighbors and played with their children, and made frequent trips to Cambridge to serve as moderator and examiner in the mathematical tripos.

In 1871 a chair in experimental physics was founded at Cambridge. It is hard to realize that at the time no courses in heat, electricity and magnetism were being taught there, and no laboratory was available for the pursuit of these arcane matters. The University, as a contemporary scholar delicately observed, "had lost touch with the great scientific movements going on outside her walls." A committee of the faculty began to bestir itself, a report was issued, and the lamentable facts fell under the gaze of the Duke of Devonshire, Chancellor of the University. He offered the money for the building and furnishing of the famous Cavendish Laboratory. Thomson, it was known, would not leave his post at Glasgow to take the new chair, and Maxwell, though at first reluctant to leave Glenlair, yielded to the urging of his friends to offer himself as a candidate. He was promptly elected.

He now devoted himself to the task of designing and superintending the erection of the laboratory. His aim was to make it the best institution of its kind, with the latest apparatus and the most effective arrangements for research. He inspected Thomson's laboratory at Glasgow and Clifton's at Oxford to learn the desirable features of both and embody them in the Cavendish. He presented to the laboratory all the apparatus in his own possession and supplemented the Duke's gift by generous money contributions. With so many details to be taken care of, the structure and its appointments were not completed until 1874. The delay, while inevitable, was inconvenient. "I have no place," wrote Maxwell, "to erect my chair, but move about like the cuckoo, depositing my notions in the Chemical Lecture Room in the first term, in the Botannical in Lent and in the Comparative Anatomy in Easter." His "notions" were the courses he gave, beginning in 1871, on heat, electricity and electromagnetism, a schedule maintained throughout the tenure of his chair. And though the audiences were often small, some of the best students were soon attracted to his lectures, which contained much important original work. The renaissance that followed in physical science at Cambridge was the direct result of his influence.

Maxwell's classic *Matter and Motion*, "a small book on a great subject," was published in 1876. About this time he contributed articles on various subjects — "Atom," "Aether," "Attraction," "Faraday," among others — to the famous ninth edition of the *Encyclopaedia Britannica*. His public lectures include a charming discourse "On the Telephone," which, though delivered when he was already very ill, is not only as clear as his best expositions but filled with gay, amusing asides. Speaking, for example, of "Professor Bell's invention," he comments on "the perfect symmetry of the whole apparatus — the wire in the middle, the two telephones at the ends of the wire, and the two gossips at the ends of the telephones. . . ." A task that occupied him for five years, almost to the very end of his life, was editing twenty packets of unpublished scientific papers of Henry Cavendish, who was great-uncle to the Duke of Devonshire. This splendid two-volume work, published in 1879, did much to fix the reputation of an immensely gifted investigator, whose important work on electricity was unknown to his contemporaries because the results were confided only to his manuscripts. Maxwell repeated Cavendish's experiments and showed that he had anticipated major discoveries in electricity, including electrostatic capacity, specific inductive capacity and Ohm's law.

As Maxwell grew older, friends remarked on his "ever-increasing soberness" of spirit. This must not be taken to mean he was invariably melancholy or withdrawn or that his nice sense of fun — about himself no less than about others — had vanished. He continued to see his many friends, to write light verse and parodies, to promenade with his dog Toby, who was at Maxwell's side even in the laboratory, to play small practical, but never mean, jokes, to engage in what was called "humorous mystification" by advancing preposterous scientific ideas in conversation while keeping a straight face. All things, he once remarked, are "full of jokes," though they are also "quite full of solemn matters," and he was as likely to stress their light as their grave aspect.

But it is true he became somewhat more reticent with the passing years, and more and more concealed his feelings and reflections beneath an ironical shell. The tough, rational, Scotch common-sense cord of his nature had always been intertwined with threads of mysticism. Often plain, even blunt, in his address, he also had an allusive way of speaking and showed a fondness for parables. He had faith in science, yet he was at bottom skeptical as to how much could be learned from science alone about nature and meaning. It was all very well, he felt, to have "ideal aspirations"; on the other hand, "It's no use thinking of the chap ye might have been." His contemporaries remember him as both modest and intellectually scornful, tentative in his scientific opinions and dogmatic when others seemed to him to be immoderately self-assured.

"No one knows what is meant by" so-and-so was his way of answering a cocksure formulation of a scientific "truth."

The most striking of Maxwell's traits was his gentleness. "His tenderness for all living things was deep and instinctive; from earliest childhood he could not hurt a fly." An extraordinary selflessness characterized his relationship to those close to him. When his brother-in-law came to London to undergo an operation, Maxwell gave up the ground floor of his house to patient and nurse and left himself with a room so small that he frequently breakfasted on his knees because there was no room for a chair at the table. Mrs. Maxwell had a serious and prolonged illness in the last years of Maxwell's life, and he insisted on nursing her. On one occasion it is reported that he did not sleep in a bed for three weeks. But his work went on as usual, and he was as cheerful as if he enjoyed the ordeal — which may indeed have been the case. Nor did he give the slightest sign of being downcast or show self-pity when his own fatal illness seized him.

In the spring of 1877 he began to be troubled with pain and a choking sensation on swallowing. For some strange reason he consulted no one about his symptoms for almost two years, though his condition grew steadily worse. His friends at Cambridge observed that he was failing, that the spring had gone out of his step. When he went home to Glenlair for the summer of 1879, he was so obviously weakening that he called for medical help. He was in terrible pain, "hardly able to lie still for a minute together, sleepless, and with no appetite for the food which he so required." He understood thoroughly that his case was hopeless, yet his main concern seemed to be about the health of his wife. In October he was told he had only a month to live. On November 5 he died. "No man," wrote his physician, Dr. Paget, "ever met death more consciously or more calmly." When Maxwell was buried in Parton Churchyard at Glenlair, the world had not yet caught up with his ideas. Even today it has not fully explored the kingdom created by his imagination.

Oersted established a connection between electric currents and magnetism; Faraday found the connection between magnetic fields and induced electric currents. But it was Maxwell who synthesized and extended these two results.

14 On the Induction of Electric Currents

James Clerk Maxwell

An excerpt from his *Treatise on Electricity and Magnetism* published in 1873.

528.] THE discovery by Örsted of the magnetic action of an electric current led by a direct process of reasoning to that of magnetization by electric currents, and of the mechanical action between electric currents. It was not, however, till 1831 that Faraday, who had been for some time endeavouring to produce electric currents by magnetic or electric action, discovered the conditions of magneto-electric induction. The method which Faraday employed in his researches consisted in a constant appeal to experiment as a means of testing the truth of his ideas, and a constant cultivation of ideas under the direct influence of experiment. In his published researches we find these ideas expressed in language which is all the better fitted for a nascent science, because it is somewhat alien from the style of physicists who have been accustomed to established mathematical forms of thought.

The experimental investigation by which Ampère established the laws of the mechanical action between electric currents is one of the most brilliant achievements in science.

The whole, theory and experiment, seems as if it had leaped, full grown and full armed, from the brain of the 'Newton of electricity.' It is perfect in form, and unassailable in accuracy, and it is summed up in a formula from which all the phenomena may be deduced, and which must always remain the cardinal formula of electro-dynamics.

The method of Ampère, however, though cast into an inductive form, does not allow us to trace the formation of the ideas which guided it. We can scarcely believe that Ampère really discovered the law of action by means of the experiments which he describes. We are led to suspect, what, indeed, he tells us himself*, that he

* *Théorie des Phenomènes Electrodynamiques*, p. 9.

discovered the law by some process which he has not shewn us, and that when he had afterwards built up a perfect demonstration he removed all traces of the scaffolding by which he had raised it.

Faraday, on the other hand, shews us his unsuccessful as well as his successful experiments, and his crude ideas as well as his developed ones, and the reader, however inferior to him in inductive power, feels sympathy even more than admiration, and is tempted to believe that, if he had the opportunity, he too would be a discoverer. Every student therefore should read Ampère's research as a splendid example of scientific style in the statement of a discovery, but he should also study Faraday for the cultivation of a scientific spirit, by means of the action and reaction which will take place between the newly discovered facts as introduced to him by Faraday and the nascent ideas in his own mind.

It was perhaps for the advantage of science that Faraday, though thoroughly conscious of the fundamental forms of space, time, and force, was not a professed mathematician. He was not tempted to enter into the many interesting researches in pure mathematics which his discoveries would have suggested if they had been exhibited in a mathematical form, and he did not feel called upon either to force his results into a shape acceptable to the mathematical taste of the time, or to express them in a form which mathematicians might attack. He was thus left at leisure to do his proper work, to coordinate his ideas with his facts, and to express them in natural, untechnical language.

It is mainly with the hope of making these ideas the basis of a mathematical method that I have undertaken this treatise.

529.] We are accustomed to consider the universe as made up of parts, and mathematicians usually begin by considering a single particle, and then conceiving its relation to another particle, and so on. This has generally been supposed the most natural method. To conceive of a particle, however, requires a process of abstraction, since all our perceptions are related to extended bodies, so that the idea of the *all* that is in our consciousness at a given instant is perhaps as primitive an idea as that of any individual thing. Hence there may be a mathematical method in which we proceed from the whole to the parts instead of from the parts to the whole. For example, Euclid, in his first book, conceives a line as traced out by a point, a surface as swept out by a line, and a solid as generated by a surface. But he also defines a surface as the boundary of a solid, a line as the edge of a surface, and a point as the extremity of a line.

In like manner we may conceive the potential of a material system as a function found by a certain process of integration with respect to the masses of the bodies in the field, or we may suppose these masses themselves to have no other mathematical meaning than the volume-integrals of $\frac{1}{4\pi}\nabla^2\Psi$, where Ψ is the potential.

In electrical investigations we may use formulae in which the quantities involved are the distances of certain bodies, and the electrifications or currents in these bodies, or we may use formulae which involve other quantities, each of which is continuous through all space.

The mathematical process employed in the first method is integration along lines, over surfaces, and throughout finite spaces, those employed in the second method are partial differential equations and integrations throughout all space.

The method of Faraday seems to be intimately related to the second of these modes of treatment. He never considers bodies as existing with nothing between them but their distance, and acting on one another according to some function of that distance. He conceives all space as a field of force, the lines of force being in general curved, and those due to any body extending from it on all sides, their directions being modified by the presence of other bodies. He even speaks of the lines of force belonging to a body as in some sense part of itself, so that in its action on distant bodies it cannot be said to act where it is not. This, however, is not a dominant idea with Faraday. I think he would rather have said that the field of space is full of lines of force, whose arrangement depends on that of the bodies in the field, and that the mechanical and electrical action on each body is determined by the lines which abut on it.

The magnetic properties of certain materials and the electric effects produced by friction were both known in ancient days. Oersted's experiment with electric current and a compass showed that electricity and magnetism are related. Maxwell found the connection between the two phenomena in his electromagnetic equations.

15 The Relationship of Electricity and Magnetism

D. K. C. MacDonald

Excerpt from his book, *Faraday, Maxwell, and Kelvin,* published in 1964.

We know that an electric current can produce forces on a magnet in its vicinity, or, in other words, an electric current produces a magnetic "field." Faraday had shown, moreover, that a changing magnetic field (produced either by moving a magnet or by varying an electric current in a coil) could induce an electric current in a neighboring, but separate, coil of wire. Thus, through these fundamental experiments of Oersted, Ampère, and particularly Faraday, various vital facts had been discovered about how electric currents and magnets could interact with one another and, as we have said earlier, these discoveries were already leading to exciting practical developments such as the electric telegraph and the submarine cables. But, in broad terms, what James Clerk Maxwell tried to do was to build up a more *general* picture of these interactions between electric and magnetic effects (or "fields") without worrying so much about actual coils of wire with electric currents in them, or about how in practice one actually produced the magnetic fields. Following Faraday's general lead in concentrating on the "lines of force" or the "fields," Maxwell tried to work out directly and quantitatively the interaction in space of the electric field on the magnetic field, and vice versa, wherever they might exist. In his mind Maxwell invented, or designed, various semi-mechanical models to build up his theory, but in the end he could discard this mental scaffolding and give a complete mathematical description of electromagnetic behavior which holds good to this day.

Consider the production of a magnetic field by a current of electricity in a coil. We know that such a current always involves a movement of electric charge, so from the electrical point of view we may say that something is changing all the time. One of the things Maxwell did was to generalize this discovery boldly, saying in essence: [I] *"A Changing Electric Field Will Always Produce a Magnetic Field."*

But, on the other hand, Faraday had shown that the movement of a magnet could produce an electric current, as we have already seen; so on the same lines this can be generalized to say: [II] *"A Changing Magnetic Field Can Produce an Electric Field."*

The ultimate result of James Clerk Maxwell's work was, in effect, that he expressed these two basic ideas in precise, quantitative terms, and he came out finally with what are now known as *Maxwell's Equations,* which, as I already have said, remain today the standard method of predicting how electricity and magnetism will behave under any given conditions. The acme of Maxwell's work, however, was his discovery that when applied in free, empty space his equations took on a form which is equally descriptive of any undamped wave motion propagating itself freely from place to place. Thus, if you drop a stone into a large pond of water a ripple or wave will proceed out from that place, and some of the energy from the falling stone will radiate outward in the wave from the splash. If you shout to somebody else some distance away, then it is a vibration or wave in the air around you which carries the sound to the distant person; or if you rig up a long, tight rope or string between two points, and then "twang" the rope, you can see a wave running along the rope, and this wave carries some of the energy that you put in the "twang." Again, if there is a violent storm at sea, the energy from this storm gets carried over long distances by waves in the ocean; the waves which smash on the rocks of Newfoundland may well be getting their energy from a storm a thousand miles or more out in the Atlantic Ocean. In each of these latter examples the waves will be damped to some degree or other. For example, waves traveling on the surface of the sea lose some energy by dragging deeper layers of water, by the very fact that water is not entirely free to move by itself, but has a viscosity or "stickiness," which means that the waves ultimately suffer losses by friction.

The particularly remarkable, and unique, feature of *electromagnetic* waves is the fact that they can propagate themselves quite freely without damping through empty space where no matter whatsoever is present, but it is not difficult to see from the two italicized statements above that a self-propelled wave motion of the electromagnetic field might be possible.

Imagine that we have electric and magnetic fields present in a small region of space, and that the fields are changing suitably with time. As the electric field changes at some point in space it will produce a magnetic field in the neighborhood, and if things are right this magnetic field will then reinforce the magnetic field in some regions, and in turn the over-all changing magnetic field will produce again a fresh electric field in its neighborhood. What Maxwell's equations showed was that this process, perhaps somewhat reminiscent of an endless game of leapfrog, could indeed be self-maintained, with the energy constantly radiating outward from where the waves started.

But this was not all. Maxwell was able to predict from this theory, moreover, the *speed* with which such an electromagnetic wave should travel in space. This speed was simply determined by the ratio of two measurements which could be made on electric and magnetic quantities in the laboratory, and it turned out that the speed predicted in this way was very close to the already known speed of light (about 300,000 km/sec $\approx$ 186,000 miles/sec). Furthermore, it is also a well-known characteristic of light that it too can propagate through empty space, as witness the light of day which reaches us unfailingly from the sun across about a hundred million miles of empty space. So Maxwell could finally say with confidence that, physically speaking, light must be a form of electromagnetic radiation.

Some years after Maxwell's death, Heinrich Hertz (1857–94) was able to show experimentally, using electrical apparatus, the direct generation and detection of the electromagnetic waves predicted by Maxwell. These "Hertzian waves" are the great-grandfather of the waves which carry all our radio and television broadcasts today, and in fact radio waves, television waves, light waves, X-rays, and gamma rays, are all members of one and the same family—electromagnetic waves. In free space they all travel with identically the same speed, which for convenience we always refer to as "the velocity of light." What distinguishes one type of wave from another is simply its rate of vibration, or the corresponding wave length (i.e., the distance between two successive "crests" or "troughs" of a wave). A typical radio wave vibrates at, or has a frequency (f) of, about a million times a second ($f = 10^6$ cycles/sec = 1 M c/s), and has a wave length (λ) of about 300 meters. For those who do not mind an equation, the relationship is very simple, namely $f\lambda = c$, where c denotes, as always in physical science, the velocity of light. At the other end of the scale, a gamma ray might have a wave length of only about one ten-billionth part of a centimeter ($\lambda = 10^{-10}$ cm), and a corresponding frequency of vibration of about three hundred billion billion cycles/sec ($f = 3 \times 10^{20}$ c/s).

Electromagnetic Waves

Maxwell's electromagnetic theory also led to intense discussion later about the fundamental nature of the electromagnetic waves involved. Many physicists felt that in order to have a wave at all there had to be "something" to do the waving or vibrating, and they invented a sort of all-pervading, universal, thin soup or consommé which they called the "aether." But whether it is more reasonable to talk about electromagnetic waves in free space (which still worries some people for the same sort of reason that "action at a distance" worried people), or whether it is better to try to think about an all-permeating, vibrating "aether" is not a very burning issue today. What matters now is that Maxwell's Equations are a generally accepted foundation for discussing electromagnetic behavior under the widest range of possible situations, and also that Maxwell's lead in analyzing electromagnetism by means of the electric and magnetic fields has led more generally to the concept of discussing other forms of interaction through some appropriate "field." Indeed, Maxwell himself was at first very inclined to believe that *gravitational* attraction must also be propagated in this way, but he ran up against difficulties with the energy involved which seemed to him then insurmountable.

We have seen that, starting from the picture of "action at a distance" between charges of electricity, Maxwell, following Faraday's lead, could reformulate the problem in terms of a field acting through, and at all points of, space of which the charged particles are, so to speak, now just the "terminals" or "end points." The discovery that this electromagnetic field would vibrate in free space was a great step toward identifying light as an electromagnetic wave, since the wave phenomenon of light (interference, diffraction, etc.) had been known for a long time. At the same time there had always been some persistent reasons for regarding light alternatively as a corpuscular phenomenon, and Einstein was to show, half a century later, that Maxwell's vibrating electromagnetic aether, when coupled with Planck's quantum theory first proposed around 1900, could also then be regarded in a more or less corpuscular manner. What Planck and Einstein showed was that the energy in the electromagnetic field could only exist in certain minimum-sized bundles or "quanta" dependent in magnitude on the frequency of vibration and the newly discovered Planck's constant. These "bundles" of light, or more technically "quanta" of the electromagnetic field, are generally known today as photons. So now we can think of electromagnetic interactions as either conveyed by the vibrating aether or equivalently as conveyed by streams of photons which will to some extent behave like particles. In dealing with many kinds of interactions, including those which hold an atomic nucleus together, modern physics finds it most valuable to be able to think in both these terms without being bound to regard one picture as more necessarily "real" than the other.

The formulation of Maxwell's equations opened the new area of science called electromagnetism, with far-reaching consequences.

16 The Electromagnetic Field

Albert Einstein and Leopold Infeld

Excerpt from their book entitled the *Evolution of Physics* published in 1938 and 1961.

THE REALITY OF THE FIELD

The quantitative, mathematical description of the laws of the field is summed up in what are called Maxwell's equations. The facts mentioned so far led to the formulation of these equations but their content is much richer than we have been able to indicate. Their simple form conceals a depth revealed only by careful study.

The formulation of these equations is the most important event in physics since Newton's time, not only because of their wealth of content, but also because they form a pattern for a new type of law.

The characteristic features of Maxwell's equations, appearing in all other equations of modern physics, are summarized in one sentence. Maxwell's equations are laws representing the *structure* of the field.

Why do Maxwell's equations differ in form and character from the equations of classical mechanics? What does it mean that these equations describe the structure of the field? How is it possible that, from the results of Oersted's and Faraday's experiments, we can form a new type of law, which proves so important for the further development of physics?

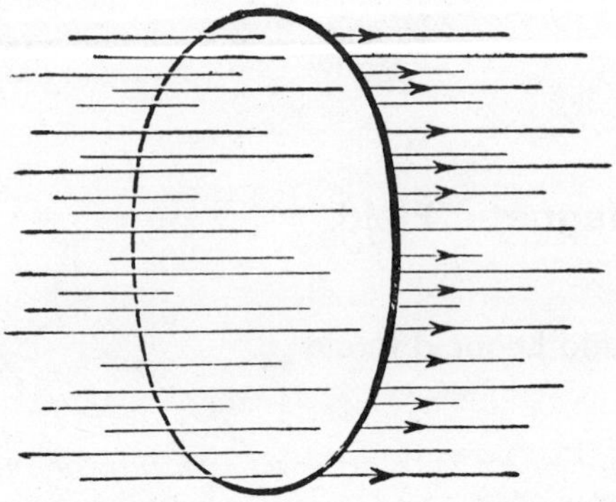

We have already seen, from Oersted's experiment, how a magnetic field coils itself around a changing electric field. We have seen, from Faraday's experiment, how an electric field coils itself around a changing magnetic field. To outline some of the characteristic features of Maxwell's theory, let us, for the moment, focus all our attention on one of these experiments, say, on that of Faraday. We repeat the drawing in which an electric current is induced by a changing magnetic field. We already know that an induced current appears if the number of lines of force, passing the surface bounded by the wire, changes. Then the current will appear if the magnetic field changes or the circuit is deformed or moved: if the number of magnetic lines passing through the surface is changed, no matter how this change is caused. To take into account all these various possibilities, to discuss their particular influences, would necessarily lead to a very complicated theory. But can we not simplify our problem? Let us try to eliminate from our considerations everything which refers to the shape of the circuit, to its length, to the surface enclosed by the wire. Let us imagine that the circuit in our last drawing becomes smaller and smaller, shrinking gradually to a very small circuit enclosing a certain point in space. Then everything concerning shape and size is quite irrelevant. In this limiting process where the closed curve shrinks to a point, size and shape automatically vanish from our considerations and we obtain laws connecting changes of magnetic and electric field at an arbitrary point in space at an arbitrary instant.

Thus, this is one of the principal steps leading to Maxwell's equations. It is again an idealized experiment performed in imagination by repeating Faraday's experiment with a circuit shrinking to a point.

We should really call it half a step rather than a whole one. So far our attention has been focused on Faraday's experiment. But the other pillar of the field theory, based on Oersted's experiment, must be considered just as carefully and in a similar manner. In this experiment the magnetic lines of force coil themselves around the current. By shrinking the circular magnetic lines of force to a point, the second half-step is performed and the whole step yields a connection between the changes of the magnetic and electric fields at an arbitrary point in space and at an arbitrary instant.

But still another essential step is necessary. According to Faraday's experiment, there must be a wire testing the existence of the electric field, just as there must be a magnetic pole, or needle, testing the existence of a magnetic field in Oersted's experiment. But Maxwell's new theoretical idea goes beyond these experimental facts. The electric and magnetic field, or in short, the *electromagnetic* field is, in Maxwell's theory, something real. The electric field is produced by a changing magnetic field, quite independently, whether or not there is a wire to test its existence; a magnetic field is produced by a changing electric field, whether or not there is a magnetic pole to test its existence.

Thus two essential steps led to Maxwell's equations. The first: in considering Oersted's and Rowland's experiments, the circular line of the magnetic field coiling itself around the current and the changing electric field, had to be shrunk to a point; in considering Faraday's experiment, the circular line of the electric field coiling itself around the changing magnetic field had to be shrunk to a point. The second step consists of the realization of the field as something real; the electromagnetic field once created exists, acts, and changes according to Maxwell's laws.

Maxwell's equations describe the structure of the electromagnetic field. All space is the scene of these laws and not, as for mechanical laws, only points in which matter or charges are present.

We remember how it was in mechanics. By knowing the position and velocity of a particle at one single instant, by knowing the acting forces, the whole future path of the particle could be forseen. In Maxwell's theory, if we know the field at one instant only, we can deduce from the equations of the theory how the whole field will change in space and time. Maxwell's equations enable us to follow the history of the field, just as the mechanical equations enabled us to follow the history of material particles.

But there is still one essential difference between mechanical laws and Maxwell's laws. A comparison of Newton's gravitational laws and Maxwell's field laws will emphasize some of the characteristic features expressed by these equations.

With the help of Newton's laws we can deduce the motion of the earth from the force acting between the sun and the earth. The laws connect the motion of the earth with the action of the far-off sun. The earth and the sun, though so far apart, are both actors in the play of forces.

In Maxwell's theory there are no material actors. The mathematical equations of this theory express the laws governing the electromagnetic field. They do not, as in Newton's laws, connect two widely separated events; they do not connect the happenings *here* with the conditions *there*. The field *here* and *now* depends on the field in the *immediate neighborhood* at a time *just past*. The equations allow us to predict what will happen a little further in space and a little later in time, if we know what happens here and now. They allow us to increase our knowledge of the field by small steps. We can deduce what happens here from that which happened far away by the summation of these very small steps. In Newton's theory, on the contrary, only big steps connecting distant events are permissible. The experiments of Oersted and Faraday can be regained from Maxwell's theory, but only by the summation of small steps each of which is governed by Maxwell's equations.

A more thorough mathematical study of Maxwell's equations shows that new and really unexpected conclusions can be drawn and the whole theory submitted to a test on a much higher level, because the theoretical consequences are now of a quantitative character and are revealed by a whole chain of logical arguments.

Let us again imagine an idealized experiment. A small sphere with an electric charge is forced, by some external influence, to oscillate rapidly and in a rhythmical way, like a pendulum. With the knowledge we already have of the changes of the field, how shall we describe everything that is going on here, in the field language?

The oscillation of the charge produces a changing electric field. This is always accompanied by a changing magnetic field. If a wire forming a closed circuit is placed in the vicinity, then again the changing magnetic field will be accompanied by an electric current in the circuit. All this is merely a repetition of known facts, but the study of Maxwell's equations gives a much deeper insight into the problem of the oscillating electric charge. By mathematical deduction from Maxwell's equations we can detect the character of the field surrounding an oscillating charge, its structure near and far from the source and its change with time. The outcome of such deduction is the *electromagnetic wave*. Energy radiates from the oscillating charge traveling with a definite speed through space; but a transference of energy, the motion of a state, is characteristic of all wave phenomena.

Different types of waves have already been considered. There was the longitudinal wave caused by the pulsating sphere, where the changes of density were propagated through the medium. There was the jelly-like medium in which the transverse wave spread. A deformation of the jelly, caused by the rotation of the sphere, moved through the medium. What kind of changes are now spreading in the case of an electromagnetic wave? Just the changes of an electromagnetic field! Every change of an electric field produces a magnetic field; every change of this magnetic field produces an electric field; every change of . . . , and so on. As field represents energy, all these changes spreading out in space, with a definite velocity, produce a wave. The electric and magnetic lines of force always lie, as deduced from the theory, on planes perpendicular to the direction of propagation. The wave produced is, therefore, transverse. The original features of the picture of the field we formed from Oersted's and Faraday's experiments are still preserved, but we now recognize that it has a deeper meaning.

The electromagnetic wave spreads in empty space. This, again, is a consequence of the theory. If the oscillating charge suddenly ceases to move, then, its field becomes electrostatic. But the series of waves created by the oscillation continues to spread. The waves lead an independent existence and the history of their changes can be followed just as that of any other material object.

We understand that our picture of an electromagnetic wave, spreading with a certain velocity in space and changing in time, follows from Maxwell's equations only because they describe the structure of the electromagnetic field at any point in space and for any instant.

There is another very important question. With what speed does the electromagnetic wave spread in empty space? The theory, with the support of some data from simple experiments having nothing to do with the actual propagation of waves, gives a clear answer: *the velocity of an electromagnetic wave is equal to the velocity of light.*

Oersted's and Faraday's experiments formed the basis on which Maxwell's laws were built. All our results so far have come from a careful study of these laws, expressed in the field language. The theoretical discovery of an electromagnetic wave spreading with the speed of light is one of the greatest achievements in the history of science.

Experiment has confirmed the prediction of theory. Fifty years ago, Hertz proved, for the first time, the existence of electromagnetic waves and confirmed experimentally that their velocity is equal to that of light. Nowadays, millions of people demonstrate that electromagnetic waves are sent and received. Their apparatus is far more complicated than that used by Hertz and detects the presence of waves thousands of miles from their sources instead of only a few yards.

Instruments borne aloft by artificial satellites and probes report that our planet is encircled by two zones containing high-energy radiation against which space travelers will have to shield themselves.

17 Radiation Belts Around the Earth

James Van Allen

An article published in *Scientific American* in 1959.

So far, the most interesting and least expected result of man's exploration of the immediate vicinity of the earth is the discovery that our planet is ringed by a region—to be exact, two regions—of high-energy radiation extending many thousands of miles into space. The discovery is of course troubling to astronauts; somehow the human body will have to be shielded from this radiation, even on a rapid transit through the region. But geophysicists, astrophysicists, solar astronomers and cosmic-ray physicists are enthralled by the fresh implications of these findings. The configuration of the region and the radiation it contains bespeak a major physical phenomenon involving cosmic rays and solar corpuscles in the vicinity of the earth. This enormous reservoir of charged particles plays a still-unexplained role as middleman in the interaction of earth and sun which is reflected in magnetic storms, in the airglow and in the beautiful displays of the aurora.

The story of the investigation goes back to 1952 and 1953, before any of us could think realistically about the use of earth satellites to explore the environment of the earth. Parties from our laboratory at the State University of Iowa spent the summers of those years aboard Coast Guard and naval vessels, cruising along a 1,500-mile line from the waters of Baffin Bay, near the magnetic pole in the far northwestern corner of Greenland, southward to the North Atlantic off the coast of Newfoundland. Along the way we launched a series of rocket-carrying balloons—"rockoons." (The balloon lifts a small rocket to an altitude of 12 to 15 miles, whence the rocket carries a modest payload of instruments to a height of 60 to 70 miles.) Our objective was to develop a profile of the cosmic-ray intensities at high altitudes and latitudes, and thus to learn the nature of the low-energy cosmic rays which at lower altitudes and latitudes are deflected by the earth's magnetic field or absorbed in the atmosphere.

Most of the readings radioed down from the rockets were in accord with plausible expectations. Two rockoons sent aloft in 1953, however, provided us with a puzzle. Launched near Newfoundland by Melvin Gottlieb and Leslie Meredith, they encountered a zone of radiation beginning at an altitude of 30 miles that was far stronger than we had expected. At first we were uneasy about the proper operation of our instruments. But critical examination of the data convinced us that we had unquestionably encountered something new in the upper atmosphere.

Significantly these measurements were made in the northern auroral zone. In this zone, which forms a ring some 23 degrees south of the north geomagnetic pole, the incidence of visible auroras reaches its maximum. Since rockets fired north and south of the zone had revealed nothing unusual, we speculated that the strong radiation played some part in the aurora. Showers of particles from the sun, it was thought, come plunging into the atmosphere along magnetic lines of force and set off these displays [see "Aurora and Airglow," by C. T. Elvey and Franklin E. Roach; SCIENTIFIC AMERICAN, September, 1955]. But the theory underlying this explanation did not explain satisfactorily why the aurora and the high-intensity radiation we had detected should occur in the auroral zone and not in the vicinity of the geomagnetic pole itself. Nor could it account for the high energies required to carry the solar particles through the atmosphere to such relatively low altitudes.

The mystery deepened when we found in later studies that the radiation persists almost continuously in the zone above 30 miles, irrespective of visible auroral displays and other known high-altitude disturbances. More discriminating detectors established that the radiation contains large numbers of electrons. Our original observations had detected X-rays only; now it turned out that the X-rays had been generated by the impact of electrons on the skin of the instrument package (as if it had been the "target" in an X-ray tube) and on the sparse atoms of the upper atmosphere itself. Sydney Chapman and Gordon Little at the University of Alaska suggested that such a process might well account for the attenuation of radio signals in the lower ionosphere of the auroral zones.

The International Geophysical Year gave us our first opportunity to investigate the "auroral soft radiation" on a more comprehensive scale. During the

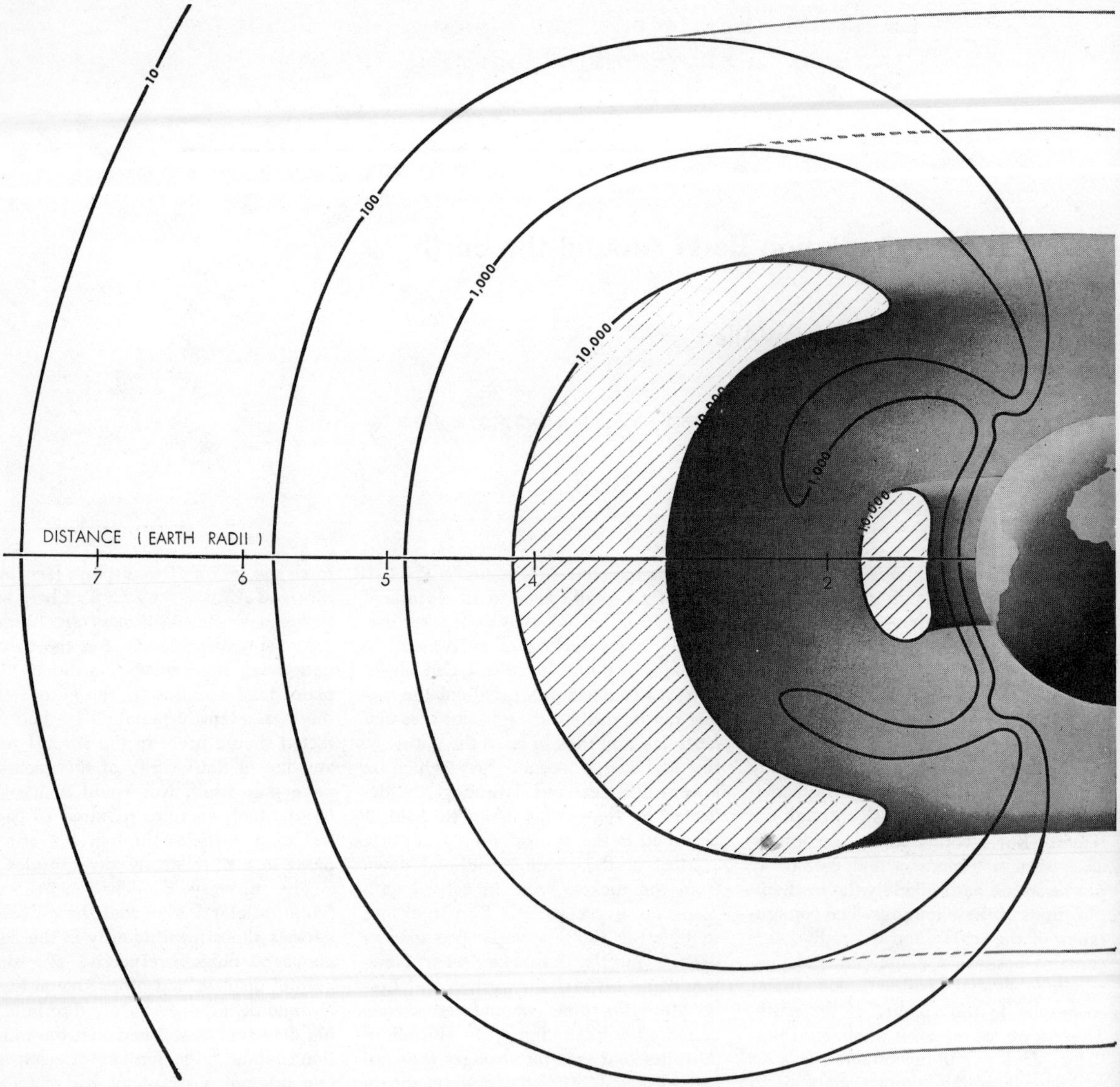

STRUCTURE OF RADIATION BELTS revealed by contours of radiation intensity (*black lines*) is shown schematically by shading (*left*); dots (*right*) suggest distribution of particles in the two belts. Contour numbers give counts per second; horizontal scale

summer and fall of 1957 Laurence Cahill and I launched a number of rockoons off the coast of Greenland and also got off one successful flight in Antarctica. The latter flight established that the radiation exists in the southern as well as the northern auroral zone. In February, 1958, Carl McIlwain fired a series of two-stage rockets through visible auroras above Fort Churchill in Canada, and discovered that the radiation includes energetic protons (hydrogen nuclei) as well as electrons.

Meanwhile all of us had been pushing a new development that greatly expanded the possibilities for high-altitude research. During the summer of 1955 the President and other Government authorities were finally persuaded that it might be feasible to place artificial satellites in orbit, and authorized an I. G. Y. project for this purpose. In January, 1956, a long-standing group of high-altitude experimentalists, called the Rocket and Satellite Research Panel, held a symposium to consider how the satellites could be most fruitfully employed. At that meeting our group proposed two projects. One was to put a satellite into an orbit nearly pole-to-pole to survey the auroral radiation in both the north and south auroral zones. Such orbits, however, did not appear to be

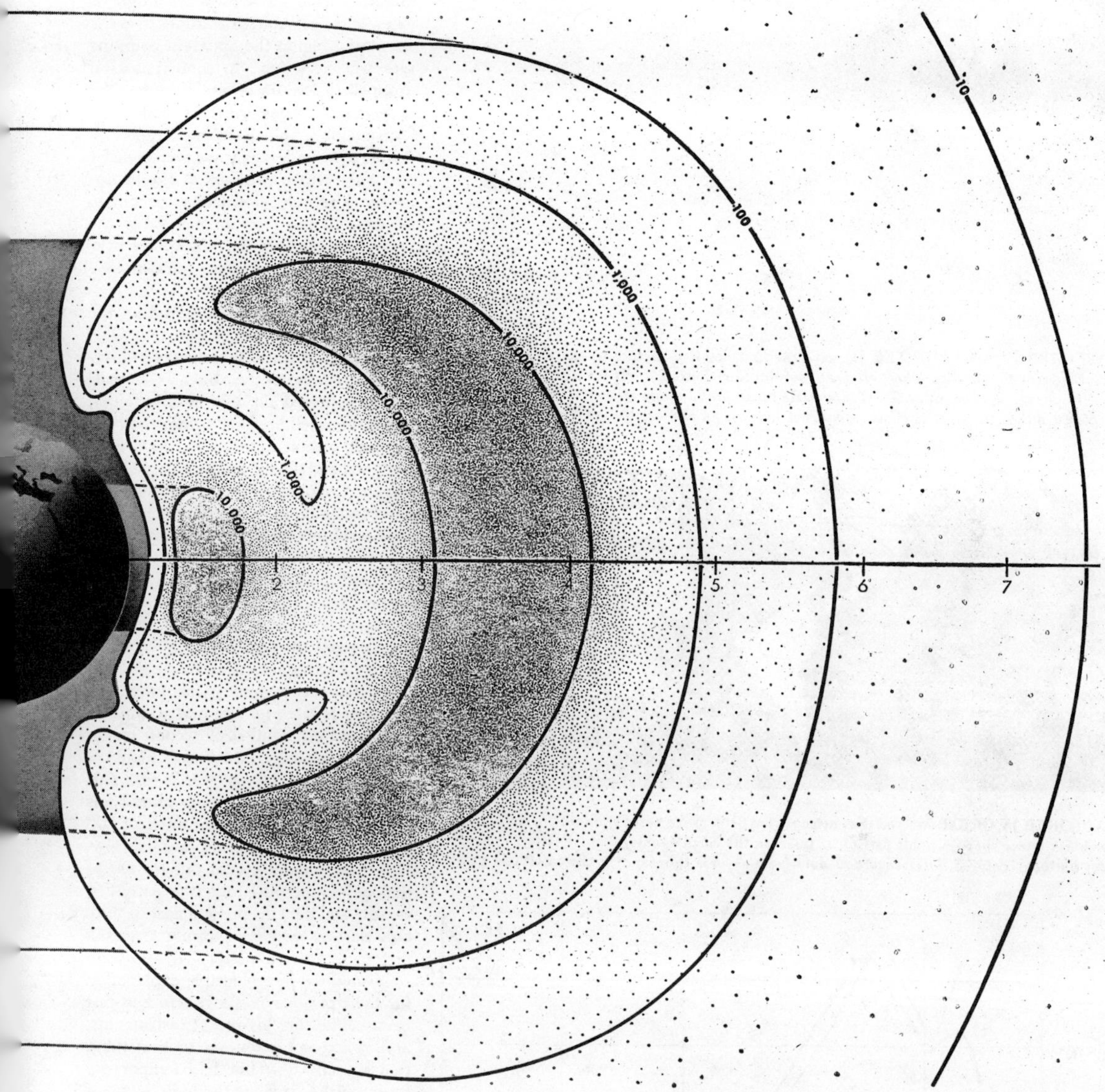

shows distance in earth radii (about 4,000 miles) from the center of the earth. Particles in the inner belt may originate with the radioactive decay of neutrons liberated in the upper atmosphere by cosmic rays; those in the outer belt probably originate in the sun.

technically feasible in the immediate future. For the time being we were forced to abandon the use of a satellite to probe farther into the auroral soft radiation. We also suggested that a satellite orbiting over the lower latitudes of the earth might usefully be employed in a comprehensive survey of cosmic-ray intensities over those regions. This project was adopted, and we were authorized to prepare suitable experimental apparatus [see "The Artificial Satellite as a Research Instrument," by James A. Van Allen; Scientific American, November, 1956]. It was planned to place this apparatus on one of the early Vanguard vehicles.

The difficulties and failures of the Vanguard are now history. Sputnik I stimulated some high government officials to accept a proposal that a number of us had been urging for more than a year: to use the proven Jupiter C rocket as a satellite-launching vehicle. As a result on January 31, 1958, Explorer I went into orbit carrying our simple cosmic-ray detector and a radio to broadcast its readings.

In the first reports from stations located in the U. S. the intensity of radiation increased with altitude along the expected curve. Several weeks later, however, we began to get tapes from stations in

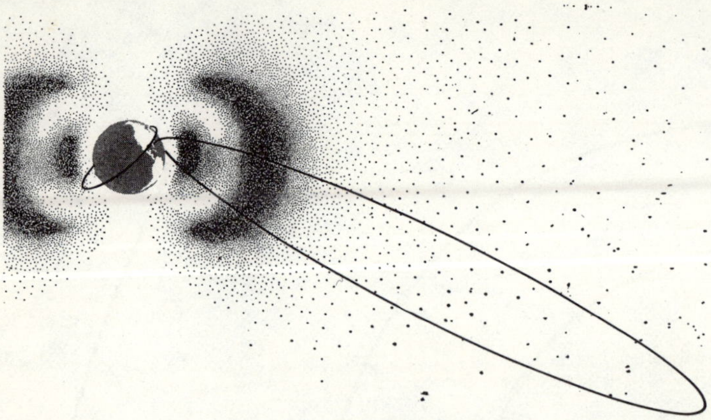

EXPLORER IV AND PIONEER III gave the first detailed picture of the radiation belts. The Explorer IV satellite (*short ellipse*) monitored radiation levels for nearly two months at altitudes up to 1,300 miles. The Pioneer III lunar probe (*long ellipse*) provided data out to 65,000 miles. Its orbit is shown distorted because of the earth's rotation during flight.

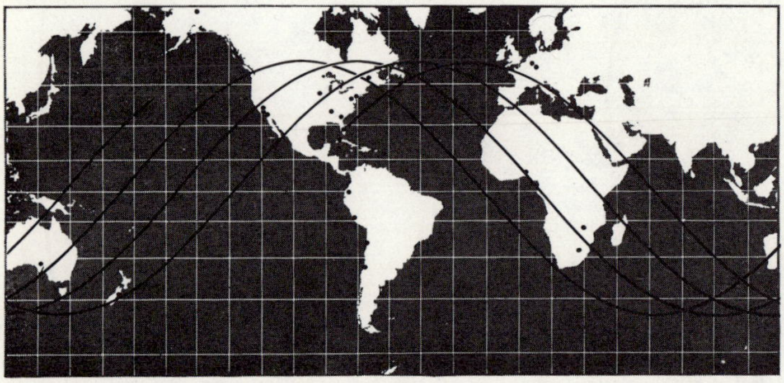

EXPLORER IV ORBIT covered the entire region 51 degrees north and south of the equator; the black curve shows a small part of its trace on the earth's surface. More than 25 observation stations (*colored dots*) recorded data from several thousand of the satellite's passes.

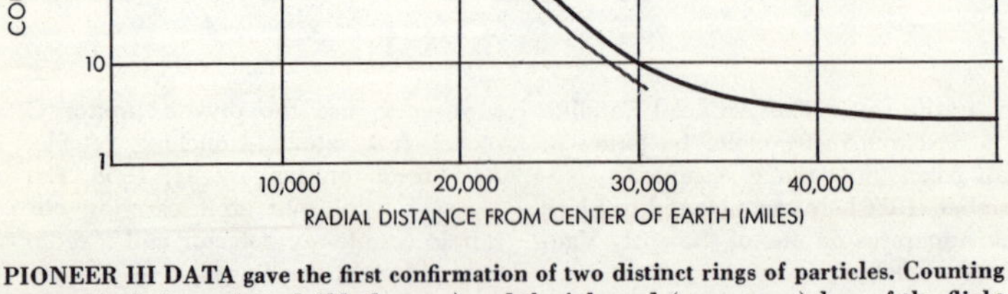

PIONEER III DATA gave the first confirmation of two distinct rings of particles. Counting rates on both the outbound (*black curve*) and the inbound (*gray curve*) legs of the flight showed two peaks. The two curves differ because they cover different sections of the belts.

South America and South Africa which gave us counting rates for much higher altitudes, due to the eccentricity of the satellite's orbit. These records brought us a new surprise. At high altitudes over the equatorial region the apparent counting rate was very low; in some passes it dropped to zero for several minutes. Yet at lower altitudes the rate had quite "reasonable" values—from 30 to 50 counts a second. Again we were uneasy about the trustworthiness of the instruments. The only alternative seemed to be that cosmic rays do not strike the uppermost layers of the atmosphere over the tropics, and we were quite unable to accept this conclusion.

Our uneasiness was increased by the incompleteness of our early data. The Explorer I apparatus broadcast its observations continuously, but its signals could be picked up only intermittently, when the satellite came within range of a ground station. Our original apparatus, designed and developed by George Ludwig for the Vanguard satellites, included a magnetic-tape recorder which could store its observations for a complete orbit around the earth and then report them in a "burst" on radio command from the ground.

By early February, working with the Jet Propulsion Laboratory, we had converted this apparatus for use in the Explorer II satellite. The first attempt to get it into orbit failed. A second rocket placed Explorer III, carrying identical apparatus, in orbit on March 26. This satellite fully confirmed the anomalous results of Explorer I. At altitudes of 200 to 300 miles the counting rate was low. When the satellite went out to 500 to 600 miles, the apparent rate ascended rapidly and then dropped almost to zero. One day, as we were puzzling over the first tapes from Explorer III, McIlwain suggested the first plausible explanation for their peculiar readings. He had just been calibrating his rocket instruments, and called our attention to something that we all knew but had temporarily forgotten: A sufficiently high level of radiation can jam the counter and send the apparent counting rate to zero. We had discovered an enormously high level of radiation, not a lack of it. As Ernest Ray, a member of our group, inaccurately but graphically exclaimed: "Space is radioactive!"

During the next two months Explorer III produced a large number of playback records, every one of which showed the same effect. At low altitudes the counting rate was reasonably attributable to

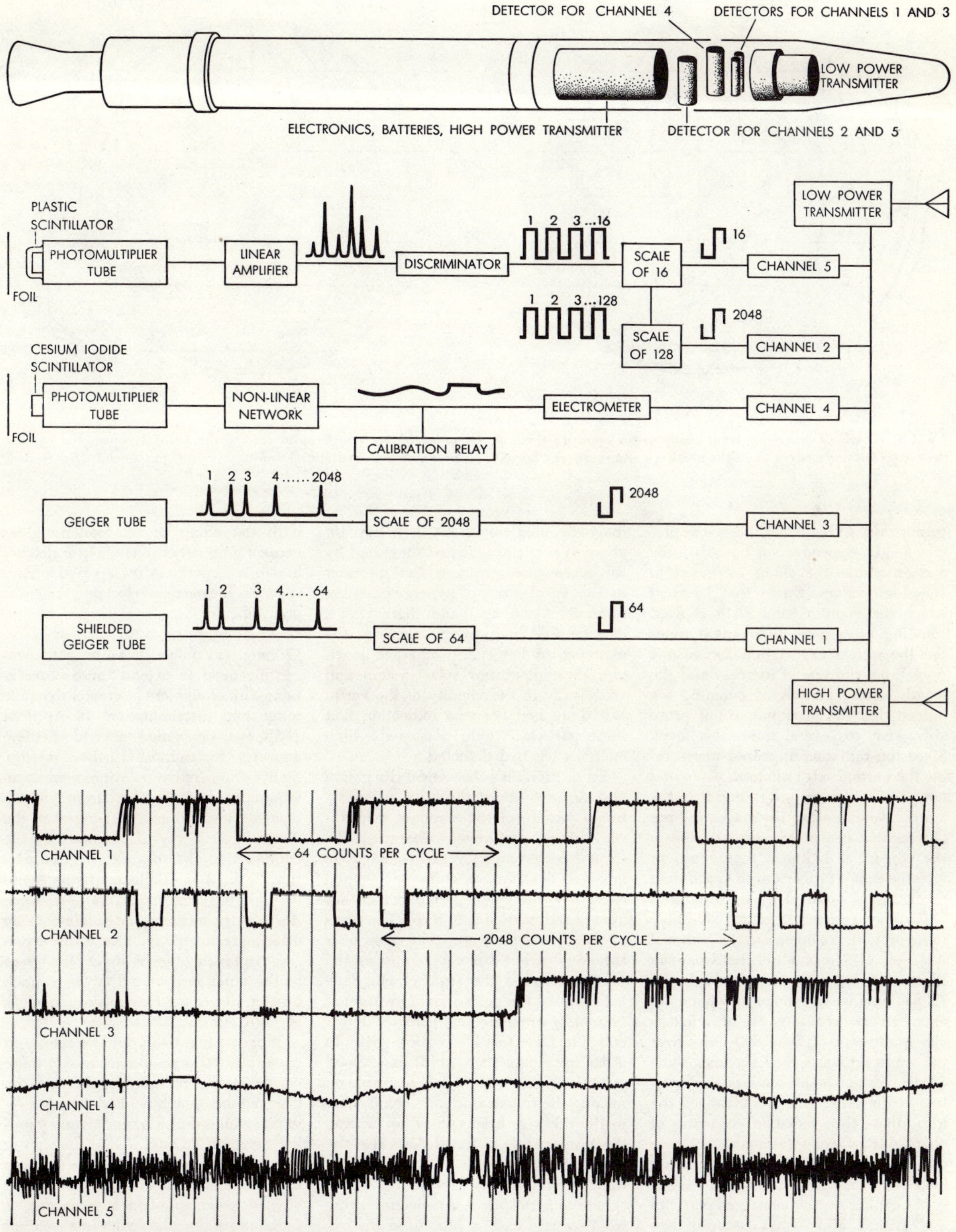

EXPLORER IV INSTRUMENTS were designed to give a detailed picture of the nature and intensity of the radiation. Plastic scintillator counted only charged particles above certain energies; two different scaling factors adapted it to both high and low counting rates. Cesium-iodide scintillator measured the total energy input rather than individual particles. Shielded and unshielded Geiger tubes could be compared to estimate the penetrability of the radiation. Radio signals suggested by the red curves in upper drawing were recorded by ground stations and later played through a multichannel oscillograph to yield records like that shown below.

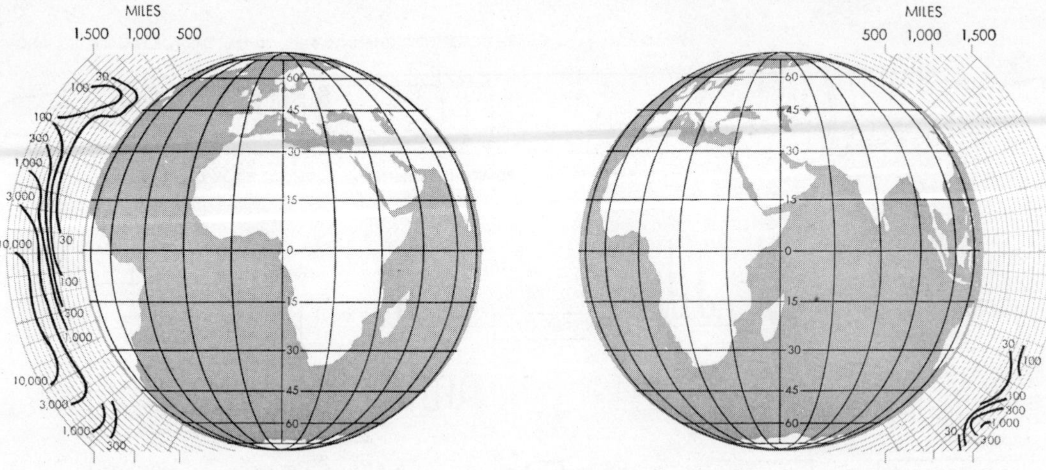

TWO SETS OF CONTOURS from readings on opposite sides of the earth (*left and center*) show the northern and southern "horns" of radiation, which point toward the auroral zones; the contour numbers show radiation intensity in counts per second. The "tipped"

cosmic rays. At higher altitudes—the precise height depended on both latitude and longitude—the count increased to very high values. Up to the points at which the counter jammed, it showed counting rates more than 1,000 times the theoretical expectation for cosmic rays. From the rate of increase and the length of the periods of jamming we judged that the maximum count probably went to several times this level. Since the radiation appeared to resemble the auroral soft radiation, we would not have been surprised to find it in the auroral zone or along the magnetic lines of force that connect these zones. But in the equatorial latitudes these lines of force lie much farther out in space than the altitudes attained by the satellites.

On May 1 of last year we were able to report with confidence to the National Academy of Sciences and the American Physical Society that Explorers I and III had discovered a major new phenomenon: a very great intensity of radiation above altitudes of some 500 miles over the entire region of their traverse, some 34 degrees north and south of the equator. At the same time we advanced the idea that the radiation consists of charged particles—presumably protons and electrons—trapped in the magnetic field of the earth.

We could rule out uncharged particles and gamma and X-rays because they would not be confined by the magnetic field, and so would be observed at lower altitudes. The possibility that the earth's magnetic field might act as a trap for charged particles was first suggested by the Norwegian physicist Carl Störmer in a classical series of papers beginning some 50 years ago, and there was a considerable body of evidence for the existence of low-energy charged particles throughout our solar system and specifically in the vicinity of the earth. But there had been no indication that these particles would possess the high energies we had detected.

From Störmer's theoretical discussion and our own observations we evolved a rough picture of the trapping mechanism. When a fast-moving charged particle is injected into the earth's magnetic field, it describes a corkscrew-shaped trajectory, the center line of which lies along a magnetic line of force. The turns of the helical path are quite open over the equator but become tighter as the particle reaches the stronger magnetic field toward the poles [*see illustration at bottom of opposite page*]. At the lower end of its trajectory the particle goes into a flat spiral and then winds back along a similar path to the other hemisphere, making the transit from one hemisphere to the other in a second or so. During this time its line of travel shifts slightly, so that the particle drifts slowly around the earth as it corkscrews from hemisphere to hemisphere. An electron drifts from west to east; a proton, in the opposite direction. At each end of its path the particle descends into regions of higher atmospheric density; collisions with the atoms of atmospheric gases cause it gradually to change its trajectory and to lose energy. After a period of days or weeks the particle is lost into the lower atmosphere.

There was obviously an urgent scientific need to extend these observations with equipment of greater dynamic range and discrimination. In April of 1958 we persuaded several Federal agencies to support further satellite flights of our radiation equipment as an adjunct to the I. G. Y. program, and we received the enthusiastic support of the National Academy of Sciences for the continuation of our work. We also persuaded the Army Ballistic Missile Agency and the Cape Canaveral Air Force Base to try to place the satellite in an orbit more steeply inclined to the equator; at an inclination of about 50 degrees to the equator it would cover a much greater area of earth and skim the edges of both auroral zones.

Working night and day, we set out at once to build new apparatus of a more discriminating nature. We retained the Geiger tube, which we had used in previous satellites, as a basic "simple-minded" detector. To be ready for the highest intensities of radiation, however, we used a much smaller tube that would yield a lower count in a given flux of radiation, and we hooked it into a circuit that would scale down its count by a much larger factor. To obtain a better idea of the penetrability of the radiation

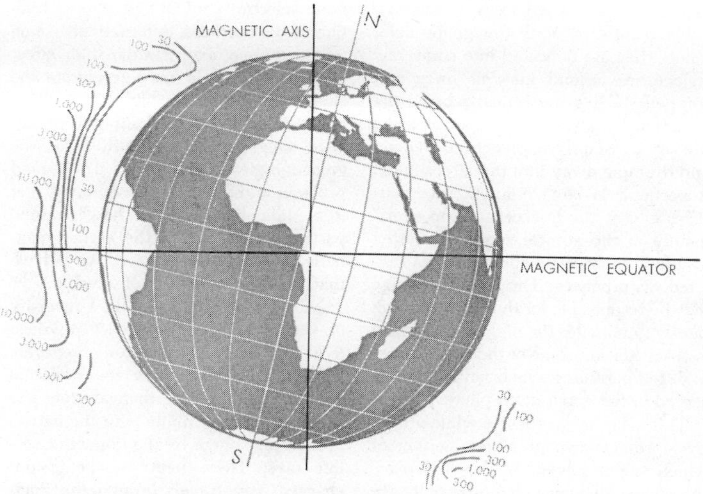

drawing at right shows the essential symmetry of the radiation around the earth's magnetic axis. The structure of the radiation zone was built up from hundreds of observed points.

we shielded a similar Geiger tube with a millimeter of lead. As a more discriminating particle detector we adopted a plastic scintillator and photomultiplier tube to respond to electrons with an energy of more than 650,000 electron volts and to protons of more than 10 million electron volts. Finally we glued a thin cesium-iodide crystal to the window of another photomultiplier tube; the light emitted by the crystal when it was irradiated would measure the over-all input of energy rather than the arrival of individual particles. To keep out light when the crystal faced the sun, we shielded it with thin, opaque nickel foil. A special amplifier gave this detector a large dynamic range extending from about .1 erg per second to 100,000 ergs per second.

Explorer IV carried this apparatus into orbit on July 26, and sent down data for almost two months. Magnetic tapes from some 25 observing stations flowed in steadily from late July to late September; altogether we obtained some 3,600 recorded passes of the satellite. A typical pass was readable for several minutes; some of the best were readable for up to 20 minutes, a large fraction of the time required for the satellite to make a turn around the earth. We are still analyzing this mass of data, but the preliminary results have already proved to be enlightening.

The readings have confirmed our earlier estimates of the maximum levels of radiation. Moreover, we have extended our observations to more than 50 degrees north and south of the equator and have been able to plot the intensity of the radiation at various latitudes and longitudes for altitudes up to 1,300 miles. The intensity contours follow the shape of the earth in the equatorial region, but as they approach high northern and southern latitudes they swing outward, then inward and sharply outward again to form "horns" reaching down toward the earth near the auroral zones [*see illustrations at the top of these two pages*]. The entire picture so far is completely consistent with the magnetic-trapping theory.

It was clear from the contours that Explorers I, III and IV penetrated only the lower portion of the radiation belt. As early as last spring we began to make hypothetical extensions of the observed contours out to a distance of several thousand miles. One of these speculative diagrams showed a single, doughnut-shaped belt of radiation with a ridge around the northern and southern edges of its inner circumference, corresponding to the horns of the contours. Another showed two belts—an outer region with a banana-shaped cross section that extended from the northern to the southern auroral zone and an inner belt over the equator with a bean-shaped cross section [*see illustration on pages 40 and 41*]. The latter diagram seemed to fit the contours better. In our seminars and after-hour discussions McIlwain held out for the two-belt theory. The rest of us tended to agree with him but preferred to stay with the single "doughnut" because of its simplicity.

To take the question out of the realm of speculation we had to secure measurements through the entire region of radiation. In May, therefore, I arranged to have one of our radiation detectors carried aboard the lunar probes planned for the fall of 1958. On October

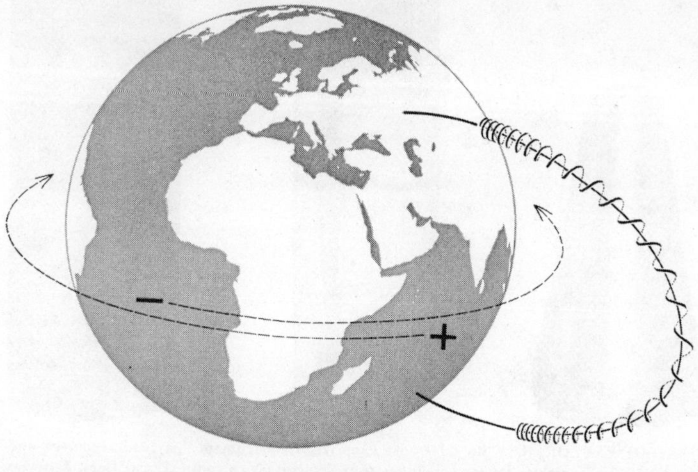

TRAPPED PARTICLES spiral rapidly back and forth along a corkscrew-shaped path whose center is a magnetic line of force. At the same time they drift slowly around the earth (*broken arrows*). Electrons (*negative*) and protons (*positive*) drift in opposite directions.

11, 12 and 13 Pioneer I, the first lunar probe, carried our instruments nearly 70,000 miles out from the earth. Though its readings were spotty, they confirmed our belief that the radiation extended outward for many thousands of miles, with its maximum intensity no more than 10,000 miles above the earth.

The next attempted moon shot, Pioneer II, was a fizzle. Pioneer III, however, went off beautifully on December 6. Although this rocket was intended to reach the vicinity of the moon, we were almost as pleased when it failed to do so, for it gave us excellent data on both the upward and downward legs of its flight, cutting through the radiation region for 65,000 miles in two places.

The observations on both legs showed a double peak in intensity [*see illustration at bottom of page 42*], establishing that there are indeed two belts rather than one. The inner belt reaches its peak at about 2,000 miles from the earth, the outer one at about 10,000 miles. Beyond 10,000 miles the radiation intensity diminishes steadily; it disappears almost completely beyond 40,000 miles. The maximum intensity of radiation in each belt is about 25,000 counts per second, equivalent to some 40,000 particles per square centimeter per second.

Most of us believe that this great reservoir of particles originates largely in the sun. The particles are somehow injected into the earth's magnetic field, where they are deflected into corkscrew trajectories around lines of force and trapped. In this theoretical scheme the radiation belts resemble a sort of leaky bucket, constantly refilled from the sun and draining away into the atmosphere. A particularly large influx of solar particles causes the bucket to "slop over," mainly in the auroral zone, generating visible auroras, magnetic storms and related disturbances. The normal leakage may be responsible for the airglow which faintly illuminates the night sky and may also account for some of the unexplained high temperatures which have been observed in the upper atmosphere.

This solar-origin theory, while attractive, presents two problems, neither of which is yet solved. In the first place the energy of many of the particles we have observed is far greater than the presumed energy of solar corpuscles. The kinetic energy of solar corpuscles has not been measured directly, but the time-lag between a solar outburst and the consequent magnetic disturbances on earth indicates that the particles are slow-moving and thus of relatively low energy. It may be that the earth's magnetic field traps only a high-energy fraction of the particles. Alternatively, some unknown magnetohydrodynamic effect of the earth's field may accelerate the sluggish particles to higher velocities. Some such process in our galaxy has been suggested as responsible for the great energies of cosmic rays. The second problem in the solar-origin theory is that it is difficult to explain how charged particles can get into the earth's magnetic field in the first place. We believe that neither problem is unsolvable.

Nicholas Christofilos of the University of California and the Soviet physicist S. N. Vernov have suggested an entirely different theory of how the radiation originates. They note that neutrons are released in large numbers in the earth's upper atmosphere by the impact of cosmic rays. These neutrons, being uncharged, can travel through the magnetic field without deflection. In due course some of them decay there into electrons and protons, which are trapped.

Our group agrees that particle-injection of this sort is going on, and at a rate which can be easily calculated; but we feel for a number of reasons that it cannot be the main source of radiation-belt particles. If we are right in supposing that the radiation belts provide the "reservoir" for the aurora, the neutron hypothesis cannot account for more than one 10,000th of the auroral energy output. Even if the association between the radiation belts and the aurora turns out to be fortuitous, preliminary indications both from our work and from the Russian experience with Sputnik III suggest that most of the particles in the radiation belt have much lower energies than those of particles that would be produced by neutron decay. A full knowledge of the energy distribution of the particles will aid greatly in clarifying their origin.

Neither theory explains why there should be two belts rather than one. It is tempting to combine the two theories and suppose that the inner belt originates with "internal injection"—*i.e.*, neutron-decay products—and the outer one with "external injection" of solar corpuscles. The two-belt configuration may of course be a transitory phenomenon, though the data from Explorer IV and Pioneer III indicate that the separate belts persisted in essentially the same form for at least five months. We should bear in mind, however, that 1958 was a year of great solar activity. Three years

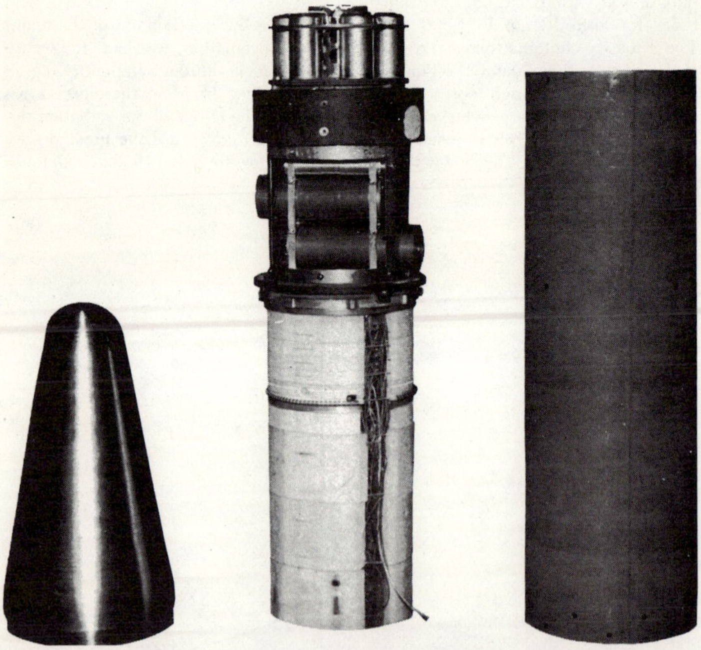

HEAD OF EXPLORER IV includes nose cone (*left*), instrument "payload" (*center*) and protective shell (*right*). Payload includes four detectors, two radio transmitters, batteries and associated electronic circuity. The outer shell is approximately six inches in diameter.

FOUR-STAGE ROCKET launched the Pioneer III moon probe on December 6, 1958. Though the flight failed to reach the moon, its outbound leg gave a continuous record of radiation out to 65,000 miles; the inbound leg gave data between 30,000 and 10,000 miles.

from now we may well find a much lower over-all intensity and perhaps a different structure altogether.

In addition to these possible long-term changes, there may be short-term fluctuations in the belts. While we feel sure that the influx and leakage of particles must balance in the long run, a major solar outbreak may temporarily increase the intensity of the radiation many-fold. If we were to detect such fluctuations and were to find that they coincide with solar outbursts on the one hand and with terrestrial magnetic disturbances on the other, we would have a plain lead to the origin of the particles. Before long we hope to launch a satellite that will monitor radiation levels for at least a year.

Our measurements show that the maximum radiation level as of 1958 is equivalent to between 10 and 100 roentgens per hour, depending on the still-undetermined proportion of protons to electrons. Since a human being exposed for two days to even 10 roentgens would have only an even chance of survival, the radiation belts obviously present an obstacle to space flight. Unless some practical way can be found to shield space-travelers against the effects of the radiation, manned space rockets can best take off through the radiation-free zone over the poles. A "space station" must orbit below 400 miles or beyond 30,000 miles from the earth. We are now planning a satellite flight that will test the efficacy of various methods of shielding.

The hazard to space-travelers may not end even when they have passed the terrestrial radiation belts. According to present knowledge the other planets of our solar system may have magnetic fields comparable to the earth's and thus may possess radiation belts of their own. The moon, however, probably has no belt, because its magnetic field appears to be feeble. Lunar probes should give us more definite information on this point before long.

How does the brain work? Part of the answer lies in electrophysiology, the study of the relation between electricity and nervous stimulation.

18 A Mirror for the Brain

W. Grey Walter

A chapter from his book *The Living Brain* published in 1963.

THE GREEKS had no word for it. To them the brain was merely "the thing in the head," and completely negligible. Concerned as so many of them were about man's possession of a mind, a soul, a spiritual endowment of the gods, it is strange they did not anticipate our much less enterprising philosophers of some score of centuries later, and invent at least a pocket in the head, a sensorium, to contain it. But no, the Greeks, seeking a habitation for the mind, could find no better place for it than the midriff, whose rhythmic movements seemed so closely linked with what went on in the mind.

The Hebrews also attributed special dignity to that part of the body; thence Jehovah plucked man's other self. Old ideas are not always as wide of the mark as they seem. The rhythm of breathing is closely related to mental states. The Greek word for diaphragm, *phren,* appears in such everyday words as *frenzy* and *fran*tic, as well as in the discredited *phreno*logy and the erudite schizo*phreni*a.

Above the midriff the classical philosophers found the vapours of the mind; below it, the humours of the feelings. Some of these ideas persisted in physiological thought until the last century and survive in the common speech of today. Hysteric refers by derivation to the womb. The four basic human temperaments were: choleric, referring to the gall bladder; phlegmatic, related to inflammation; melancholic, black bile; and sanguine, from the blood. This classification of temperaments was revived by a modern physiologist, Pavlov, to systematize his observations of learning.

As in nearly all notions that survive as long as these fossils of language have survived, there is an element of truth, of observation, in them. States of mind are certainly related to the organs and liquors designated, and may even be said in a sense to originate in them. The philosopher, William James, was responsible with Lange for a complete theory of emotion which invoked activity in the viscera as the essential precursor of deep feeling. Some of the most primitive and finest phrases in English imply this dependence of sincere or deep emotion on heart or bowels. But communication of thought is so rapid that the Greeks overlooked the existence or need of a relay station. And no doubt it is for the same reason that we all seem particularly given to the same error of over-simplification when we first begin, or refuse to begin, to consider how the mind works. We know what makes us happy or unhappy. Who, in the throes of sea-sickness, would think of dragging in the brain to account for his melancholy state?

More curious still is Greek negligence of the brain, considering their famous oracular behest, "Know thyself." Here indeed was speculation, the demand for a mirror, insistence upon a mirror. But for whom, for what? Was there, among the mysteries behind the altar, concealed perhaps in the Minerva myth, a suspicion of something more in the head than a thing, and that the organ which had to do the knowing of itself must be an organ of reflection?

The brain remained for more than two thousand years in the dark after its coming of age. When it was discovered by the anatomist, he explored it as a substance in which might be found the secret dwelling of intelligence; for by that time the mind had moved from the diaphragm to the upper story, and Shakespeare had written of the brain, "which some suppose the soul's frail dwelling-house." Dissection was high adventure in those days. Most people believed what an ironical writer today was "astonished to learn," that "it is possible for anger, envy, hatred, malice, jealousy, fear and pride, to be confined in the same highly perishable form of matter with life, intelligence, honesty, charity, patience and truth." The search for such prize packets of evil and virtue in the brain tissue, dead or alive, could only lead to disappointment. The anatomist had to be satisfied with weighing the "grey matter"—about 50 ounces for man and 5 less for woman—and making sketches of the very complicated and indeed perishable organisation of nerves and cells which his knife revealed. He could do little more. It should enlighten us at once as to the essential character of brain activity, that there was no possible understanding of the mechanism of the brain until the key to it, the electrical key, was in our hands.

There were some flashes of foresight, sparks in the scientific dark, before Galvani put his hand on the key. What generated all the speculations of the day was a new notion in science, the conception of physical motion which began to acquire importance with Galileo and continued with Newton and into our own times with Rutherford and Einstein. First among these imaginative flashes may be mentioned the novel proposal made by the 16th Century philosopher, Hobbes, when disputing the dualist theory of Descartes. The French philosopher contemplated a non-spatial mind influencing the body through the brain, and suggested the pineal gland as the rendezvous for mind and matter. The proposal advanced by Hobbes, in rejecting this popular theory, was that thought should be regarded as being produced by bodies in motion. Hobbes was born in the year of the Spanish Armada; the Royal Society had received its charter seventeen years before he died in 1679.

The controversy about the residential status of the mind is almost as much out of date as that in which the non-existence of motion seemed to be proved by the hare and tortoise fable. But the value of Hobbes' speculation was enduring; the observation and correlation of mental and physical phenomena are today a routine of physiological research.

More specific than the speculation of Hobbes was that of Dr. David Hartley about a century later. Hartley in 1749 anticipated by two hundred years the kind of theory of mental function for which evidence has been found in the last year or two. His "Observations on Man, his Frame, his Duty and his Expectations" is a milestone in the history of English thought. Hartley, a contemporary of Newton and Hume, was a pioneer of what he termed the "doctrine of mechanism." According to this, he suggested, mental phenomena are derived from rhythmic movements in the brain—vibrations, he called them; upon these is superimposed a fine structure of "vibratiuncles" which give thought and personality their subtle shades and variations. Hartley realised quite well the value of the plastic and compact virtues such a system might have. He was also the first to develop the theory of "association of ideas" in a rigorous form, relating this to his "vibratiuncles" in a manner which we should now consider strictly scientific in the sense that it is susceptible to experimental test. It is difficult for us to appreciate the originality of his notions, the gist of which is now a commonplace of electrophysiology.

Hartley wrote nearly half a century before Galvani (1737–1798) and with him we might say farewell to fancy. But to pass over the famous Galvani-Volta controversy with the bald statement that the one claimed to have discovered electricity in animals and the other its generation by metals, would be unfair to any reader who may not know how strangely truth came out of that maze of error.

The incident began with an experiment made by Luigi and Lucia Galvani in the course of their long and patient study of that still fresh mystery, electricity. The word had been in use since William Gilbert coined it in the 16th century from *elektron,* meaning amber, another pretty semantic shift; and Henry Cavendish had already, eight years before the incident, determined the identity of its dynamic laws with those of gravitation. Everybody in high society was familiar with the effects of discharges from Leyden jars upon the lifeless muscles of executed criminals; and Louis XV had, in the words of Silvanus Thomson, "caused an electric shock from a battery of Leyden jars to be administered to 700 Carthusian monks joined hand to hand, with prodigious effect." But in Bologna in 1790 the professor of anatomy had a notion that it was atmospheric electricity which acted upon the muscle tissues of animals. On a stormy evening, one version of the story goes, he and his wife had the curious idea of testing this point by tying a dead frog to the top of the iron balustrade of the court-yard, apparently using copper wire to hold it by the leg. They expected that, as the storm approached, the frog would be convulsed by electric shocks. And, as they watched the thunder cloud come near, so indeed it happened; the dead frog, hanging against the iron grill, twitched in repeated convulsions.

Further experiments convinced the Galvani that they had witnessed a form of electricity derived from living processes, not merely from the atmosphere. He published a famous account of his experiments on the relation of animal tissue to electricity: *De viribus Electricitatis in Motu Musculari Commentarius* (1791). Volta seized upon this to refute the whole of Galvani's thesis, repeating his experiment not only without the storm but without the frog, proving that the electricity in question could be generated by copper and zinc sheets. This "current electricity" as it was called, was therefore metallic, and no nonsense about any animal variety. So ended a controversy and a friendship. So began the science of electrical engineering.

Eppur, the Galvani might have repeated, *si muove.* For their discredited experiment had truly revealed, not indeed what they supposed, but something more wonderful. What had happened was that, swaying in the wind, the suspended frog had come into contact with the iron bars, between which and the copper wire a current had been generated, activating its muscles. The Galvani had demonstrated the electrical aspect of nervous stimulation.

This was an event as important to the physiologist as its counter-event was to the physicist; it was the starting-point of that branch of the science with which we are concerned here, electrophysiology.

Volta's counter-demonstration led directly to the invention of the electric battery, and economic opportunity evoked electrical engineering from the Voltaic pile. There was no such incentive for research when, a generation later, the existence of animal electricity was proved. Instead, the discovery was exploited by the academic dilettante and the quack. The Aristotelian doctors of the period, assuming that where there is electricity there is magnetism, saw in it proof also of Mesmer's "*Propositions*" which had been published in his "*Mémoire sur la Découverte du Magnétisme Animal*" in 1779, floundering deeper into mystification than Dr. Mesmer himself, who had at least declared in his "*Mémoire*" that he used the term analogically, and that he "made no further use of electricity or the magnet from 1776 onwards."

There is still controversy about the origin and nature of animal electricity. Nobody who has handled an electric eel will question the ability of an animal to generate a formidable voltage; and the current is demonstrably similar in effect to that of a mineral dry cell. On the other hand, there is no evidence that the electric energy in nerve cells is generated by electro-magnetic induction or by the accumulation of static charge. The bio-chemist finds a complicated substance, acetylcholine, associated with electric changes; it would be reasonable to anticipate the presence of some such substance having a role at least as important as that of the chemicals in a Leclanché cell.

We know that living tissue has the capacity to concentrate

potassium and distinguish it from sodium, and that neural electricity results from the differential permeability of an inter-face, or cell-partition, to these elements, the inside of a cell being negatively charged, the outside positively. Whether we call this a chemical or an electrical phenomenon is rather beside the point. There would be little profit in arguing whether a flash-lamp is an electrical or chemical device; it is more electrical than an oil lamp, more chemical than a lightning flash. We shall frequently refer to changes of potential as electrical rhythms, cycles of polar changes, more explicitly electro-chemical changes. We shall be near the truth if we keep in mind that electrical changes in living tissue, the phenomena of animal electricity, are signs of chemical events, and that there is no way of distinguishing one from the other in the animal cell or in the mineral cell. The current of a nerve impulse is a sort of electro-chemical smoke-ring about two inches long travelling along the nerve at a speed of as much as 300 feet per second.

The neglect and mystification which obscured Galvani's discovery, more sterile than any controversy, forced electrophysiology into an academic backwater for some decades. A few experiments were made; for example, by Biedermann, who published a 2-volume treatise called *Electrophysiology,* and by Dubois-Reymond, who introduced Michael Faraday's induction coil into the physiological laboratory and the term faradisation as an alternative to galvanisation into the physiotherapist's vocabulary. Faraday's electrical and electrifying research began in 1831, the date also of the foundation of the British Association for the Advancement of Science; but physiology long remained a backward child of the family.

Hampered though these experimenters were by lack of trustworthy equipment—they had to construct their own galvanometers from first principles—they gradually accumulated enough facts to show that all living tissue is sensitive in some degree to electric currents and, what is perhaps more important, all living tissue generates small voltages which change dramatically when the tissue is injured or becomes active.

These experiments were not concerned with the brain; they were made on frog's legs, fish eggs, electric eels and flayed vermin. Nor could the brain be explored in that way.

> Following life through creatures you dissect,
> You lose it in the moment you detect.

It took a war to bring the opportunity of devising a technique for exploring the human brain—and two more wars to perfect it. Two medical officers of the Prussian army, wandering through the stricken field of Sedan, had the brilliant if ghoulish notion to test the effect of the Galvanic current on the exposed brains of some of the casualties. These pioneers of 1870, Fritsch and Hitzig, found that when certain areas at the side of the brain were stimulated by the current, movements took place in the opposite side of the body.

That the brain itself produces electric currents was the discovery of an English physician, R. Caton, in 1875.

This growing nucleus of knowledge was elaborated and carried further by Ferrier in experiments with the "Faradic current." Toward the end of the century there was a spate of information which suggested that the brain of animals possessed electrical properties related to those found in nerve and muscle. Prawdwicz-Neminski in 1913 produced what he called the "electro-cerebrogram" of a dog, and was the first to attempt to classify such observations.

The electrical changes in the brain, however, are minute. The experiments of all these workers were made on the exposed brains of animals. There were no means of amplification in those days, whereby the impulses reaching the exterior of the cranium could be observed or recorded, even if their presence had been suspected. On the other hand, the grosser electrical currents generated by the rhythmically contracting muscles of the heart were perceptible without amplification. Electro-cardiography became a routine clinical aid a generation before the invention of the thermionic tube made it possible to study the electrical activity of the intact human brain.

From an unexpected quarter, at the turn of the century, came an entirely new development. Turn up the section on the brain in a pre-war textbook of physiology and you will find gleanings from clinical neuro-anatomy and—Pavlov. Almost as if recapitulating the history of physiological ideas, Pavlov's work began below the midriff. He found that the process of digestion could not be understood without reference to the nervous system, and so commenced his laborious study of learning in animals.

In the gospel according to Stalin, Pavlov founded not merely a branch of physiology as Galvani had done, but a whole new science—Soviet physiology. His work indeed was original; it owed nothing to Galvani, lying quite outside electrophysiology, to which it was nevertheless eventually, though not in Pavlov's day, to contribute so much in the way of understanding.

For nearly two generations Pavlov's experiments were the major source of information on brain physiology. Workers in the English laboratories had not permitted themselves to explore further than the top of the spinal cord. One took an anatomical glance at the brain, and turned away in despair. This was not accountable to any peculiar weakness of physiological tradition but to the exigencies of scientific method itself. A discipline had been building up through the centuries which demanded that in any experiment there should be only one variable and its variations should be measurable against a controlled background. In physiology this meant that in any experiment there should be only one thing at a time under investigation—one single function, say, of an organ—and that the changes of material or function should be measurable. There seemed to be no possibility of isolating one single variable, one single mode of activity, among the myriad functions of the brain. Thus there was something like a taboo against the study of the brain. The success of Pavlov in breaking this taboo early in the century was due to his contrivance for isolating his experimental animals from all but two stimuli; his fame rests on his measurement of responses to the stimuli.

There was no easy way through the academic undergrowth of traditional electrophysiology to the electrical mechanisms underlying brain functions. The Cambridge school of electrophysiology, under a succession of dexterous and original experimenters beginning toward the end of the last century, developed its own techniques in special fields of research, particularly in the electrical signs of activity in muscles, nerves and sense organs. At the same time, the Oxford school under the leadership of Sherrington was beginning to unravel some of the problems of reflex function of the spinal cord. In both these schools the procedure adopted, to comply with the traditional requirements of scientific method, was to dissect out or isolate the organ or part of an organ to be studied. This was often carried to the extreme of isolating a single nerve fibre only a few thousandths of a millimetre in diameter, so as to eliminate all but a single functional unit.

Imagine, then, how refreshing and tantalizing were the reports from Pavlov's laboratory in Leningrad to those engaged on the meticulous dissection of invisible nerve tendrils and the analysis of the impulses which we induced them to transmit. After four years spent working literally in a cage and chained by the ankle—not for punishment but for electrical screening—enlargement came when my professor of that date, the late Sir Joseph Barcroft, assigned me to establishing a laboratory in association with a visiting pupil of Pavlov, Rosenthal. We spent a year or so on mastering the technique and improving it by the introduction of certain electronic devices. The Russian results were confirmed. To do more than this would have required staff and equipment far beyond the resources of the Cambridge laboratory.

Meanwhile, another major event in the history of physiology had taken place. Berger, in 1928, at last brought Hartley's vibrations into the laboratory and with them a method which seemed to hold out the promise of an investigation of electrical brain activity as precise as were the reflex measurements of Pavlov. When Pavlov visited England some time after we heard of this, as the English exponent of his work I had the privilege of discussing it with him on familiar terms. Among other things, I asked him if he saw any relation between the two methods of observing cerebral activity, his method and Berger's. The latter, I was even then beginning to suspect, might in some way provide a clue to *how* the conditioning of a reflex was effected in the brain. But Pavlov showed no desire to look behind the scenes. He was not in the least interested in the mechanism of cerebral events; they just happened, and it was the happening and its consequence that interested him, not how they happened. Soviet physiology embalmed the body of this limited doctrine as mystically as the body of Lenin, for the foundations of their science. The process of conditioning reflexes has a specious affinity with the Marxian syllogism. Others have found in the phenomena sufficient substantiation for a gospel of Behaviourism.

Pavlov was before his time. He would have been a greater man, his work would have been more fertile in his lifetime, and Russian science might have been spared a labyrinthine deviation, had the work of Berger come to acknowledgement and fruition in his day. But again there was delay; Berger waved the fairy wand in 1928; the transformation of Cinderella was a process of years.

There were reasons for this delay. For one thing, Berger was not a physiologist and his reports were vitiated by the vagueness and variety of his claims and the desultory nature of his technique. He was indeed a surprisingly unscientific scientist, as personal acquaintance with him later confirmed.

The first occasion on which the possibilities of clinical electroencephalography were discussed in England was quite an informal one. It was in the old Central Pathological Laboratory at the Maudsley Hospital in London, in 1929. The team there under Professor Golla was in some difficulty about electrical apparatus; they were trying to get some records of the "Berger rhythm," using amplifiers with an old galvanometer that fused every time they switched on the current. Golla was anxious to use the Matthews oscillograph, then the last word in robust accuracy, to measure peripheral and central conduction times. I was still working at Cambridge under the watchful eye of Adrian and Matthews and was pleased to introduce this novelty to him and at the same time, with undergraduate superiority, put him right on a few other points. When, at lunch around the laboratory table, he referred to the recent publication of Berger's claims, I readily declared that anybody could record a wobbly line, it was a string of artefacts, even if there were anything significant in it there was nothing you could measure, and so on. Golla agreed with milder scepticism, but added: "If this new apparatus is as good as you say, it should be easy to find out whether Berger's rhythm is only artefact; and if it isn't, the frequency seems remarkably constant; surely one could measure that quite accurately." And he surmised that there would be variations of the rhythm in disease.

Cambridge still could not accept the brain as a proper study for the physiologist. The wobbly line did not convince us or anybody else at that time. Berger's "elektrenkephalograms" were almost completely disregarded. His entirely original and painstaking work received little recognition until in May, 1934, Adrian and Matthews gave the first convincing demonstration of the "Berger rhythm" to an English audience, a meeting of the Physiological Society at Cambridge.

Meanwhile, Golla was reorganising his laboratory, and his confidence in the possibilities of the Berger method was growing. When he invited me to join his research team as physiologist at the Central Pathological Laboratory, my first task was to visit the German laboratories, including particularly that of Hans Berger.

Berger, in 1935, was not regarded by his associates as in the front rank of German psychiatrists, having rather the reputation of being a crank. He seemed to me to be a modest and dignified person, full of good humour, and as unperturbed by lack of recognition as he was later by the fame it eventually brought him. But he had one fatal weakness: he was completely ignorant of the technical and physical basis of his method. He knew nothing about mechanics or electricity. This handicap made it impossible for him to correct serious shortcomings in his experiments. His method was a simple adaptation of the electrocardiographic technique by which the electrical impulses generated by the heart are recorded. At first he inserted silver wires under the subject's scalp; later he used silver foil bound to the head with a rubber bandage. Nearly always he put one electrode over the forehead and one over the back of the head; leads were taken from these to an Edelmann galvanometer, a light and sensitive "string" type of instrument, and records were taken by an assistant photographer. A potential change of one-ten-thousandth of a volt—a very modest sensitivity by present standards—could just be detected by this apparatus. Each record laboriously produced was equivalent to that of two or three seconds of modern continuous pen recording. The line did show a wobble at about 10 cycles per second. (See Figure 3.) He had lately acquired a tube amplifier to drive his galvanometer, and his pride and pleasure in the sweeping excursions of line obtained by its use were endearing.

Berger carried the matter as far as his technical handicap permitted. He had observed that the larger and more regular

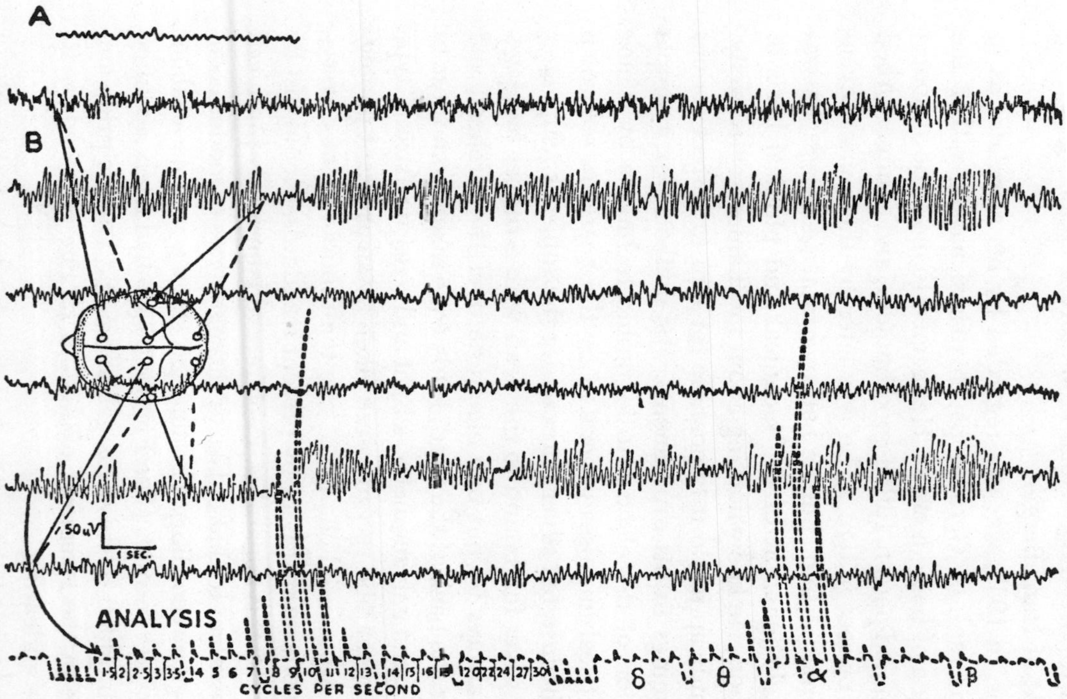

Figure 3. "The line did show a wobble at about 10 cycles per second." (a) A tracing from one of Berger's earliest records. (b) Record from a modern laboratory showing consistency of automatic analysis over 10-second periods.

rhythms tended to stop when the subject opened his eyes or solved some problem in mental arithmetic. This was confirmed by Adrian and Matthews with leads from electrodes on Adrian's head attached to a Matthews amplifier and ink-writing oscillograph. This superior apparatus, and a more careful location of electrodes, enabled them to go a step further and prove that the 10 cycles per second rhythm arises in the visual association areas in the occiput and not, as Berger supposed, from the whole brain.

Only some years later was it realised what an important step this was. Its significance could not be recognised while so little was known about the components of the "wobbly line," the electroencephalogram or, abbreviated, EEG. Unavoidably at the time, the significance of the salient character of the normal EEG was overlooked; it was found, in Adrian's phrase, "disappointingly constant." The attention of many early workers in electroencephalography therefore turned from normal research to the study of nervous disease. In immediate rewards this has always been a rich field. In this instance, a surprising state was soon reached wherein what might be called the electropathology of the brain was further advanced than its electrophysiology.

In the pathological laboratory, Golla's earlier surmise, that there would be variations of the rhythmic oscillation in disease, was soon verified. A technique was developed there by which the central point of the disturbance in the tissue could be accurately determined. For surgery, the immediate result of perfecting this technique was important; it made possible the location of tumours, brain injuries, or other physical damage to the brain. It was helpful in many head cases during the war as well as in daily surgical practice.

The study of epilepsy and mental disorders also began to occupy the attention of many EEG workers. The difficulties encountered in these subjects threw into prominent relief the essential complexity of the problem as compared with those of classical physiology. The hope of isolating single functions had now been abandoned; those who entered this field were committed to studying the brain as a whole organ and through it the body as a whole organism. They were therefore forced to multiply their sources of information.

It is now the general EEG practice, not only for clinical purposes, but in research, to use a number of electrodes simultaneously, indeed as many as possible and convenient. The standard make of EEG recorder has eight channels. Eight pens are simultaneously tracing lines in which the recordist, after long experience, can recognise the main components of a complex graph. The graphs can also be automatically analysed into their component frequencies. A more satisfactory method of watching the electrical changes in all the main areas, as in a moving picture, a much more informative convention than the drawing of lines, has been devised at the Burden Neurological Institute. This will be described after a simple explanation of what is meant by the rhythmic composition of the normal EEG; for its nature, rather than the methods of recording and analysing it, is of first importance for understanding what follows.

If you move a pencil amply but regularly up and down on a paper that is being drawn steadily from right to left, the result will be a regular series of curves. If at the same time the paper is moving up and down, another series of curves will be added to the line drawn. If the table is shaking, the vibration will be added to the line as a ripple. There will then be three components integrated in the one wavy line, which will begin to look something like an EEG record. The line gives a coded or conventional record of the various frequencies and amplitudes of different physical movements. In similar coded or integrated fashion the EEG line reports the frequencies and amplitudes of the electrical changes in the different parts of the brain tapped by the electrodes on the scalp, their minute currents being relayed by an amplifier to the oscillograph which activates the pens.

All EEG records contain many more components than this; some may show as many as 20 or 30 at a time in significant sizes. Actually there may be tens of thousands of impulses woven together in such a manner that only the grosser combinations are discernible.

A compound curve is of course more easily put together than taken apart. (See Figure 4.) The adequate analysis of a few inches of EEG records would require the painstaking computation of a mathematician—it might take him a week or so. The modern automatic analyser in use in most laboratories writes out the values of 24 components every 10 seconds, as well as any averaging needed over longer periods.

The electrical changes which give rise to the alternating currents of variable frequency and amplitude thus recorded arise in the cells of the brain itself; there is no question of any other power supply. The brain must be pictured as a vast aggregation of electrical cells, numerous as the stars of the Galaxy, some 10 thousand million of them, through which surge the restless tides of our electrical being relatively thousands of times more potent than the force of gravity. It is when a million or so of these cells repeatedly fire together that the rhythm of their discharge becomes measureable in frequency and amplitude.

What makes these million cells act together—or indeed what causes a single cell to discharge—is not known. We are still a long way from any explanation of these basic mechanics of the brain. Future research may well carry us, as it has car-

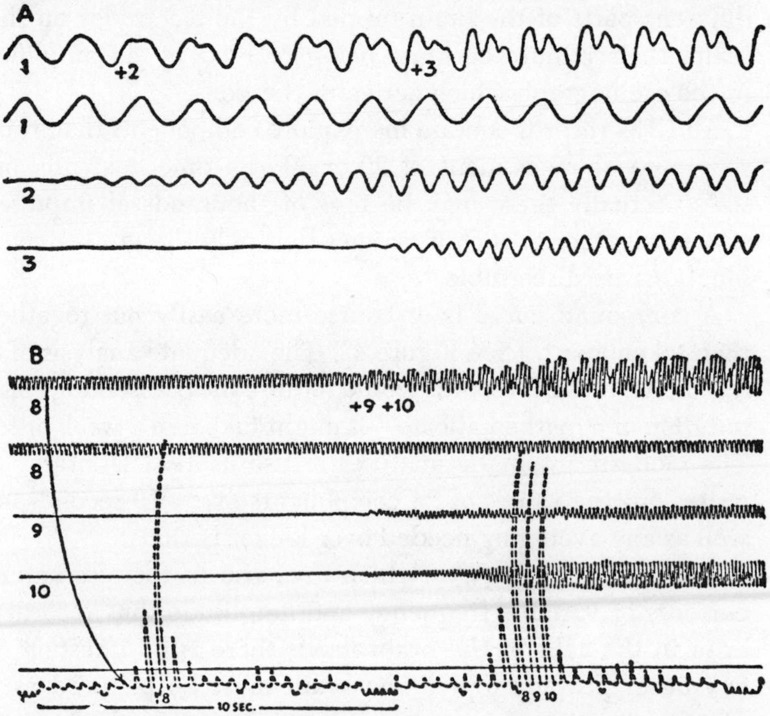

Figure 4. "A compound curve is more easily put together than taken apart." (a) A compound curve in which the three components can be detected by visual inspection, ratios 1:2 and 2:3. (b) The three components (ratios 8:9, 9:10) of this compound curve cannot be determined at sight. The bottom line shows their frequencies automatically recorded every 10 seconds. Note the accidental similarity of this curve to the EEG record of alpha rhythms in Figure 3 (b).

ried the physicist in his attempt to understand the composition of our atomic being, into vistas of ever increasing enchantment but describable only in the convention of mathematical language. Today, as we travel from one fresh vista to another, the propriety of the language we use, the convention we adopt, becomes increasingly important. Arithmetic is an adequate language for describing the height and time of the tides, but if we want to predict their rise and fall we have to use a different language, an algebra, with its special notation and theorems. In similar fashion, the electrical waves and tides in the brain can be described adequately by counting, by arithmetic; but there are many unknown quantities when we come to the more ambitious purposes of understanding and predicting brain behaviour—many *x*'s and *y*'s; so it will have to have its algebra. The word is forbidding to some people; but, after all, it means no more than "the putting together of broken pieces."

EEG records may be considered, then, as the bits and pieces of a mirror for the brain, itself *speculum speculorum*. They must be carefully sorted before even trying to fit them together with bits from other sources. Their information comes as a conventional message, coded. You may crack the code, but that does not imply that the information will necessarily be of high significance. Supposing, for instance, you pick up a coded message which you think may be about a momentous political secret. In the first stage of decoding it you might ascertain that the order of frequency of the letters was *ETAONI*. This does not sound very useful information; but reference to the letter-frequency tables would assure you at least that it was a message in English and possibly intelligible. Likewise, we watch the frequencies as well as the amplitude and origin of the brain rhythms, knowing that many earnest seekers for the truth have spent lifetimes trying to decipher what they thought were real messages, only to find that their horoscopes and alembics contained gibberish. The scientist is used to such hazards of research; it is only the ignorant and superstitious who regard him, or think he regards himself, as a magician or priest who is right about everything all the time.

Brain research has just about reached the stage where the letter frequencies of the code indicate intelligibility and their grouping significance. But there is this complication. The ordinary coded message is a sequence in time; events in the brain are not a single sequence in time—they occur in three-dimensional space, in that one bit of space which is more crowded with events than any other we can conceive. We may tap a greater number of sectors of the brain and set more pens scribbling; but the effect of this will only be to multiply the number of code signals, to the increasing embarrassment of the observer, unless the order and inter-relation of the signals can be clarified and emphasised. Redundancy is already a serious problem of the laboratory.

The function of a nervous system is to receive, correlate, store and generate many signals. A human brain is a mechanism not only far more intricate than any other but one that has a long individual history. To study such a problem in terms of frequency and amplitude as a limited function of time—in wavy lines—is at the best over-simplification. And the redundancy is indeed enormous. Information at the rate of about 3,600 amplitudes per minute may be coming through each of the eight channels during the average recording period of 20 minutes; so the total information in a routine record may be represented by more than half a million numbers; yet the usual description of a record consists only of a few sentences. Only rarely does an observer use more than one-hundredth of one per cent of the available information.

"What's in a brain that ink may character . . . ?"

For combining greater clarity with greater economy, many elaborations of methods have been adopted in clinic and laboratory. They still do not overcome the fundamental embarrassment of redundancy and the error of over-simplification, both due to the limitations of a time scale. A promising alternative is a machine that draws a snapshot map instead of a long history, projecting the electrical data visually on a spatial co-ordinate system which can be laid out so as to represent a simple map or model of the head. This moving panorama of the brain rhythms does approximate to Sherrington's "enchanted loom where millions of flashing shuttles weave a dissolving pattern, always a meaningful pattern though never an abiding one." (Figure 5.)

We have called the apparatus which achieves this sort of effect at the Burden Institute a toposcope, by reason of its display of topographic detail. The equipment was developed by Harold Shipton, whose imaginative engineering transformed the early models from entertainment to education. Two of its 24 channels are for monitoring the stimuli; the others, instead of being connected with pens, lead the electrical activity of the brain tapped by the electrodes for display on the screens of small cathode-ray tubes. So instead of wavy lines on a moving paper, the observer sees, to quote Sherrington again, "a sparkling field of rhythmic flashing points with trains of travelling sparks hurrying hither and thither." Assembled in the display console, 22 of the tubes give a kind of Mercator's projection of the brain. Frequency, phase and time relations of the rhythms are shown in what at first appears to be a completely bewildering variety of patterns in each tube and in their ensemble. Then, as the practised eye gains familiarity with the scene, many details of brain activity are seen for the first time. A conventional pen machine is simultaneously at the disposal of the observer, synchronised so that, by turning a switch, a written record of the activity seen in any five of the tubes can be made. Another attachment is a camera with which at the same time permanent snapshot records of the display can be obtained. (Figure 6.)

Thus, from Berger's crude galvanometer to this elaborate apparatus requiring a whole room of its own, electroencephalography has progressed from a technique to a science. Its clinical benefits, by-products of free research, are acknowledged; they can be gauged by the vast multiplication of EEG laboratories. From Berger's lone clinic have sprung several hundred EEG centres—more than 50 in England alone. Literally millions of yards of paper have been covered with frantic scribblings. In every civilized country there is a special learned society devoted to the discussion of the records and to disputation on technique and theory. These societies are banded together in an International Federation, which publishes a quarterly Journal and organises international congresses.

For a science born, as it were, bastard and neglected in infancy, this is a long way to have travelled in its first quarter of a century. If it is to provide the mirror which the brain requires to see itself steadily and whole, there is still a long road ahead. The following chapters give the prospect as seen from the present milestone, assuming that such studies are allowed to continue. Looking back, we realise that the present scale of work as compared with previous physiological research is elaborate and expensive. But our annual cost of conducting planned investigations of a fundamental nature into man's supreme faculties is less than half that of one medium tank, and the money spent on brain research in all England is barely one-tenth of one per cent of the cost of the national mental health services.

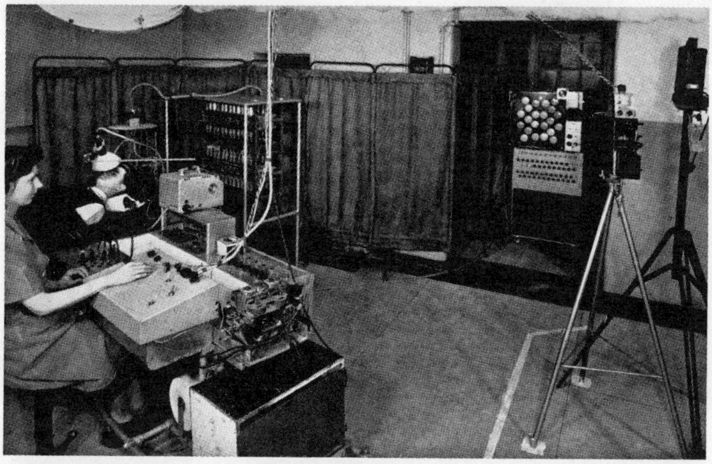

Figure 5. ". . . a moving panorama of the brain rhythms." The Toposcope Laboratory. The subject's couch and triggered stroboscope (flicker) reflector at extreme left beyond desk of 6-channel pen recorder with remote control panel. The 22-channel toposcope amplifier is in the background, the display panel at right centre, camera and projector at extreme right.

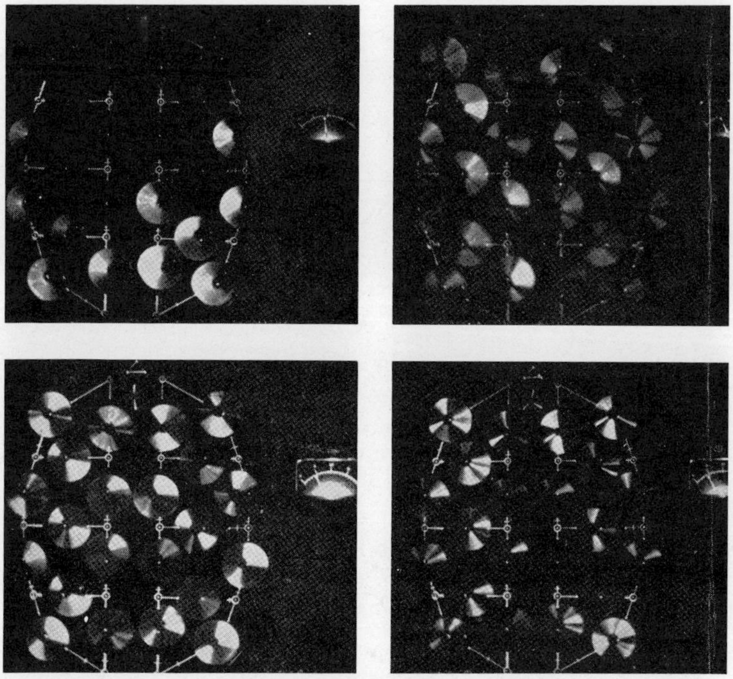

Figure 6. ". . . always a meaningful pattern though never an abiding one." Snapshots of the "sparkling field of rhythmic flashing points." Each of the tube screens, which form a chart of the head seen from above with nose at top, shows by the flashing sectors of its disc the activity of the corresponding area of the brain. (*Top left*) Resting alpha rhythms. (*Top right*) Theta rhythms in anger. (*Bottom left*) Wide response to double flashes of light. (*Bottom right*) Spread of response to triple flashes.

Physics is full of concepts of which we cannot form simple pictures. Therefore the authors, like most modern scientists, recommend taking a "mathematical view."

---

19 Scientific Imagination

Richard P. Feynman, Robert B. Leighton, and Matthew Sands

Excerpt from *The Feynman Lectures on Physics,* Volume II, 1964.

I have asked you to imagine these electric and magnetic fields. What do you do? Do you know how? How do *I* imagine the electric and magnetic field? What do *I* actually see? What are the demands of scientific imagination? Is it any different from trying to imagine that the room is full of invisible angels? No, it is not like imagining invisible angels. It requires a much higher degree of imagination to understand the electromagnetic field than to understand invisible angels. Why? Because to make invisible angels understandable, all I have to do is to alter their properties *a little bit*—I make them slightly visible, and then I can see the shapes of their wings, and bodies, and halos. Once I succeed in imagining a visible angel, the abstraction required—which is to take almost invisible angels and imagine them completely invisible—is relatively easy. So you say, "Professor, please give me an approximate description of the electromagnetic waves, even though it may be slightly inaccurate, so that I too can see them as well as I can see almost invisible angels. Then I will modify the picture to the necessary abstraction."

I'm sorry I can't do that for you. I don't know how. I have no picture of this electromagnetic field that is in any sense accurate. I have known about the electromagnetic field a long time—I was in the same position 25 years ago that you are now, and I have had 25 years more of experience thinking about these wiggling waves. When I start describing the magnetic field moving through space, I speak of the ***E***- and ***B*** fields and wave my arms and you may imagine that I can see them.

I'll tell you what I see. I see some kind of vague shadowy, wiggling lines—here and there is an *E* and *B* written on them somehow, and perhaps some of the lines have arrows on them—an arrow here or there which disappears when I look too closely at it. When I talk about the fields swishing through space, I have a terrible confusion between the symbols I use to describe the objects and the objects themselves. I cannot really make a picture that is even nearly like the true waves. So if you have some difficulty in making such a picture, you should not be worried that your difficulty is unusual.

Our science makes terrific demands on the imagination. The degree of imagination that is required is much more extreme than that required for some of the ancient ideas. The modern ideas are much harder to imagine. We use a lot of tools, though. We use mathematical equations and rules, and make a lot of pictures. What I realize now is that when I talk about the electromagnetic field in space, I see some kind of a superposition of all of the diagrams which I've ever seen drawn about them. I don't see little bundles of field lines running about because it worries me that if I ran at a different speed the bundles would disappear. I don't even always see the electric and magnetic fields because sometimes I think I should have made a picture with the vector potential and the scalar potential, for those were perhaps the more physically significant things that were wiggling.

Perhaps the only hope, you say, is to take a mathematical view. Now what is a mathematical view? From a mathematical view, there is an electric field vector and a magnetic field vector at every point in space; that is, there are six numbers associated with every point. Can you imagine six numbers associated with each point in space? That's too hard. Can you imagine even *one* number associated with every point? I cannot! I can imagine such a thing as the temperature at every point in space. That seems to be understandable. There is a hotness and coldness that varies from place to place. But I honestly do not understand the idea of a *number* at every point.

So perhaps we should put the question: Can we represent the electric field by something more like a temperature, say like the displacement of a piece of jello? Suppose that we were to begin by imagining that the world was filled with thin jello and that the fields represented some distortion—say a stretching or twisting—of the jello. Then we could visualize the field. After we "see" what it is like we could abstract the jello away. For many years that's what people tried to do. Maxwell, Ampere, Faraday, and others tried to understand electromagnetism this way. (Sometimes they called the abstract jello "ether.") But it turned out that the attempt to imagine the electromagnetic field in that way was really standing in the way of progress. We are unfortunately limited to abstractions, to using instruments to detect the field, to using mathematical symbols to describe the field, etc. But nevertheless, in some sense the fields are real, because after we are all finished fiddling around with mathematical equations—with or without making pictures and drawings or trying to visualize the thing—we can still make the instruments detect the signals from Mariner II and find out about galaxies a billion miles away, and so on.

The whole question of imagination in science is often misunderstood by people in other disciplines. They try to test our imagination in the following way. They say, "Here is a picture of some people in a situation. What do you imagine will happen next?" When we say, "I can't imagine," they may think we have a weak imagination. They overlook the fact that whatever we are *allowed* to imagine in science must be *consistent with everything else we know:* that the electric fields and the waves we talk about are not just some happy thoughts which we are free to make as we wish, but ideas which must be consistent with all the laws of physics we know. We can't allow ourselves to seriously imagine things which are obviously in contradiction to the known laws of nature. And so our kind of imagination is quite a difficult game. One has to have the imagination to think of something that has never been seen before, never been heard of before. At the same time the thoughts are restricted in a strait jacket, so to speak, limited by the conditions that come from our knowledge of the way nature really is. The problem of creating something which is new, but which is consistent with everything which has been seen before, is one of extreme difficulty.

While I'm on this subject I want to talk about whether it will ever be possible to imagine *beauty* that we can't *see*. It is an interesting question. When we look at a rainbow, it looks beautiful to us. Everybody says, "Ooh, a rainbow." (You see how scientific I am. I am afraid to say something is beautiful unless I have an experimental way of defining it.) But how would we describe a rainbow if we were blind? We *are* blind when we measure the infrared reflection coefficient of sodium chloride, or when we talk about the frequency of the waves that are coming from some galaxy that we can't see—we make a diagram, we make a plot. For instance, for the rainbow, such a plot would be the intensity of radiation vs. wavelength measured with a spectrophotometer for each direction in the sky. Generally, such measurements would give a curve that was rather flat. Then some day, someone would discover that for certain conditions of the weather, and at certain angles in the sky, the spectrum of intensity as a function of wavelength would behave strangely; it would have a bump. As the angle of the instrument was varied only a little bit, the maximum of the bump would move from one wavelength to another. Then one day the physical review of the blind men might publish a technical article with the title "The Intensity of Radiation as a Function of Angle under Certain Conditions of the Weather." In this article there might appear a graph such as the one in Fig. 20–5. The author would perhaps remark that at the larger angles there was more radiation at long wavelengths, whereas for the smaller angles the maximum in the radiation came at shorter wavelengths. (From our point of view, we would say that the light at 40° is predominantly green and the light at 42° is predominantly red.)

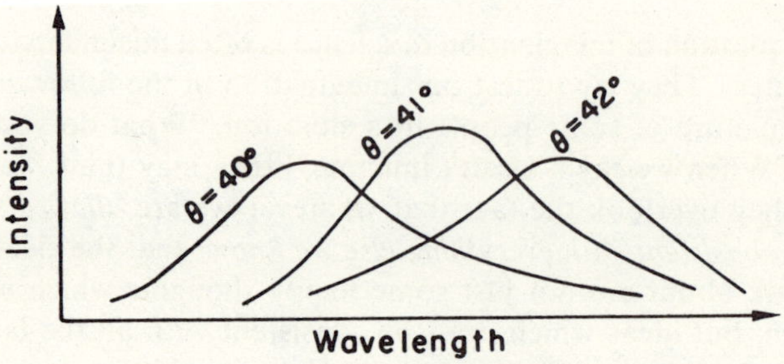

Fig. 20–5. The intensity of electromagnetic waves as a function of wavelength for three angles (measured from the direction opposite the sun), observed only with certain meteorological conditions.

Now do we find the graph of Fig. 20–5 beautiful? It contains much more detail than we apprehend when we look at a rainbow, because our eyes cannot see the exact details in the shape of a spectrum. The eye, however, finds the rainbow beautiful. Do we have enough imagination to see in the spectral curves the same beauty we see when we look directly at the rainbow? I don't know.

But suppose I have a graph of the reflection coefficient of a sodium chloride crystal as a function of wavelength in the infrared, and also as a function of angle. I would have a representation of how it would look to my eyes if they could see in the infrared—perhaps some glowing, shiny "green," mixed with reflections from the surface in a "metallic red." That would be a beautiful thing, but I don't know whether I can ever look at a graph of the reflection coefficient of NaCl measured with some instrument and say that it has the same beauty.

On the other hand, even if we cannot see beauty in particular measured results, we *can* already claim to see a certain beauty in the equations which describe general physical laws. For example, in the wave equation (20.9), there's something nice about the regularity of the appearance of the x, the y, the z, and the t. And this nice symmetry in appearance of the x, y, z, and t suggests to the mind still a greater beauty which has to do with the four dimensions, the possibility that space has four-dimensional symmetry, the possibility of analyzing that and the developments of the special theory of relativity. So there is plenty of intellectual beauty associated with the equations.

$$\frac{\partial^2\psi}{\partial x^2} + \frac{\partial^2\psi}{\partial y^2} + \frac{\partial^2\psi}{\partial z^2} - \frac{1}{c^2}\frac{\partial^2\psi}{\partial t^2} = 0 \qquad (20.9)$$

Magnifying glasses, spectacles, cameras, projectors, eyes, microscopes, telescopes—they all work on the same simple principles.

20 Lenses and Optical Instruments

Physical Science Study Committee

From the textbook *Physics* by Physical Science Study Committee, 1962.

OPTICAL instruments — cameras, projectors, telescopes, and microscopes — usually are built with lenses; that is, with pieces of refractive materials to converge or diverge light according to our design. A whole industry is devoted to the design and production of such instruments or their components. All these instruments are understood and designed in terms of Snell's law. The whole field of applications rests on the simple summary of refraction that we reached in the last chapter, $n_1 \sin \theta_1 = n_2 \sin \theta_2$. Most optical technology stems from this little bit of physics.

In this chapter, we want to learn how the laws of refraction are related to the construction of lenses and optical systems. An extensive treatment of the design of optical systems is, however, beyond the purpose of this chapter.

14–1. The Convergence of Light by a Set of Prisms

We found in Chapter 12 that we could control and redirect light beams by the use of curved mirrors. Devices that can accomplish similar purposes through refraction, instead of reflection, are called lenses. To understand how a lens operates, let us examine the behavior of light in passing through the combination of a plate of glass with parallel sides and the two triangular prisms shown in Fig. 14–1 (a). If a parallel beam of light falls on this system from the left, so that it is normally incident on the plate of glass, it will behave as indicated by the rays shown in the figure. The light that passes through the plate in the center will continue along its original direction, since the angle of incidence is 0°. Light striking the upper prism will be deviated downward by an amount depending on the opening angle of the prism and on its index of refrac-

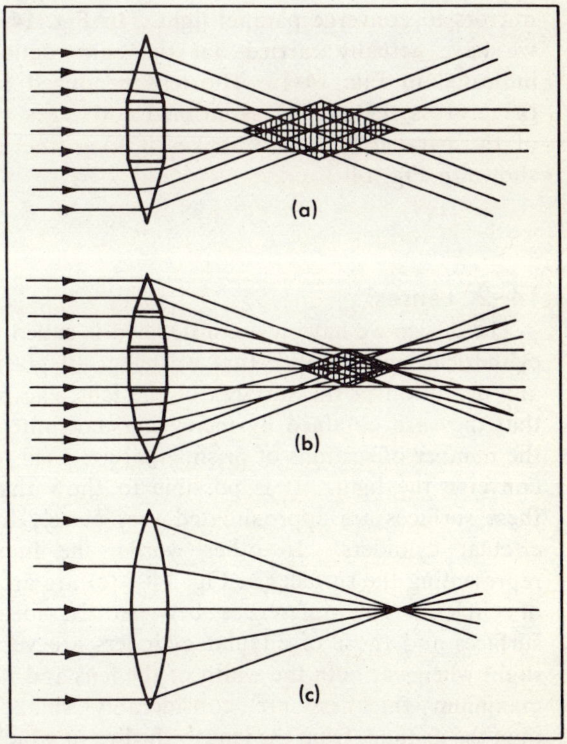

14–1. Construction of a lens by the process of subdividing prismatic sections.

tion. Similarly, light striking the lower prism will be deviated upward. As a result, there is a region, shown shaded in the figure, through which passes almost all of the light that falls on the plate and the prisms.

The convergence of a parallel beam of light into a limited region by this system resembles the convergence of a similar beam by a set of mirrors. (See Section 12–6.) While working with mirrors, we decreased the size of the region into which the light was converged by using an increased number of mirrors, each smaller than the original one. Let us try the same scheme here. Fig. 14–1 (b) shows parts of the central plate and of the two prisms cut away and replaced by pieces of new prisms. The size of the shaded region is clearly smaller than it was before.

If we continue the process of removing parts of the prisms and replacing them by sections having smaller opening angles, we come closer and closer to a piece of glass with the smoothly curved surface shown in Fig. 14–1 (c). This device is the limit that is approached as we increase the number of prisms indefinitely, just as the parabolic mirror of Fig. 12–16 was the limit approached as we used more and more plane mirrors to converge parallel light. In Fig. 14–2 we have actually carried out the construction indicated in Fig. 14–1. The lens produced by the process that we have outlined converges all of the parallel light that strikes it to a line as shown in Fig. 14–3.

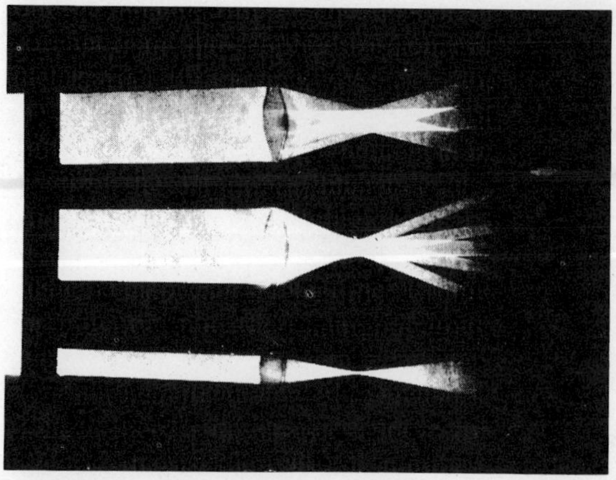

14–2. The experiments diagramed in Fig. 14–1.

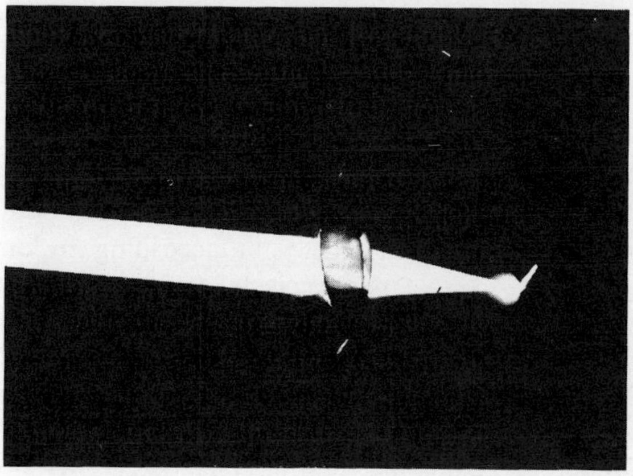

14–3. Convergence of light by a cylindrical lens like the one shown in Fig. 14–1. Note that the light is brought to a focus along a line.

14–2. Lenses

The device we have just constructed is called a cylindrical lens. Notice that we have not given any definition of the surfaces of the lens, except that they are obtained by increasing indefinitely the number of sections of prisms that are used to converge the light. It is possible to show that these surfaces are approximated very closely by circular cylinders. In other words, the lines representing the surfaces in Fig. 14–1 (c) are arcs of circles. The differences between the ideal surfaces and those of circular cylinders are very slight whenever both the width of the lens and its maximum thickness are considerably smaller than the distance from the lens to the line at which parallel light is converged.

Cylindrical lenses bring the light from a distant point source of light to a focus along a line. For most purposes we prefer that the light from a point source should be focused at a point. This focusing can be accomplished by constructing a lens whose surfaces curve equally in all directions. Such surfaces are portions of spheres. Almost all lenses are bounded by two spherical surfaces.

The line passing through the center of the lens and on which the centers of the two spheres are located is called the *axis* of the lens. The point on this axis at which incident parallel rays focus or converge is the *principal focus*, F. The distance of the principal focus from the center of the lens is known as the *focal length*, f.

The two surfaces of a lens do not always have the same radius. For example, the lens shown in Fig. 14–4 has a spherical surface of much larger

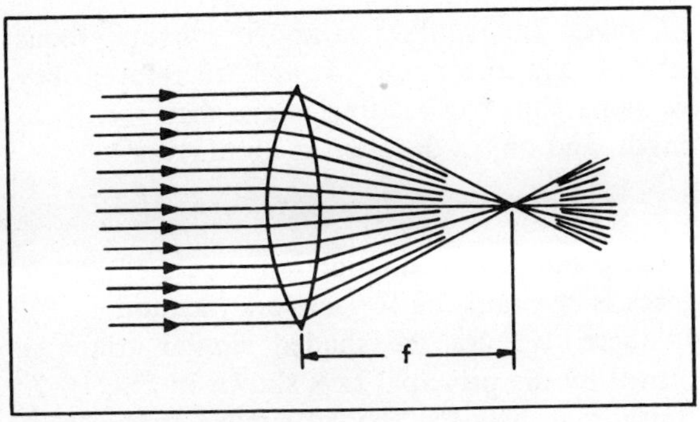

14–4. A lens with surfaces of unequal radii.

radius at its right-hand boundary than it has at the left.

If a lens is thin compared to its focal length, it makes no difference which side of the lens the light enters, the focal length is always the same. This symmetry is obvious if the lens is itself symmetric. That it is true for all thin lenses can easily be shown by an experiment in which a lens is used to focus the parallel rays of the sun to a point on a piece of paper or cardboard. If the lens is then flipped over, the focus occurs at the same distance from the lens (Fig. 14–5).

This result is also predicted by a detailed application of Snell's law from which we find

$$\frac{1}{f} = (n - 1)\left(\frac{1}{R_1} + \frac{1}{R_2}\right),$$

where R_1 and R_2 are the radii of the opposing spherical surfaces.* We see that interchanging R_1 and R_2, which is equivalent to turning the lens over, does not change the calculated value of f.

From this equation, we can also see that when R_1 and R_2 are small, the lens will have a short focal length. This is illustrated in Fig. 14–6 where we see that the paths of light rays through the lens in (b) are bent more sharply, so that the focal length is shorter than in (a).

14–3. Real Images Formed by Lenses

We have thus far concentrated our attention on the focusing of light by a lens when the light comes from a very distant object. In the practical use of lenses, we are commonly interested in the light coming from near-by objects and we all know that lenses do form images of such objects. We can locate the images with the help of the knowledge that we have gained about the behavior of initially parallel rays.

Fig. 14–7 shows a lens, an object H_o, and its image H_i. To find the location of this image, we draw the two principal rays from the top of the object, one ray parallel to the axis and the other through the principal focus F_2. The ray parallel to the axis is bent by the lens so as to pass through the principal focus F_1. We also know that rays coming from the right and parallel to the axis would be deviated to pass through the other principal focus F_2. It follows from the reversibility of light paths that the ray from the top of H_o that passes through F_2 from the left must travel parallel to the axis after it has passed through the lens. All rays starting from the top of H_o will converge very close to the point at which these two bent rays intersect. This point is therefore the real image of the top of H_o.

We could have chosen any other point on the object and located its image in the same way. Had we done so for a number of points, we would have found that the image, H_i, falls along the line that is shown in the figure.

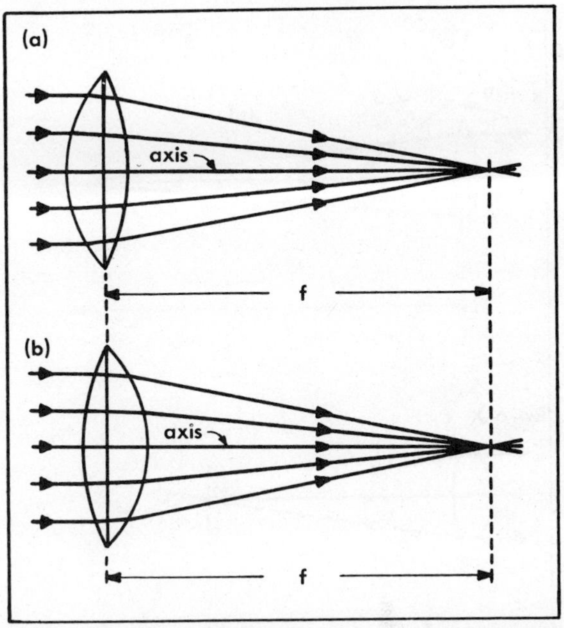

14–5. The principal foci of a lens. For thin lenses the focal distance is the same for parallel light entering either the side with a small radius (a) or the side with a larger radius (b).

*We shall not give the proof of this "lens maker's" formula here. Although no new physics is involved, the proof is a long-winded application of trigonometry and Snell's law. Later, however, we can use the results of further study to get the formula more easily. It is therefore discussed at the end of Part II (see pages 302–303).

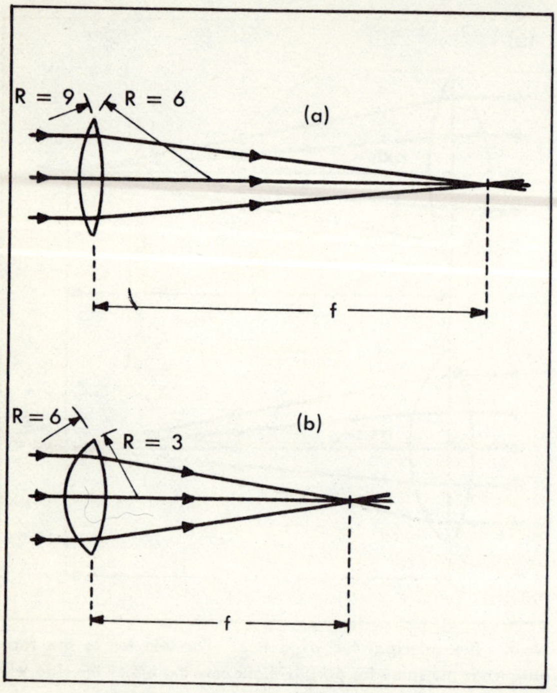

14–6. The shorter the radius of the surface of a lens, the shorter the focal length.

You probably have noticed that, in constructing the two principal rays, we have not considered the exact path of the ray within the lens, but have broken it sharply. This approximate construction is good enough for our present purposes because our location of the two principal foci is accurate only if the lens thickness (at its center) is small compared with the focal length. The only lenses to which our construction accurately applies are therefore *thin lenses*. For the purposes of ray diagrams, we may consider such lenses to be circular plates perpendicular to the axis.

Convex lenses, like parabolic mirrors, focus parallel rays to a point. Lenses, therefore, obey the same equation relating image distance, focal length, and object distance as do mirrors:

$$S_i S_o = f^2.$$

The proof of this equation in the case of convex lenses is the same as for mirrors (Section 12–9). As there, we use the shaded similar triangles formed by the principal rays shown in Fig. 14–7. Considering first the shaded similar triangles to the left of the lens, we see that $H_i/H_o = f/S_o$. The shaded triangles to the right of the lens give $H_i/H_o = S_i/f$. Combining the two equations, we have

$$S_o S_i = f^2.$$

14–4. Camera, Projector, and Eye

Produce an image of the sun with a convex lens. Since the sun is far away, the image is formed practically at the principal focus and you can see it there on a piece of paper. Images of closer objects lie beyond the principal focus; and, in order to capture them on paper or on a photographic film, we have to change the distance between lens and film. To make a photographic camera, then, we usually make a light-tight box with a bellows that allows us to move the lens. By adjusting the length of the bellows, we can place a sharp image on the photographic film. With some cameras we can place a piece of ground glass where the film is later inserted.

14–7. The formation of a real image by a converging lens.

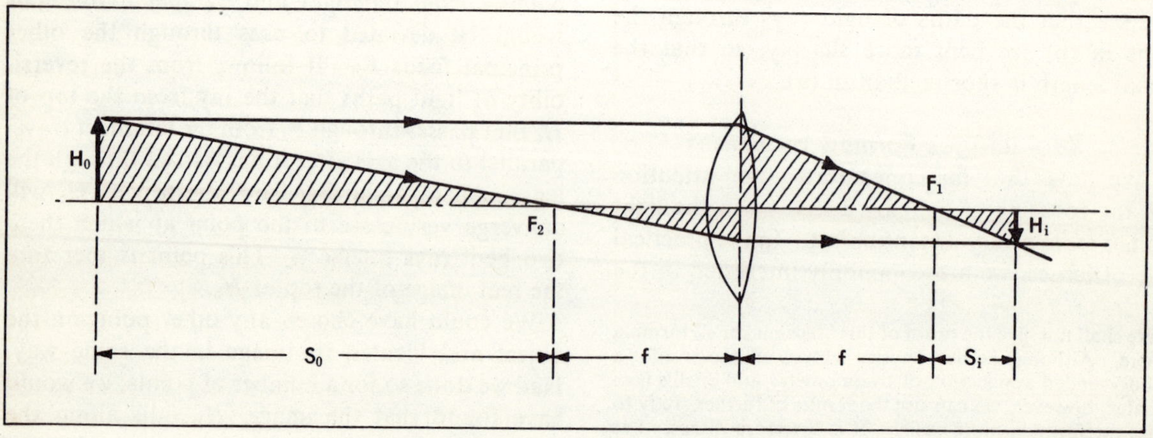

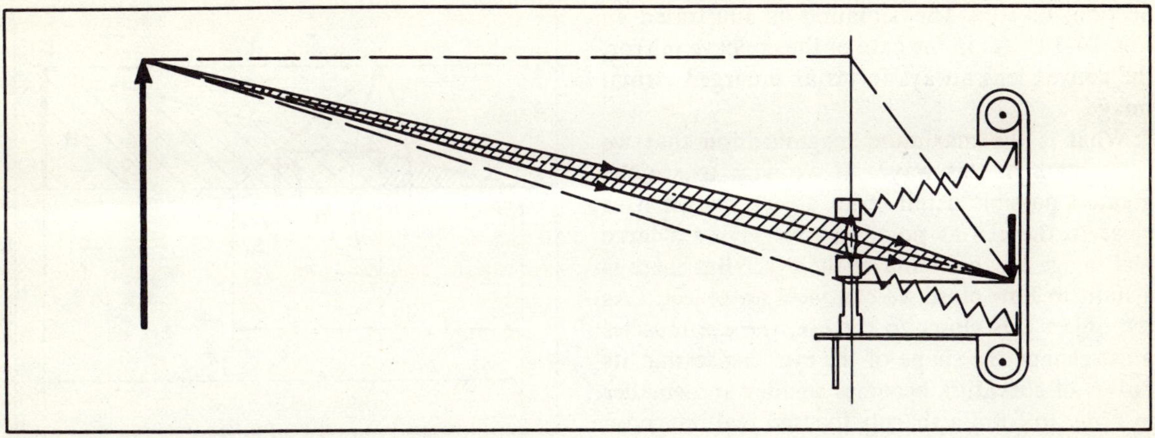

14–8. A camera. The rays of light that form the image of the head of the arrow are indicated.

We can then view the image directly and focus sharply on the particular object we want to photograph (Fig. 14–8).

As long as the object is more than twice the focal distance from the lens, so that S_o is longer than f, the image size is smaller than the object, as $H_i/H_o = f/S_o$ shows. When a small object is brought closer to the principal focus, the image moves to distances behind the lens that are large compared with the focal length; also, the image becomes bigger than the object. Consequently, to photograph small objects, a lens of short focal length is useful.

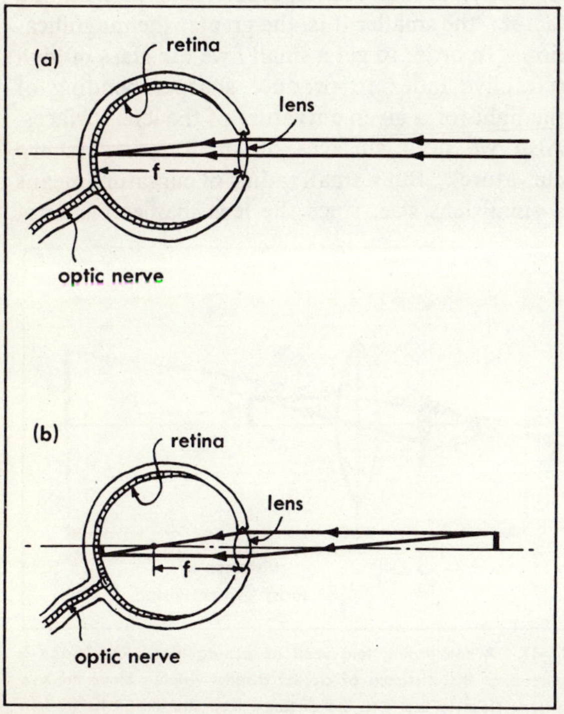

14–9. The lens of an eye adjusted to focus the light from a distant object (a) and from one near by.

A projector is just a camera worked backwards. You can make one by taking the back off a camera, mounting the slides or film where the film usually goes, and shining a bright light through the film and out through the lens. The lens then forms an enlarged image well in front of the camera, where you can place a screen.

In cameras, projectors, and other man-made optical instruments, images are always brought into focus by changing the position of a lens with respect to the object. The eye, on the other hand, is unusual: it focuses images on the retina by changing its curvature and hence the focal length of its lens. When an object is at a very large distance from the eye, the rays entering the eye are nearly parallel and an image is formed at the principal focus as shown in Fig. 14–9 (a). When a close-by object is viewed, the image is formed beyond the focal point, and eye muscles form the elastic eye lens into a sharper curve, decreasing its focal length so that a real image will form on the retina [Fig. 14–9 (b)].

14–5. The Magnifier or Simple Microscope

Let us go back to the small object that we brought close to the principal focus of a lens. As the object is moved through the principal focus the real image moves infinitely far away on the other side of the lens; and when the object is between the lens and the principal focus a virtual image is formed behind the object just as in the case of a concave mirror that we discussed in

Section 12–10. The situation is illustrated in Fig. 14–10. As in the case of the concave mirror, the convex lens always forms an enlarged virtual image.

What is the maximum magnification that we can obtain in this way? If we wish to see the greatest possible detail in an object, we get it as close to the eye as possible, thus giving a large real image on the retina of the eye. But there is a limit to how close we can view an object. As the object gets closer to the eye, the eye muscles must change the shape of the eye lens so that its radius of curvature becomes smaller and smaller in order to form a sharply focused real image on the retina. Soon a limit is reached; the adult eye cannot accommodate to an object closer than about 25 cm. This object distance is called the distance of most distinct vision. Try bringing a pencil closer and closer to your eye. You will see more and more detail until finally, with a great straining of your eye muscles, you can no longer keep a sharp image. Is your distance of most distinct vision greater or less than the average of 25 cm?

A convex lens helps us to see more detail by forming an enlarged virtual image which we can place at a comfortable distance from the eye. We notice in Fig. 14–10 that no matter where the object is placed between the lens and F_2, the top of the image always lies on the line F_1D, and $H_i = \frac{H_o}{f} S_i$ as usual.

Consequently, to make the image look as large as possible we should bring our eye right up to the lens as in Fig. 14–11; and in addition we should move the object (or the lens and our eye) until the image gets as close as we can clearly accommodate. This is the way to get the largest angle between the rays entering our eye from the top and from the bottom of the object; and since this light is what the eye works with, it is the way to make the object (or its virtual image) look largest.

Now for our own comfort we place the image at the distance of most distinct vision d, so the image distance S_i (measured from F_1) is approximately given by $S_i = d + f$. Therefore

$$H_i = \frac{H_o}{f} S_i = \frac{H_o}{f}(d + f) = H_o\left(\frac{d}{f} + 1\right).$$

Furthermore, since we are looking at this image

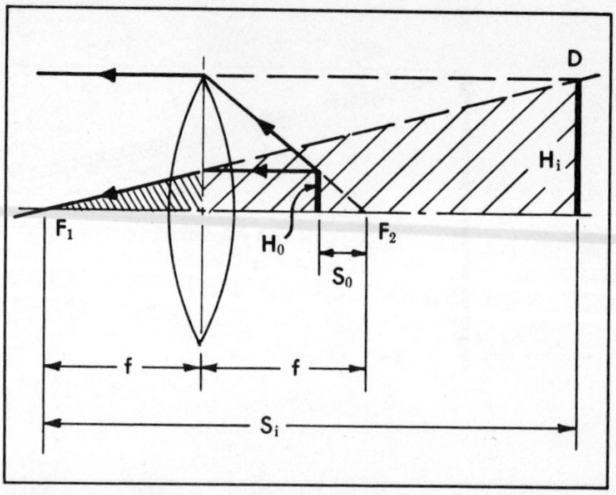

14–10. Formation of a virtual image by a converging lens.

from the distance of most distinct vision just as we could best look at the object without the aid of the lens, the magnification of the image we see is H_i/H_o. That is, maximum magnification is

$$\frac{H_i}{H_o} = \frac{d}{f} + 1.$$

This equation tells us the greatest magnification of a simple microscope. What, then, determines how great a magnification we can get? The focal length, f, of the convex lens is the determining factor; the smaller it is, the greater the magnification. In order to get a small f we use glass of high refractive index to produce sharper bending of the light for a given curvature of the lens surfaces. Also we need surfaces of small radius (sharp curvature). But a small radius of curvature means a small lens size, since the lens diameter cannot

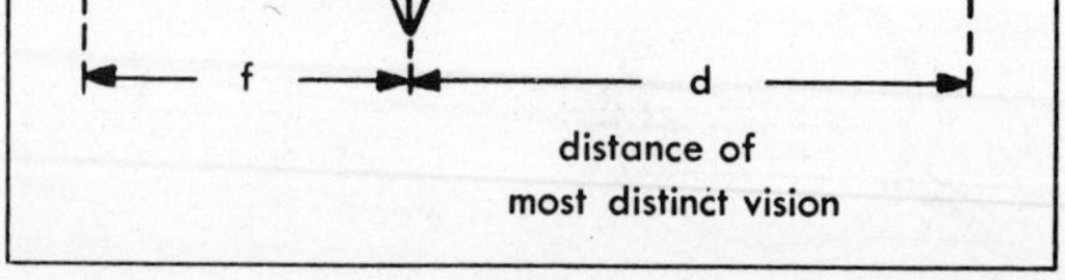

14–11. A converging lens used as a magnifier. The image is placed at the distance of closest distinct vision. Since the eye is very close to the lens, the distance from the image to the lens is about the same as that to the eye.

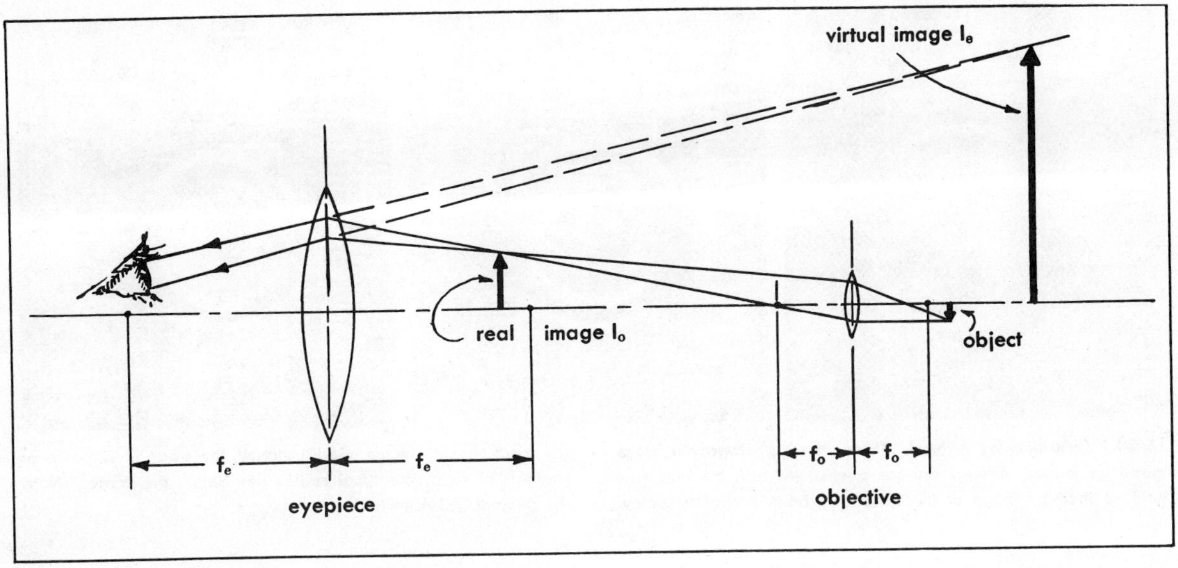

14–12. Compound microscope. The eyepiece and objective are located at the two ends of a tube. The diagram illustrates the formation of the images.

be greater than twice the radius of curvature. So we see that a high magnification means a small lens such as a jeweler wears over his eye while examining the works of your watch. (A big magnifying glass gives a large field of view but little magnification.)

As with all instruments, a simple microscope has its limitations. A small lens means little light is intercepted and even with strong illumination of the object, at high magnifying powers, the image is too dim to see. The compound microscope, described in the next section, circumvents this difficulty and gives sharper and brighter images, but it, too, has its limitations.

In spite of the limits of the simple microscope, it can give magnifications of 100 or so times; and in one of the most exciting discoveries in the history of science, Leeuwenhoek, a Dutch scientist of the seventeenth century, used such a microscope with a lens only a millimeter or two in diameter to look for the first time upon the teeming life of microscopic organisms invisible to the unaided eye. He made his lens, as you can make yours, by heating a glass rod until soft and then pulling it, like taffy, into a long, threadlike filament. He then slowly fed a piece of the filament into a flame until a spherical, molten globule of the right size formed on its end. This sphere he then used as his microscope.

14–6. The Compound Microscope; Telescopes

The most common and the most useful type of optical microscope today is the compound microscope. Perhaps you have used one in a biology course. It consists of a long tube fitted with converging lenses at each end (Fig. 14–12). The bottom lens, called the objective lens, is of short focal length and the microscope is adjusted until the object is just beyond the principal focus of the objective lens so that it forms an enlarged real image, I_o. This image is at such a distance from the eyepiece (slightly nearer the eyepiece than its principal focus) that a virtual image, I_e, is formed at the distance of most distinct vision. The eyepiece is thus used as a simple microscope to examine the enlarged real image formed by the objective.

You can make a crude compound microscope with two cheap, short-focal-length lenses mounted in a cardboard tube. Of course, to take full advantage of the potential of such an instrument, very carefully ground multiple lenses must be used for both eyepiece and objective.

Refracting telescopes are much like microscopes, but to increase the magnification and to gather more light, they have a large objective lens of long focal length. (Recall that $H_i/H_o = f/S_o$ and note that S_o is now large.) As in the compound microscope, the real image can be examined through an eyepiece.

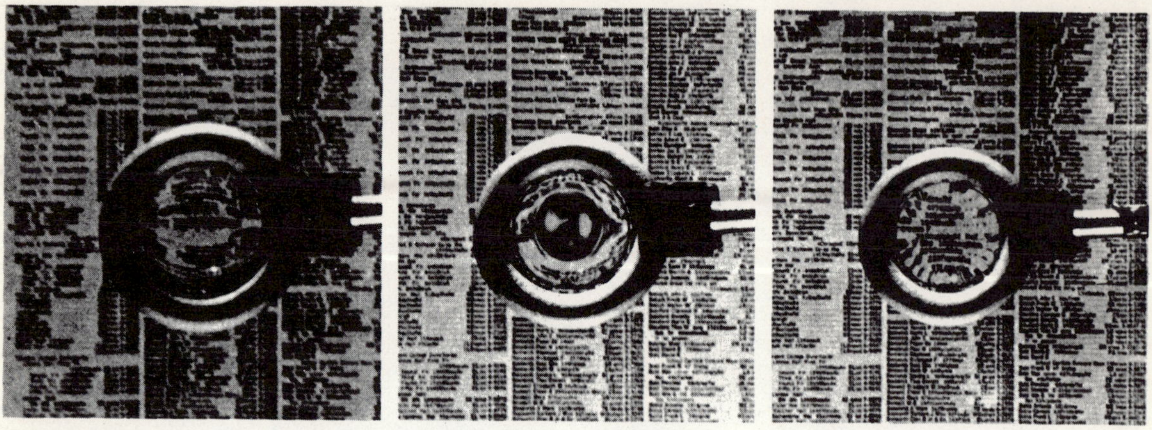

14–13. Distortion by a lens. These three photographs were made by looking through the same lens. At left, the lens was held so that the page of the telephone book is slightly below the focal region; in the middle picture, the page is in the focal region; at right, the focal region lies below the page. Note the geometrical distortions.

14–7. Limitations of Optical Instruments

If you hold a magnifying glass close to this page, you can see a clear, undistorted and slightly magnified image of the print. Now slowly raise the glass from the paper and at the same time increase the distance between your eyes and the lens. At some positions the image appears distorted. If your eyes are far enough from the glass, you may also detect some rainbow colors when looking at a corner of the page. In Figure 14–13 we see three pictures. They were made by looking through the same lens. In the first picture the lens was held so that the page of the telephone book lies entirely (but slightly) below the focal region; in the second picture the page is in the focal region; and in the last the focal region lies below the page. Clearly, in each case the image of the page looks quite different from the page itself. Part of the game of designing really good optical instruments is to minimize the geometrical distortions so apparent in these pictures.

What are the origins of these defects in images? First, we know even for mirrors (Chapter 12) that a surface designed to bring light from one small object to a sharp focus is not the correct surface to bring light to an exact focus from an object at a different place. The same is true for lens surfaces. Some blurring of the image therefore results. In addition, when we look through different parts of a lens the images are at different positions (and the magnification is different). The image therefore is distorted. In photography distortion and blurring are often cut down by using a "stop," a barrier with a small hole in it so that we use only a selected portion of the lens.

The colored edges of images usually arise because of the dispersion of the light that passes through a lens. The focal length of a lens is slightly longer for red light than it is for blue light, because the blue light is refracted more strongly than the red. This undesirable effect is called *chromatic aberration.* It can be greatly reduced by using a weakly diverging lens made of glass for which the index of refraction changes greatly with color, in conjunction with a strongly

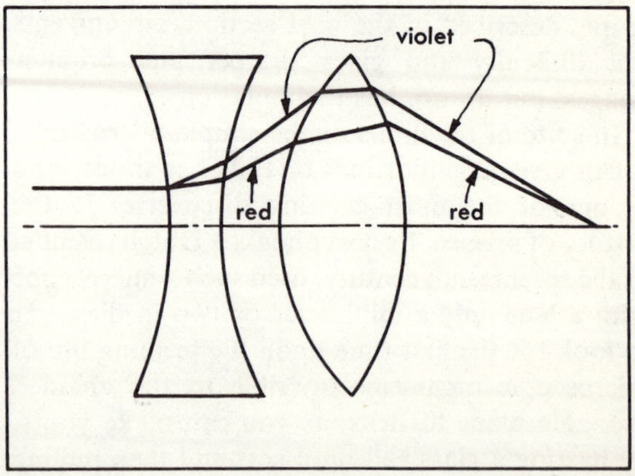

14–14. A lens built of two pieces to minimize the different focal properties of different colors. Such doublets are often made with one common surface and glued together. They are called achromatic lenses.

converging lens of glass for which the index of refraction changes less with color (Fig. 14–14). This trick makes the focal properties of the whole system of lenses nearly the same for all colors.

The problem of designing a system of lenses with the smallest amount of distortions and aberrations is a very complicated one. But the complications arise only in the detailed applications of the laws of refraction; they involve no new principle. Disentangling these complications will not enrich our understanding of basic optical phenomena, and therefore we shall not do it here.

There is one limiting factor affecting optical magnifiers which causes a blur in the image and is of a fundamental nature. This is the inevitable diffraction which results from the limited size of the objective lens through which the light must pass. At high magnification it is this blurring that prevents us from seeing finer and finer detail. We shall learn more about diffraction in Chapter 19.

FOR HOME, DESK, AND LAB

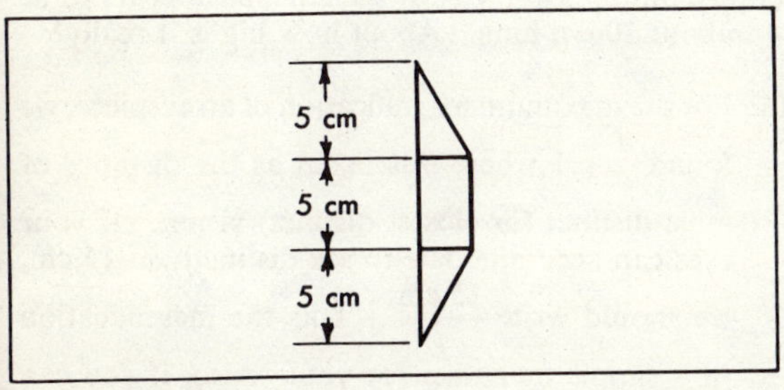

14–15. For Problem 1.

1. A crude converging lens can be constructed by placing two 30°–60°–90° glass prisms together with a glass block as shown in Fig. 14–15.

(a) What is the focal length of this "lens" to one significant figure?

(b) Would such a lens form a clear image? Explain.

2. If two 45° prisms of glass (index = 1.50) are arranged as in Fig. 14–16 they will not converge parallel light. Why not? What will happen to the light?

3. Some lighthouses and light buoys mark the positions of dangerous rocks and shoals. The light must be concentrated at a low angle with respect to the horizon (light directed upward is wasted) and must be equally visible from all points of the compass.

(a) Can you design a "lens" which will do this?

(b) Instead of using a continuous curved surface, such lights often use a lens made of sections of prisms. Can you draw a diagram of such a lens? It is called a *Fresnel lens* after the French physicist who first devised such a lens.

(c) Automobile headlights are constructed to give a wide, flat, horizontal beam. Parabolic reflectors are made to give a narrow beam which passes through a Fresnel lens in the front of the headlight. Examine an automobile headlight and see if you can understand how it gives broad, horizontal beams.

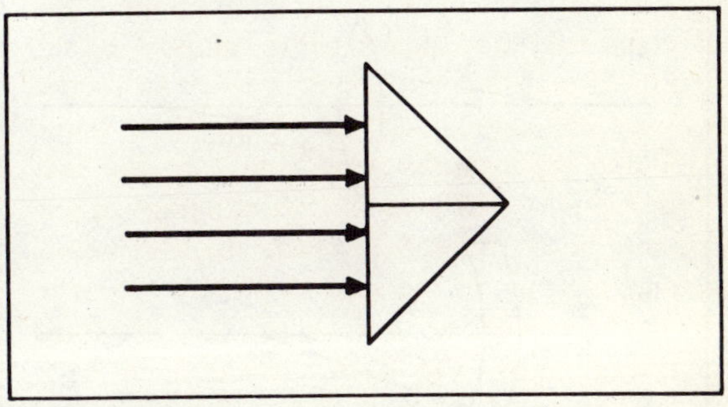

14–16. For Problem 2.

4. Use the Lens Maker's Formula

$$\frac{1}{f} = (n - 1)\left(\frac{1}{R_1} + \frac{1}{R_2}\right)$$

to find the focal length of a glass lens (n = 1.50) with one flat surface and one with a radius of 10 cm. (Such a lens is called a *plano-convex lens.*)

5. (a) What are the focal lengths of the two lenses shown in Fig. 14–17? (Index of glass = 1.50.)

(b) How does the focal length of (b) compare with a flat block of glass?

6. A lens (index = 1.50) has a focal length in air of 20.0 cm.

(a) Is its focal length in water greater or less than in air?

(b) What is its focal length in water?

Hint: Notice that every individual refraction depends on the relative index of refraction.

7. A lens whose focal length is 10 cm is used in a slide projector to give a real image on a screen at a distance of 6.0 meters.
(a) What will be the magnification?
(b) How far is the lens placed from the slide?

8. Prove that if two identical converging lenses of focal length 10 cm are placed 40 cm apart, the combination will form an upright image of an object that is 20 cm away from the first lens and the magnification will be 1.

9. (a) Prove that the size of the image of the sun produced by a convex lens is proportional to the focal length. What is the constant of proportionality?
(b) How large an image of the sun (diameter 1.4×10^9 m) will be formed by a lens of focal length 1.0 meter?
(c) What will be the ratio of the size of the images of the sun formed by a lens of 10 cm focal length and a lens of 10 m focal length?

10. How large an image will be formed of an artificial satellite (53 cm in diameter) passing at an altitude of 500 miles, if it is photographed with a camera whose focal length is 10 cm? Would you expect an actual photograph to show a larger or smaller image than the size you have calculated?

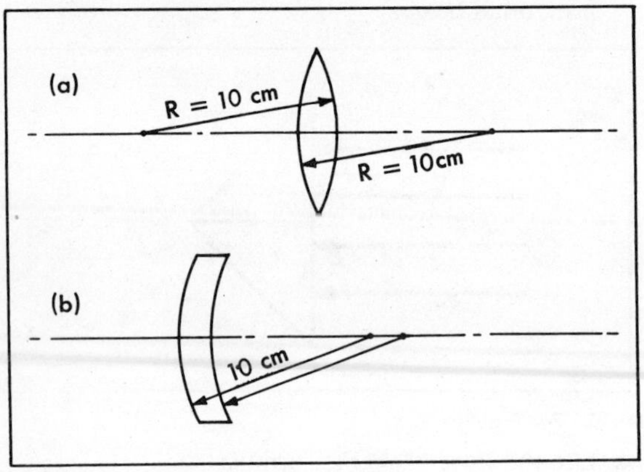

14–17. For Problem 5.

11. (a) What is the focal length of the lens in Fig. 14–18? (Index of the glass is 1.50.)
(b) By sketching the paths of some light rays, show what the lens does to incident light parallel to its axis.
(c) From the ray diagram of Fig. 14–18 (b), show that $S_oS_i = f^2$. Notice from which focal points S_o and S_i are measured.
(d) What happens as you move the object toward the lens? Can S_i ever get bigger than S_o? Is the image ever bigger than the object?
(e) How would you find (experimentally) the focal length of a diverging lens?

12. Assume your distance of most distinct vision is 15 cm. What is the maximum magnification that can be obtained with each of the following convex lenses when used as a magnifying glass or simple microscope?
(a) $f = 30$ cm,
(b) $f = 10$ cm,
(c) $f = 1$ cm,
(d) $f = 1$ mm.
(e) Graph the maximum magnification as a function of "f."

13. Assume your distance of most distinct vision is 25 cm. A compound microscope has an eyepiece of 2.0 cm focal length and an objective of 4.0 mm focal length. The distance between objective and eyepiece is 22.3 cm. What is its magnification to two significant figures?

14. Using the microscope of Problem 13, with the same adjustment, we see an amoeba. With a ruler, we measure the size of the virtual image by looking at it with one eye and at the ruler with the other. On the ruler the amoeba appears to be about 10 cm long. About how big is it really?

15. For the maximum magnification of an eyepiece, we found $\frac{d}{f} + 1$ where d is taken as the distance of most distinct (or closest distinct) vision. If your eyes can accommodate to see distinctly at 15 cm, we should write $\frac{15 \text{ cm}}{f} + 1$ as the magnification of a simple magnifier for you. Also, if you can

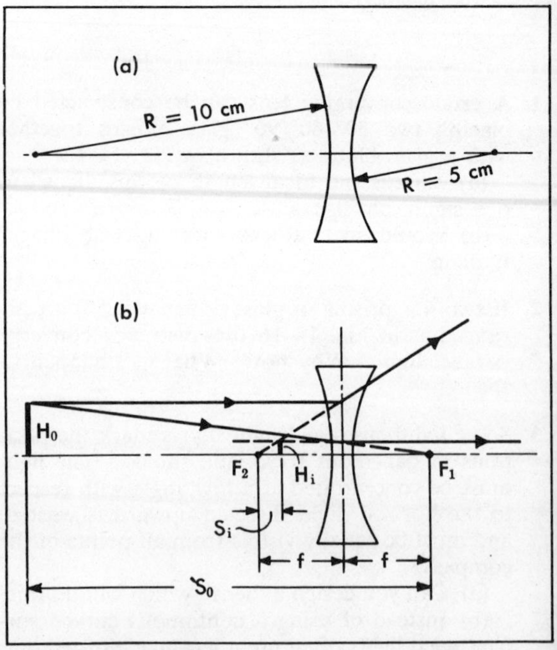

14–18. For Problem 11.

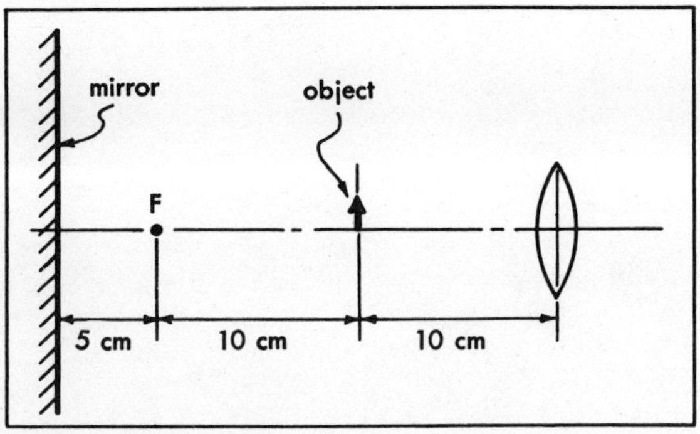

14–19. For Problem 19.

accommodate to images no closer than 35 cm, $\frac{35\text{ cm}}{f} + 1$ would apply. Why does the magnification go up for someone who accommodates poorly at small distances? Does he see more detail than someone who can accommodate closer? Be prepared to discuss this question in class.

16. Two lenses both have a focal length of 20 cm, but one has a diameter four times that of the other. Draw sketches of the two lenses and tell how the images they form differ.

17. (a) What is the ratio of the focal lengths of a crown-glass lens for violet light and for red light? (The index of refraction for various colors is given in Table 4, Chapter 13.)
(b) Is the ratio the same for all kinds of glass?

18. A lens of focal length 20 cm is placed 30 cm from a plane mirror and an object is placed on the axis 10 cm from the mirror. Where will the image of the object be found?

19. Where are the images of the object in Fig. 14–19? Can you see all the images if you look through the lens
(a) with your eye near the lens?
(b) with your eye far from the lens?

FURTHER READING

ROGERS, FRANCES, *Lens Magic.* Lippincott, 1957. A history of the development of lenses, and a description of their many applications.

TEXEREAU, JEAN, *How to Make a Telescope.* Interscience Publishing Co., 1957.

THOMPSON, ALLYN J., *Making Your Own Telescope.* Sky Publishing Co., Cambridge, Mass., 1947.

WALD, GEORGE, "Eye and Camera." *Scientific American*, August, 1950 (p. 32).

Everybody, however intelligent, has a mental block about some aspect of life. This article from a British magazine of humor, describes how electricity *ought* to behave.

---

21 Baffled!

Keith Waterhouse

Article in *Punch*, 1968.

YOU learn something new every day. With no thought of self-improvement, for example, I was reading that story of Thurber's in which he recalls his mother's belief that electricity leaks out of an empty light socket if the switch has been left on. From this I gathered—going by the general context, and the known fact that Thurber was a humorist—that it doesn't.

I picked up another piece of electrical knowledge in 1951, while working as a drama critic on the *Yorkshire Evening Post*. Wanting to imply that a certain actress had given a muted performance, I wrote that while undoubtedly she had an electric presence, on this occasion it was as if the electricity had been immersed in water. A kindly sub-editor explained to me that when electricity gets wet, by some miracle of the elements it intensifies rather than diminishes. I have never seen the sense of this, but I conceded the point and have used only gas-driven metaphor since that date.

I was never taught electricity at school, nor was it often a topic of dinner-table conversation among my parents. What I know about the subject I have mastered the hard way. Take, as an instance, television, an electrical device of awesome complexity. Unlike more privileged students, who are able to go running to m'tutor every time the framehold goes wobbly, I have had to learn in the School of Life that on the large rented model the knobs are on the front whereas on the HMV portable they are on the side. Similarly with electric irons. When I bought my first electric iron there was no plug attached, presumably in case I wanted to wind the flex around my neck and jump off Westminster Bridge with it. There was a leaflet explaining how to get the plug on, but this was of course in German, the international language of the household appliances industry. Only by putting my natural intelligence to the problem did I eventually work out the solution—find a German-speaking electrician.

And so, what with having perforce to change a light bulb

here and tune in a transistor radio there, I have picked up a pretty sound working knowledge of electrical matters. It is not comprehensive, God knows—I still can't fully understand why you can't boil an egg on an electric guitar—but when I jot down a summary of what I have learned, I marvel that I have never been asked to write for the *Electrical Journal:*

1. Most electricity is manufactured in power stations where it is fed into wires which are then wound around large drums.

2. Some electricity, however, does not need to go along wires. That used in portable radios, for example, and that used in lightning. This kind of electricity is not generated but is just lying about in the air, loose.

3. Electricity becomes intensified when wet. Electric kettles are immune to this.

4. Electricity has to be earthed. That is to say, it has to be connected with the ground before it can function, except in the case of aeroplanes, which have separate arrangements.

5. Electricity makes a low humming noise. This noise may be pitched at different levels for use in doorbells, telephones, electric organs, etc.

6. Although electricity does not leak out of an empty light socket, that light socket is nevertheless live if you happen to shove your finger in it when the switch is at the "on" position. So if it is not leaking, what else is it doing?

7. Electricity is made up of two ingredients, negative and positive. One ingredient travels along a wire covered with red plastic, and the other along a wire covered with black plastic. When these two wires meet together in what we call a plug, the different ingredients are mixed together to form electricity. Washing machines need stronger electricity, and for this a booster ingredient is required. This travels along a wire covered with green plastic.

8. Stronger electricity cannot be used for electric razors. Electric razors make a fizzing sound when attached to a power plug.

9. Electricity may be stored in batteries. Big batteries do not necessarily hold more electricity than small batteries. In big batteries the electricity is just shovelled in, while in small batteries (transistors) it is packed flat.

10. Electricity is composed of small particles called electrons, an electron weighing only 1 1·837 as much as an atom of the lightest chemical element, hydrogen, unless the *Encyclopædia Britannica* is a liar.

Incurious people are content to take all this as read. They press a switch and the light comes on, and that is all they know about the miracle in their homes. This has never done for me. I have to know how things work, and if I cannot find out from some technical handbook—the *Every Boys' Wonder Book* series does an advanced manual on electricity—then I combine such information as I already have with simple logic. Thus it is very easy to deduce that the light switch controls a small clamp or vice which grips the wires very hard, so that the electricity cannot get through. When the switch is flicked on the vice is relaxed and the electricity travels to the light bulb where a bit of wire, called the element, is left bare. Here, for the first time, we can actually *see* the electricity, in the form of a small spark. This spark is enlarged many hundreds of times by the curved bulb which is made of magnifying glass.

Why, is our next question, do these light bulbs have a limited life? As any schoolboy knows, heat converts oxygen into moisture. When all the oxygen in the light bulb has become liquified in this manner, it naturally quenches the electric spark. Some years ago a man in Birmingham invented an everlasting electric light bulb which, since it contained no oxygen, would never go out. The rights in it were bought up by the Atlas people who keep it locked in their safe.

Now we come to electricity as a source of power rather than a source of light or heat. Why, when you plug in an electric iron, does it get hot, whereas when you plug in an electric fan it does not get hot but whirrs round and round? The answer is that when light or heat is required we use bare electricity, whereas when power is required we keep the electricity covered up. The constant flow of sparks, unable to escape, is converted into energy. This energy is fed into a motor which makes things go round and round.

I have not yet touched on fuse wire. It has always amazed me that an industry which is so enterprising in most respects—the invention of colour electricity for use in traffic lights and the harnessing of negative electricity for refrigeration are two examples that come to mind—should still, two hundred years after James Watt invented the electric kettle, be manufacturing fuse wire too thin. I pass on a hint for what it is worth. There is available from hardware shops a sturdy wire used mostly for making chicken runs, and this is far more durable than the stuff sold by electricians (who must, I appreciate, make a living). By using chicken wire I now have a fuse box which—even when the spin-dryer burst into flames due to too much booster electricity having been fed into it—has for six months been as impregnable as the Bank of England.

But why have fuse wire at all? I completely understand that the fuse box is the junction at which the wires leading from the power station join, or fuse with, the wires belonging to the house, and that these two sets of wires have got to be connected with each other somehow. But what is wrong with a simple knot? Perhaps I might make this the subject of a paper for the *Electrical Journal* which, I now see from the *Writers' and Artists' Year Book*, welcomes electro-technical contributions not exceeding 3,000 words.

In some respects, I reiterate, my knowledge is imperfect. I have not yet explored the field of neon signs—how do they make the electricity move about? And the pop-up toaster—how does it know when the toast is ready? With an electronic eye, presumably—and this brings us to another fruitful area. What is the difference between electricity and electronics? Or is there a difference? Is electronics now just the smart word to use, like high-speed gas? How can an English computer speak French, which requires a different voltage? Logic would answer these questions too, and many of a more technical nature, but the light over my desk has just gone out. A valve blown somewhere, I expect.

NORMAN LEADER ALLEN

Norman Leader Allen, British physicist, was born in 1927 and received his B.Sc. from the University of Birmingham, England, in 1948 and his Ph.D. in 1951. Allen has been a staff member of Massachusetts Institute of Technology and is now a lecturer in the Electrical and Electronic Engineering Department at the University of Leeds. In addition to his book, Threshold Pressure for Arc Discharges, he has written extensively in scientific journals on arc discharges, cosmic rays and plasma physics.

ALBERT V. BAEZ

Albert V. Baez, born in Puebla, Mexico, in 1912, received his B.A. at Drew (1933), an M.A. from Syracuse (1936), and a doctorate in physics from Stanford University (1950). He has taught at Drew University, Wagner College, Stanford, and Harvard. From 1949 to 1950 he was a physicist in the aeronautical laboratory at Cornell, and from 1951 to 1958 professor of physics at the University of Redlands. He was physicist to the Film Group of the Physical Science Study Committee, and for six years headed the science teaching section at UNESCO in Paris.

STANLEY SUMNER BALLARD

Stanley S. Ballard, Professor of Physics and chairman of the department at the University of Florida, Gainesville, was born in Los Angeles in 1908. He received his A.B. from Pomona College, and M.A. and Ph.D. from the University of California. He has taught at the University of Hawaii, Tufts University, and has been a research physicist at the Scripps Institution of Oceanography. Ballard has served as president of the Optical Society of America. His specialities are spectroscopy, optical and infrared instrumentation, and properties of optical materials. Ballard is coauthor of Physics Principles.

JOHN M. CARROLL

John M. Carroll was born in Philadelphia in 1925, and educated at Lehigh University, and Hofstra. He was editor at Electronics Magazine from 1952 to 1964, became professor of industrial engineering at Lehigh in 1964, and Associate Professor of the Department of Computer Science, University of Western Ontario, London, Ontario, Canada, since 1968. His professional work is in industrial engineering and electronics.

ARTHUR C. CLARKE

Arthur C. Clarke, British scientist and writer is a Fellow of the Royal Astronomical Society. During World War II he served as technical officer in charge of the first aircraft ground-controlled approach project. He has won the Kalinga Prize, given by UNESCO for the popularization of science. The feasibility of many of the current space developments was perceived and outlined by Clarke in the 1930's. His science fiction novels include Childhoods End and The City and the Stars.

ALBERT EINSTEIN

Albert Einstein, considered to be the most creative physical scientist since Newton, was nevertheless a humble and sometimes rather shy man. He was born in Ulm, Germany, in 1879. He seemed to learn so slowly that his parents feared that he might be retarded. After graduating to the Polytechnic Institute in Zurich, he became a junior official at the Patent Office at Berne. At the age of twenty-six, and quite unknown, he published three revolutionary papers in theoretical physics in 1905. The first paper extended Max Planck's ideas of quantization of energy, and established the quantum theory of radiation. For this work he received the Nobel Prize for 1921. The second paper gave a mathematical theory of Brownian motion, yielding a calculation of the size of a molecule. His third paper founded the special theory of relativity. Einstein's later work centered on the general theory of relativity. His work had a profound influence not only on physics, but also on philosophy. An eloquent and widely beloved man, Einstein took an active part in liberal and anti-war movements. Fleeing from Nazi Germany, he settled in the United States in 1933 at the Institute for Advanced Study in Princeton. He died in 1955.

RICHARD PHILLIPS FEYNMAN

Richard Feynman was born in New York in 1918, and graduated from the Massachusetts Institute of Technology in 1939. He received his doctorate in theoretical physics from Princeton in 1942, and worked at Los Alamos during the Second World War. From 1945 to 1951 he taught at Cornell, and since 1951 has been Tolman Professor of Physics at the California Institute of Technology. Professor Feynman received the Albert Einstein Award in 1954, and in 1965 was named a Foreign Member of the Royal Society. In 1966 he was awarded the Nobel Prize in Physics, which he shared with Shinichero Tomonaga and Julian Schwinger, for work in quantum field theory.

LEOPOLD INFELD

Leopold Infeld, a co-worker with Albert Einstein in general relativity theory, was born in 1898 in Poland. After studying at the Cracow and Berlin

Universities, he became a Rockefeller Fellow at Cambridge where he worked with Max Born in electromagnetic theory, and then a member of the Institute for Advanced Study at Princeton. For eleven years he was Professor of Applied Mathematics at the University of Toronto. He then returned to Poland and became Professor of Physics at the University of Warsaw and until his death on 16 January 1968 he was Director of the Theoretical Physics Institute at the university. A member of the presidium of the Polish Academy of Science, Infeld conducted research in theoretical physics, especially relativity and quantum theories. Infeld was the author of The New Field Theory, The World in Modern Science, Quest, Albert Einstein, and with Einstein The Evolution of Physics.

K. SCOTT KINERSON

Dr. Kinerson was educated at the University of New Hampshire, Rensselaer Polytechnic Institute, and Michigan State University. After serving in the U.S. Army from 1943 to 1946, he became Instructor in Physics at the University of Massachusetts at Fort Devens, in 1946. In 1948 he joined the staff of Russell Sage College in Troy, New York as Instructor in Physics. He is presently Chairman of the Department of Physics and Mathematics at that college. He is a co-author of Introduction to Natural Sciences, Part I—The Physical Sciences, 1968.

THOMAS JEFFERSON

Thomas Jefferson, third President of the United States, was born in 1743 at Shadwell in Goochland County, Virginia. He studied Greek, Latin, and mathematics at the College of William and Mary for two years, and later became a lawyer. From 1768 to 1775 Jefferson was a member of the Virginia House of Burgesses. In 1775 he was elected to the Second Continental Congress, and in 1776 he drafted the Declaration of Independence. Jefferson felt a conflicting devotion to the tranquil pursuits of science and public service. His interests ranged over such fields as agriculture, meteorology, paleontology, ethnology, botany, and medicine. He believed in the freedom of the scientific mind and the importance of basing conclusions on observations and experiment. Jefferson demanded utility of science, hence his numerous inventions and interest in improvements and simplifications of agricultural tools and techniques, and in balloons, dry docks, submarines, even the furniture in his home (swivel chairs and music stands). Because of his prominence as a public figure, he was influential in increasing and improving science education in America. He died on July 4, 1826, the fiftieth anniversary of the Declaration of Independence.

MATTHEW JOSEPHSON

Matthew Josephson, prolific writer and magazine editor, was born in Brooklyn in 1899. He received his B.A. from Columbia University in 1920. Josephson was successively editor of the Broom, Transition, and The New Republic, which he left in 1932. In 1948 he was elected to the National Institute of Arts and Letters and also was a traveling Guggenheim fellow for creative literature. He is the author of Zola and His Time, The Robber Barons, and Portrait of the Artist as American.

ROBERT B. LEIGHTON

Robert B. Leighton, born in Detroit, Michigan in 1919, was first a student and then a faculty member at California Institute of Technology. He is a member of the International Astronomical Union, the National Academy of Science and the American Physics Society. Professor Leighton's work deals with the theory of solids, cosmic rays, high energy physics, and solar physics.

ABRAHAM S. LUCHINS

Dr. Luchins received a B.A. degree from Brooklyn College (1935), M.A. degree from Columbia University (1936), and his PhD. at New York University (1940). He was research assistant to the psychologist Max Wertheimer, clinical psychologist in the United States Army, and Director of Mental Hygiene Clinic for the Veterans' Administration. He was taught at McGill University, University of Oregon, University of Miami, and since 1962 has been Professor at the State University of New York at Albany. His publications include: Logical Foundations of Mathematics for Behavioral Scientists (1965) and Group Therapy: A Guide (1964); and he was a co-author of Introduction to Natural Science (Parts I and II), 1968 and 1970.

DAVID KEITH CHALMERS MACDONALD

David Keith Chalmers MacDonald was born in Glasgow, Scotland, in 1920 and received his M.A. in mathematics and natural philosophy from Edinburgh University in 1941. After serving with the Royal Mechanical and Electrical Engineers during World War II, he received his Ph.D. in 1946 from Edinburgh. Then he attended Oxford as a research fellow and received a Ph.D. in 1949. In 1951 Dr. MacDonald went to Canada and started a low temperature physics research laboratory for the National Research Council. MacDonald was appointed to the physics department at Ottawa University in 1955 and elected Fellow of the Royal Society of London in 1960. Aside from numerous articles in scientific journals, he was the author of Near Zero: An Introduction to Low Temperature Physics and Faraday, Maxwell, and Kelvin. MacDonald died in 1963.

JAMES CLERK MAXWELL

See J. R. Newman's articles in Readers 3 and 4.

ALAN S. MELTZER

Alan S. Meltzer was born in New York in 1932 and educated at the University of Syracuse, and at Princeton, where he received his Ph.D. in astronomy, in 1956. He was physicist at the Smithsonian Astrophysical Observatory from 1956 to 1957. Presently he is Assistant Professor of Astronomy at Rensselaer Polytechnic Institute at Troy, New York. His areas of investigation include solar and stellar spectroscopy, and solar-terrestrial relations.

ALBERT ABRAHAM MICHELSON

Precision measurement in experimental physics was the lifelong passion of A.A. Michelson (1852–1931), who became in 1907 the first American to win a Nobel Prize in one of the sciences. Born in Prussia but raised in California and Nevada, Michelson attended the U.S. Naval Academy and was teaching there in 1879 when he first improved the methods of measuring the velocity of light on earth. After a post-graduate education in Europe he returned to the United States where he taught physics at the college that became Case Institute of Technology, then at Clark University, and at the University of Chicago. While in Europe he invented the famous instrument called the Michelson interferometer and while in Cleveland at Case in 1887, he and E.W. Morley improved this device in an effort to measure the absolute velocity of the Earth as it hurtles through space. The failure of the Michelson-Morley aether-drift experiment was an important result that showed a deep flaw in 19th-century physics. Although Michelson remained a creative experimentalist in physical optics, meteorology, astrophysics and spectroscopy throughout his life, he died still believing in the wave model of the nature of light and in his "beloved aether." His experimental value of the speed of light, refined still further just before his death, remain the accepted value of one of the few "absolute" constants in physics for several decades.

JAMES ROY NEWMAN

James R. Newman, lawyer and mathematician, was born in New York City in 1907. He received his A.B. from the College of the City of New York and LL.B. from Columbia. Admitted to the New York bar in 1929, he practiced there for twelve years. During World War II he served as chief intelligence officer, U. S. Embassy, London, and in 1945 as special assistant to the Senate Committee on Atomic Energy. From 1956–57 he wassenior editor of The New Republic, and since 1948 had been a member of the board of editors for Scientific American where he was responsible for the book review section. At the same time he was a visting lecturer at the Yale Law School. J. R. Newman is the author of What is Science?, Science and Sensibility, and editor of Common Sense of the Exact Sciences, The World of Mathematics, and the Harper Encyclopedia of Science. He died in 1966.

V. LAWRENCE PARSEGIAN

V. Lawrence Parsegian studied at M.I.T., Washington University, and New York University, obtaining his Ph.D. in physics in 1948. He has been professor of nuclear science and engineering at Rensselaer Polytechnic Institute, since 1954, and holds the distinguished Chair of Rensselaer professorship. In addition to his research activities, he has chaired a curriculum development project to improve college science teaching.

PHYSICAL SCIENCE STUDY COMMITTEE (PSSC)

As one of the earliest curriculum development groups, formed in 1956 and consisting of scientists and educators, it produced materials for a new high school physics course (first published in 1962). These continue to be used by many students and teachers in the U.S., and portions of the course have been adapted also for use in other countries.

MATTHEW SANDS

Matthew Sands was born in Oxford, Massachusetts, in 1919. He attended Clark College, Rice Institute of Technology. During World War II he worked at the the Los Alamos Scientific Laboratory. He was Professor of Physics at the California Institute of Technology before joining the linear accelerator group at Stanford University. Professor Sands specializes in electronic instrumentation for nuclear physics, cosmic rays, and high-energy physics. He served as chairman of the Commission on College Physics.

WILLIAM ASAHEL SHURCLIFF

Born in Boston in 1909, William A. Shurcliff was educated at Harvard, receiving his Ph.D. in physics in 1934. During the war he served as technical aide to the Office of Scientific Research and Development, National Defense Research Committee, and Manhattan project. Then he was with the Polaroid Corporation as senior scientist and project leader. He is now a Research Fellow at the Electron Accelerator at Harvard. Shurcliff is the author of Polarized Light: Production and Use and Bombs at Bikini. His technical interests include emission spectroscopy, absorption spectrophotometry, atomic energy, gamma radiation dosimeters, microscope design, and color vision. He has headed a citizen's group to examine the deleterious effects of the planned supersonic transport planes.

JAMES ALFRED VAN ALLEN

James Alfred Van Allen, discoverer of the "Van Allen radiation belt," was born at Mt. Pleasant, Iowa, in 1914. After his undergraduate work at Iowa Wesleyan College, he received his M.S. and in 1939 his Ph.D. from the State University of Iowa, where he is now a Professor of Physics and Astronomy. He has been a Carnegie research fellow, and a research associate at Princeton, and is the recipient of numerous honorary doctorates. For his distinguished work in nuclear physics, cosmic rays and space probes, he has been awarded the Hickman Medal from the American Rocket Society, the Distinguished Civilian Service Medal of the U.S. Army, and the Hill Award of the Institute of Aerospace Science.

EDGAR VILLCHUR

Edgar Villchur is President and Director of Research of the Foundation for Hearing Aid Research in Woodstock, New York. He was born in New York City in 1917 and received a M.S.Ed. from the City College of New York. He has taught at New York University, and was President and Chief Designer of Accoustic Research, Inc., a manufacturing company in the high fidelity field.

GEORGE WALD

George Wald was born in New York in 1906 and received his education at New York University and Columbia University. He did research in biology at the Universities of Berlin, Zurich, and Chicago, and joined the faculty of Harvard University in 1935, where he now is professor of biology. He is the recipient of many honors for his work on the biochemistry of vision, including the Nobel Prize in physiology and medicine for 1967. He is also widely regarded as one of the outstanding teachers of biology.

WILLIAM GREY WALTER

William Grey Walter was born in 1911 and received his M.A. and Sc.D. (1947) from Cambridge University. He was a Rockefeller Fellow at the Maudsley Hospital in England. W. Grey Walter is a pioneer in the use of electroencephalography for translating the minute electrical currents of the human brain into physical patterns which may be studied for the information they give us on brain processes. Walter is the author of The Living Brain, Further Outlook, The Curve of the Snowflake and articles to various scientific journals.